Due Diligence an
Corporate Governance

by

Dr Linda S Spedding

Members of the LexisNexis Group worldwide

United Kingdom	LexisNexis UK, a Division of Reed Elsevier (UK) Ltd, 2 Addiscombe Road, CROYDON CR9 5AF
Argentina	LexisNexis Argentina, BUENOS AIRES
Australia	LexisNexis Butterworths, CHATSWOOD, New South Wales
Austria	LexisNexis Verlag ARD Orac GmbH & Co KG, VIENNA
Canada	LexisNexis Butterworths, MARKHAM, Ontario
Chile	LexisNexis Chile Ltda, SANTIAGO DE CHILE
Czech Republic	Nakladatelství Orac sro, PRAGUE
France	Editions du Juris-Classeur SA, PARIS
Germany	LexisNexis Deutschland GmbH, FRANKFURT AND MUNSTER
Hong Kong	LexisNexis Butterworths, HONG KONG
Hungary	HVG-Orac, BUDAPEST
India	LexisNexis Butterworths, NEW DELHI
Ireland	Butterworths (Ireland) Ltd, DUBLIN
Italy	Giuffrè Editore, MILAN
Malaysia	Malayan Law Journal Sdn Bhd, KUALA LUMPUR
New Zealand	LexisNexis Butterworths, WELLINGTON
Poland	Wydawnictwo Prawnicze LexisNexis, WARSAW
Singapore	LexisNexis Butterworths, SINGAPORE
South Africa	LexisNexis Butterworths, Durban
Switzerland	Stämpfli Verlag AG, BERNE
USA	LexisNexis, DAYTON, Ohio

A CIP Catalogue record for this book is available from the British Library.

ISBN 0 754526224

Typeset by Kerrypress Ltd, Luton, Beds www.kerrypress.co.uk

Printed and bound in Great Britain by Antony Rowe Ltd, Chippenham, Wiltshire

Visit LexisNexis UK at www.lexisnexis.co.uk

Dedication

To all of my family, especially Ajan and his father:

> 'The greatest service you can give to the world
> Is to take responsibility for yourself,
> Your relationships and the environment.'

Dr S Purna, *The Truth Will Set You Free* (1987) Element Books, p 55.

Acknowledgements

I, the Author, wish to express my thanks to all of those who have encouraged and supported the preparation of this handbook, both those who are mentioned expressly and those who are not.

This handbook could not have been completed without the contribution of reliable colleagues who have submitted the chapters referred to in the table of contents and whose profiles are mentioned separately. For these contributions I have enjoyed a purely editorial role in the main.

In addition several other colleagues have provided some material and checklists that have enabled the preparation of the remaining chapters in a timely manner. With this in mind acknowledgement and thanks to Charles Bacon, CEO, Due.Com Group of Companies, Christopher Davis and Harry Benham of City of London law firm, Davis & Co, Hemant Batra of Kesar Dass B. & Associates in Delhi, Mike Dance of Jackson Parton, Shipping Solicitors, London, Jonathan Barber and Andrew Reese of SERM, Professor Bob Lee of BRASS, David Kaye, Steven T Miano of Wolf, Block, Schorr and Solis-Cohen LLP, Zaid Hamzah, joint CEO, Lexfutura Limited, Claude S Lineberry, Jayant Bhuyan of ASSOCHAM, Shalini Agarwal and Sakate Khaitan, Partners and K Radhika, Associate ALMT Legal, Stephen Mason a barrister specialising in e-risks, e-business, IT, data protection and commercial law and William L Pence of Akerman Senterfitt, as well as to all of my friends and colleagues at SERM, EMIS Professional Publishers Ltd and the Forum for Private Business.

As regards the appendices, I would like to acknowledge the following for their input: Chapter 1: Appendix 1 – Hemant Batra of Kesar Dass B. & Associates in Delhi, Appendices 2 and 3 – Christopher Davis, Chapter 8: Appendix 1 – Jayant Bhuyan of ASSOCHAM, Shalini Agarwal and Sakate Khaitan, Partners and K Radhika, Associate ALMT Legal, Appendix 2 – Herbert Williams and Chapter 16: Appendix – William L Pence of Akerman Senterfitt.

I wish to express my appreciation of the support and editorial comment provided by Ruth Eldon, my co-director at Women In Law Ltd, as well as my colleagues Claire Gillman and Jill Ryan.

I also wish to thank my publishers, LexisNexis, for their enthusiasm and support.

Many thanks to Dr Purna, whose guidance and ongoing advice have been a constant source of inspiration.

I also acknowledge the painstaking proofreading of my mother, who has provided this service lovingly since my first publication in 1986 and who will celebrate her 80th birthday on the due publication date. In addition I acknowledge the encouragement of my father, who has respected my writing over the years.

Finally, I acknowledge the patience, encouragement and support of my family – especially my son Ajan – who have shared my time and attention as I worked on this project and generally for their ongoing belief in my work.

While I gratefully acknowledge the assistance of all of the above, myself and the contributors accept responsibility for the final publication.

Dr Linda S Spedding

Foreword

The International Chamber of Commerce (ICC), the world business organisation, promotes an open international trade and investment system and the market economy. Our conviction that trade is a powerful force for peace and prosperity dates from our earliest years, when the small group of business leaders who founded ICC called themselves 'the merchants of peace'. Increasingly, experience is demonstrating the significant benefits brought by international trade and investment in faster economic growth and significant reductions in poverty – as strongly evidenced by the current economic success of India and China resulting in significantly reduced poverty levels in both those countries. However, for the effective operation of the international trading system, good governance at country level and at company level are both essential. ICC promotes the benefit of a rules based trading system and clear self-regulation in all aspects of operation built around an effective corporate governance system. For corporate governance to be effective clear policies must be in place but also the tools, guidelines and experience must be available to translate policy into effective implementation.

It is for this reason that we welcome this new publication by Dr Linda Spedding which clearly demonstrates the importance of effective due diligence, particularly when making international acquisitions. It provides clear guidance and case studies to help all involved understand the complexity of the issues involved and to demonstrate the detailed work that is necessary both to ensure that the benefits of an acquisition can be realised and that there are no unexpected problems, for example through damage to corporate reputation that more than offsets the targeted benefits. As high profile business failures tarnish the reputation of international business it is essential that business responds by having the policies and practices in place in day-to-day operations and in particular, as this book so effectively demonstrates, when a major business development such as an acquisition is being implemented.

Dr Spedding is well known to the ICC as an international lawyer with a substantial reputation as consultant, author and speaker on significant commercial matters. She is to be congratulated on this valuable and timely contribution to the debate on due diligence and sound corporate governance.

Andrew Hope
Director
ICC United Kingdom

Foreword

[illegible]

[illegible]

[illegible]

[illegible]

[illegible]

Director

[illegible]

Preface

Times are changing: no longer do organisations aspire to profit for shareholders alone – they are increasingly answerable to other stakeholders. As a result of regulatory and media pressure, in particular transparency, openness and fair play are needed to be successful and sustainable in business. While traditional concepts and vehicles still exist and are referred to in this handbook, other meanings and structures are also developing. It is a constant evolutionary process from which few organisations can escape, regardless of size, sector or location.

The effective running of any business, large or small, is no minor undertaking. We are in an era of instant communication, media interest and increasing global management requirements and trends. As a result, enhanced business knowledge, improved commercial awareness and the appropriate technology are vital. Traditional barriers and divisions of responsibility are being dismantled and the concepts of due diligence, corporate governance and risk management must be recognised as holistic business issues.

This handbook is not intended to deal with every aspect of due diligence – the intention is to consider due diligence having regard to corporate governance and related drivers. It combines traditional business strategies with the more recent trends in due diligence and corporate governance in an accessible manner that enables business representatives and their advisers to operate in a more proactive and responsible way. The intention is that it should provide an overview of the key concerns and offer some practical tools in the form of checklists and case studies that can assist with business planning.

A preliminary overview of the traditional approach to legal due diligence is dealt with at the outset. However, the reader should appreciate that this handbook regards the due diligence process as an ongoing exercise. It extends to areas of business activity that go well beyond the transaction/deals with which it is usually associated to embrace many aspects of business operations and performance.

Corporate governance and risk management can also be seen as part of an organisation's ongoing internal due diligence. While certain matters clearly have a basis in regulation it is important to regard this handbook as a business resource rather than a legal text. As a general rule, the stated position relates to the UK unless otherwise mentioned or indicated. Wherever it is helpful, comparative and/or regional or international trends and standards are also discussed. In view of the global nature of the developments and their extensive impact, the international dimension is considered. Selected countries include India, Hong Kong, China, Australia and Japan. Bearing in mind the enormous influence of US developments in general, and the recent *Sarbanes-Oxley Act of 2002* (*SOX*) in particular, the handbook provides an overview of the American framework.

The reader should be aware that this is a vast and developing debate – indeed many of the issues are each worthy of a book in themselves. As such, the treatment in this handbook must be selective with the intention to provide a different approach or angle to a subject that is covered regularly by the media and the internet.

Finally, it should be emphasised that throughout the handbook attention is given to small business issues and to small companies including small and medium-sized enterprises (SMEs) as far as is practicable. Since they are regarded as the backbone of most economies, any comments addressed to them therefore have a broader relevance for the economy and society.

It is to be hoped that the reader enjoys the contents as much as the Author has enjoyed dealing with such an important debate.

Dr Linda S Spedding

Author and contributors

Dr Linda S Spedding, international lawyer and business adviser, SERM
Dr Linda S Spedding is a practising solicitor (England and Wales) and attorney at law (USA). She has worked at the Legal Service of the European Commission in Brussels and in matters before the European Court in Luxembourg. During her career she has been a partner and a consultant with international law firms. Currently Dr Spedding provides consultancy advice to law firms and to clients in both the private and public sectors. As an international lawyer also qualified as an Advocate in India, Dr Spedding keeps abreast of practical professional developments through her network of long-standing professional connections and membership of associations. She has practised in the area of due diligence as a consultant with Davis & Company, which specialises in international mergers and acquisitions and advises on integrating acquisition targets as well as with other colleagues overseas. She has been particularly concerned with legal business issues that affect small business and is policy adviser to the Forum for Private Business, the leading lobbying organisation in the UK for small business. As editor for *Advising Business* journal, including the *Due Diligence Quarterly,* Dr Spedding has written extensively on due diligence and related areas of business risk. In her capacity as international environmental lawyer and adviser to SERM she has developed corporate governance as a specialist area of advice for business.

Stephen David Jones and Rashmi Sharma, Jirehouse Capital (*Chapters 2 and 3*)
Stephen David Jones and Rashmi Sharma are solicitors both qualified to practise in England and Hong Kong. Jirehouse Capital is based in London with associated operations in Mauritius, Nevis and Panama. It provides a broad range of corporate financial services to its international client base, including structuring and financing businesses and transactions and the provision of trustee and corporate services and legal and tax advisory services. It also provides, through its affiliates, investment portfolio services and is a co-founder of iSentry, which provides web-based 'anti-money laundering' compliance solutions and services to professionals for the exchange of know-your-customer (KYC) information.

Olga Reese, Cass Business School (*Chapter 7*)
Olga Reese is currently completing an MSc in Insurance and Risk Management at Cass Business School (dissertation and interest in reputation risk management etc) and has already completed her MA in Politics, Security and Integration at SSEES, UCL in 2003. During the MA Olga wrote a dissertation on the Russian oil industry in the context of CSR.

Laura Washburn and David Roth, Bradley Arant Rose & White LLP (*Chapter 11*)
Founded in 1871, Bradley Arant Rose & White LLP has a large number of lawyers practising in the US. Laura Washburn is a member of Bradley Arant's Corporate and Securities Practice Group, focusing on public companies and compliance with federal and state securities laws. She is also a member of the firm's Sarbanes-Oxley Task Force, and has developed expertise advising public and private companies of the impact of the *Sarbanes-Oxley Act of 2002* on corporate governance. David Roth is a partner in Bradley Arant's Birmingham office where he focuses his practice on environmental and toxic tort matters. He represents clients on a full range of Federal and state environmental matters including those related to the Clean Air and Water Acts, CERCLA, RCRA, and the Endangered Species Act. A large part of David's practice involves providing regulatory and permitting guidance to businesses operating in the State of Alabama. He also represents clients investigating and responding to allegations of potential environmental violations – both internally and externally.

Ramni Taneja and Jitheesh Thilak, Little & Co. (*Chapter 12*)
Ramni Taneja and Jitheesh Thilak are Advocates at Little & Co, New Delhi, India. Little and Co was established in 1856 in Bombay, India and has an extensive international and all India civil practice. The areas of practice of the firm are: acquisitions and mergers; domestic and foreign banking; commercial disputes; conflict of laws; contracts; corporate governance; foreign direct investment; finance; venture capital; securities; international offerings; joint ventures; mutual funds; insurance; leasing; corporations; energy; intellectual property; shipping; air transportation; litigation before all tribunals and courts in India; and commercial arbitration. The firm advises business groups, government companies and commercial banks in India – Bombay and Delhi – and has an extensive international clientele.

Angela Wang & Co. (*Chapter 13*)
Established in 1995, Angela Wang & Co. is a business-focused Greater China legal practice with a dedicated group of local and expatriate lawyers qualified in multiple jurisdictions. Experienced lawyers in our Hong Kong and Shanghai offices advise a wide range of business clients on all aspects of corporate and commercial law, mergers and acquisitions, corporate finance, employment law and commercial litigation in Hong Kong and the PRC.

Jim Harrowell, Partner Hunt & Hunt (*Chapter 14*)
Jim Harrowell has been a partner of the national law firm Hunt & Hunt for over 20 years. He qualified in law and accountancy at the University of New South Wales and is a fellow of the Australian Institute of Company Directors and the Taxation Institute of Australia. Jim is currently the International Chairman of Interlaw, a legal network with some 65 member firms in over 100 major cities throughout the world. Jim has been appointed a foreign Arbitrator by the Chinese Arbitration Commission in Beijing and by the Shanghai Arbitration Commission in Shanghai. Jim has had considerable experience, both as a board member and legal adviser to boards on issues relating to

establishing aculture of good corporate governance and dealing with the legal issues arising from failures by organisations to implement a good corporate governance culture.

Shuichi Namba, Koichi Nakatani and Richard G Small, Momo-o, Matsuo & Namba (*Chapter 15*)
Shuichi Namba, LL.M (Columbia), is member of the Japanese Bar, New York Bar and California Bar. He became an associate at Ozaki & Momo-o in 1984. In 1986 he enrolled at Columbia Law School from which he was awarded a Master of Law (LLM) in 1987. In February 1988 he was admitted to the New York Bar. In December 1988 he was admitted to the California Bar. From June 1989 he became a founding partner of Momo-o, Matsuo & Namba. Koichi Nakatani is a member of the Japanese Bar. He joined the Legal Department of IBM (Japan) in 1997. In 2004 he became an associate at Momo-o, Matsuo & Namba. Dr Richard G. Small, LLM. (London), PhD (London), member of the New York Bar, joined Momo-o, Matsuo & Namba in 2001 following a Research Fellowship at the International Center for Comparative Law and Politics, Graduate School of Law and Politics, University of Tokyo.

Nicola Jones, Solicitor (*Chapter 18*)
Nicola Jones is a solicitor in England and Wales (1987), and was called to the Bermuda Bar in 1997. She specialised in UK taxation, corporate and trust law. She is currently working as family business and wealth adviser and contributes to professional press and international public speaking engagements.

[illegible] performance culture [illegible]

[illegible] (Chapter 10)

[illegible]

[illegible] (Chapter 13)

[illegible]

Contents

Table of International Statutes

India

Table of Statutory Instruments

Table of Cases

List of Key Abbreviations

ABI	=	Association of British Insurers
ACCA	=	Association of Chartered Certified Accountants
ADR	=	Alternative dispute resolution
ADR	=	American depository receipts
AGM	=	Annual general meeting
AIRMIC	=	The Association of Insurance and Risk Management
ALARM	=	The National Forum for Risk Management in the Public Sector
ASB	=	The Accounting Standards Board
ASSOCHAM	=	Associated Chambers of Commerce and Industry
AR&A	=	Annual Report and Accounts
ASTM	=	The American Society for Testing and Materials
BP	=	British Petroleum
BPO	=	Business process outsourcing
BS	=	British Standard
BSI	=	British Standards Institute
BTR	=	Blackout Trading Restriction
CASC	=	The China Accounting Standard Committee
CBI	=	The Confederation of British Industry
CC	=	The Charity Commission
CC	=	Climate change
CCR	=	Current customer review
CDD	=	Cultural due diligence
CEO	=	Chief Executive Officer
CERCLA	=	The Comprehensive Environmental Response, Compensation and Liability Act
CFA	=	Conditional fee agreement
CFCs	=	Chlorofluorocarbons
CFO	=	Chief Finance Officer
CFR	=	Cost and freight terms

CICPA	=	The Chinese Institute of Certified Public Accountants
CIF	=	Cost, insurance and freight
CII	=	The Confederation of Indian Industry
CIO	=	The Charitable Incorporated Organisation
CLB	=	Company Law Board
CLO	=	Chief Legal Officer
CNAO	=	The China National Audit Office
CO	=	Companies Ordinance
CPR	=	The Civil Procedure Rules
CRM	=	Cause Related Marketing
CORE	=	The Corporate Responsibility Bill
CSR	=	Corporate social responsibility
CSRC	=	The Chinese Securities Regulatory Commission
DPA	=	The Data Protection Act 1998
DTI	=	The Department of Trade and Industry
EBRD	=	The European Bank for Reconstruction and Development
EC	=	European Community
ECE	=	Economic Commission for Europe
ECJ	=	Court of Justice of the European Communities
EMAS	=	Eco-Management and Audit Regulation
EDD	=	Environmental due diligence
EDIFAR	=	Electronic Data Information Filing And Retrieval
EEA	=	European Environment Agency
EHS	=	Environment, health and safety
EPA	=	Environmental Protection Agency
EPA 1990	=	The Environmental Protection Act
EPCRA (or SARA Title III)	=	The Emergency Preparedness and Community Right to Know Act
ETI	=	Ethical Trading Initiative
EXW	=	Ex-works ex-factory terms or free on board
EU	=	European Union
FATF	=	The Financial Action Task Force
FAS	=	Free alongside ship
FCA	=	Free carrier
FDI	=	Foreign direct investment

FDCPA	=	The Fair Debt Collection Practices Act (amended 1996)
FDIC	=	The Federal Deposit Insurance Corporation
FEMA	=	The Foreign Exchange Management Act 2000
FIUs	=	The Financial Intelligence Units
FOB	=	Free on board
FPA	=	Free of particular average
FPB	=	The Forum for Private Business
FSA	=	The Financial Services Authority
GAAP	=	Generally accepted accounting principles
GAFTA	=	The Grain and Feed Trade Association
GLBA	=	Gramm-Leach-Bliley Act (Financial Modernization Act of 1999)
GDR	=	Global depositary receipts
GITIC	=	Guangdong International Trust and Investment Corporation
GM	=	Genetically-modified
GRI	=	The Global Reporting Initiative
HAZOPs	=	Hazard and operability studies
HIPAA	=	Health Insurance Portability and Accountability Act of 1996
HMG	=	The UK Government
HS&E	=	Health, Safety and Environmental
HSSE	=	Health, safety, social and environmental
IBE	=	The Institute of Business Ethics
ICAEW	=	The Institute of Chartered Accountants in England and Wales
ICC	=	The International Chambers of Commerce
ICCR	=	The Interfaith Center of Corporate Responsibility
ICSA	=	The Institute of Chartered Secretaries and Administrators
IDR Act	=	The Industries Development and Regulation Act 1951
IEC	=	The International Electro Technical Commission
IFRS	=	The International Financial Reporting Standards
IIA	=	The Institute of Internal Auditors
ILO	=	The International Labour Organisation

IMF	=	The International Monetary Fund
IMO	=	The International Maritime Organisation
IMoLIN	=	The International Money Laundering Information Network
INEDs	=	Independent non-executive directors
IOSCO	=	The International Organisation of Securities Commissions
IPRs	=	Intellectual property rights
IT	=	Information technology
ITES	=	Information technology enabled services
KYC	=	The know-your-customer
LCD	=	Lord Chancellor's Department
LLC	=	Limited Liability Corporation
LLP	=	Limited Liability Partnership
Ltd	=	Limited
M&A	=	Mergers and acquisitions
MIT	=	Ministry of Information Technology
MOF	=	The Ministry of Finance
MTRP Act	=	The Monopolies and Restrictive Trade Practices Act 1969
NASSCOM	=	National Association of Software and Service Companies
NBCC	=	The National Building Construction Corporation Limited
NEDs	=	Non executive directors
NGOs	=	Non governmental organisations
NIC	=	The National Informatics Center
NPCSC	=	The National People's Congress Standing Committee
NSDL	=	The National Securities Depository Limited
OCC	=	The Office of the Comptroller of the Currency
OEM	=	Original equipment manufacturers
OECD	=	The Organisation for Economic Cooperation and Development
OFR	=	Operating and financial review
OSHA	=	Occupational Safety and Health Administration
PCAOB	=	Public Company Accounting Oversight Board
PLC	=	Public Limited Company
PRC	=	The People's Republic of China

PwC	=	Pricewaterhouse Coopers LLP
QLLC	=	Qualified Legal Compliance Committee
R&D	=	Research and development
Sch	=	Schedule
SCEEMAS	=	The Small Company Environmental and Energy Management Assistance Scheme
SCRA	=	The Securities Contracts (Regulation) Act 1956
SEBI	=	The Securities and Exchange Board of India
SEBI Act	=	The Securities and Exchange Board of India Act 1992
SEC	=	The Securities and Exchange Commission
SETC	=	The State Economic and Trade Committee
SEED	=	Social, ethical and environmental disclosure
SHSE	=	The Shanghai Stock Exchange
SERM	=	Safety/Social and Environmental/Ethical Risk Management
SFO	=	Securities and Futures Ordinance
SICA	=	The Sick Industrial Companies (Special Provisions) Act 1985
SIR	=	The standard information return
SMEs	=	Small and medium-sized businesses
SORP	=	The Statement of Recommended Practice for Charities
SOX	=	The Sarbanes-Oxley Act of 2002
SRI	=	Sustainable and responsible investment
SRI	=	Statutory rate of interest
STR	=	Suspicious transaction report
SU	=	The Strategy Unit
SZSE	=	The Shenzhen Stock Exchange
UEAPME	=	The European representative body for small business
UK	=	The United Kingdom
UKELA	=	The United Kingdom Environmental Lawyers Association
UNDP	=	United Nations Development Programme
UNEP	=	The United Nations Environment Programme
US	=	The United States
USA	=	The United States of America
UTI	=	Unit Trust of India

VAT	=	Value Added Tax
WEEE	=	Waste Electrical and Electronic Equipment
WICEM 2	=	Second World Industry Conference on Environmental Management
WTO	=	The World Trade Organization
WWF-UK	=	The World Wide Fund for Nature or World Wildlife Fund UK

Chapter 1 Introduction and Traditional Due Diligence

Introduction 1.1

In today's business world the concepts of due diligence and corporate governance are of increasing importance. Both concepts have broadened as regards their scope and meaning. Indeed their application has also come to overlap as a result of the regulatory and voluntary frameworks that are emerging globally. From purely economic roots they have come to encompass many aspects of corporate behaviour. Moreover, in view of the corporate scandals that continue to attract media headlines and demonstrate the need for improved corporate governance, all organisations – regardless of their size or location – should regard these issues as paramount. An understanding and respect for due diligence and corporate governance make absolute business sense.

Methodology 1.2

An overview of the traditional approach to legal due diligence is dealt with in this chapter. This handbook takes the approach that the due diligence process is on ongoing exercise. It extends to areas of business activity that go well beyond the transaction/deal with which it is usually associated to embrace many aspects of business operations and performance. However, this chapter will focus in some detail on giving an account of the processes, context and typical aspects of due diligence in its traditional mould, ie that relating to corporate transactions and particularly merger and acquisition activity. As such, this chapter gives an overview of the legal issues and concepts relating to due diligence in corporate transaction, the process, documentation and some information regarding typical stumbling blocks and tactics. The chapter also begins to flag the route to a view of due diligence through a corporate governance framework (which will be seen to be necessary, given current trends). It concludes with three helpful appendices contrasting typical due diligence steps and a specimen auction or tender process letter.

Definition of due diligence 1.3

Traditionally due diligence has involved a process of discovery that is relevant in key business transactions, as well as operational activities. As is seen in more detail below, due diligence has become the norm in decision-making as regards:

- joint ventures;
- mergers and acquisitions;
- selecting appropriate partners;
- choosing the right jurisdiction or location; and
- buying and selling assets.

In the context of such transactions, highly defined methodologies have evolved, some of which may be found in the sample checklists in the **APPENDIX 1** to this chapter and the Appendix to **CHAPTER 2**. Historically the process was rather drawn out since it involved the physical examination of extensive documentation on site. More recently the development of technology has brought about effective due diligence that can be much more timely and pressured, involving both Internet and Intranet capability (see **CHAPTER 9**).

Any search of a law dictionary reveals the meaning of 'due' as payable or immediately enforceable. 'Diligence' involves care, attention and application. In Scottish law 'diligence' also means proceeding for payment. Together, therefore, the words have an interesting emphasis on the enforceability of the process. Having regard to the purpose of due diligence and the evolution of the term, it is no surprise that the legal process has become increasingly comprehensive as regulatory and other business frameworks have developed.

It is essential, however, to set some parameters around the terms that we want to review and analyse. For due diligence, there is an endless variety of related words in the dictionary that we can apply. For the present purpose it is not very useful to set out all of the available definitions, nor to suggest all the legal terms that appertain to due diligence. It is more helpful to propose a few meanings that support the premise in this chapter. For example Charles Bacon, CEO of Due.Com Group of Companies (a US group providing decision making software to support due diligence), defined traditional due diligence as:

> '...mainly a legal and financial course of action, first designed to avoid litigation and risk, second to determine the value, price and risk of a transaction, and third to confirm various facts, data and representations'.

A variation on that definition is to:

> 'Assist management to justify the price of a merger, acquisition, alliance or joint venture by verifying, validating and analysing available data.'
>
> (*Charles Bacon, CEO, Due.Com*)

Due diligence activities, as with everything, needs a starting point. If there are talks about a possible merger, each party to the transaction must be willing to commence a due diligence activity. However, this is where the definition of the activity of due diligence can become blurred. In a merger situation, there would normally be a significant amount of due diligence prior to any informal or formal conversation. The diligence required is to determine whether there is enough information that can lead to conversations about a possible merger. Thus the starting point for any due diligence activity is never one single step with a single starting point. Consequently, the actions surrounding due diligence must be adaptable within a framework that places the organisation and its owners, employees and advisers in a constant state of data collection and data organisation that can support whatever process is being started.

Therefore, a third definition of due diligence can be to:

> 'Provide a framework within which organisations can continuously confirm that their actions and transactions are supported by the policies, procedures, and management decision-making methodologies.'
>
> (*Charles Bacon, CEO, Due.Com*)

All companies will perform due diligence in some manner – informally or formally. Of course, larger organisations require a more structured formal approach. Too many people doing everything their own way can create an abundance of data with no information. On the other hand, smaller organisations may do everything informally with no notes, no documentation and a lot of ad hoc decision-making. Within each company, there needs to be an understanding of how due diligence works for the common good. Classically this is the role of management as they set the policies, procedures and culture of the company and how it conducts its business.

US origins 1.4

In the context of commercial transactions the term 'due diligence' evidently originated in the US in *section 11(b)(3)* of the *Securities Act of 1933*. This section provided a defence of due diligence to those who had made reasonable investigation into matters contained in a prospectus for the issue of securities. This process of evaluation in the US has been termed 'due diligence' since then. Moreover, its scope has been extended internationally beyond investigations into the accuracy of prospectuses to include:

- any investigation into the acquisition of a company or assets in a commercial context;

- risk analysis in financing; and
- general pre-contractual enquiries.

For further discussion of commercial due diligence in selected sectors see **CHAPTER 6.**

The clear majority of traditional due diligence projects are directed and performed by legal professionals and secondarily by tax and audit financial professionals. It is widely recognised that the legal professional's primary obligation in traditional due diligence concerns the prospective liabilities, and the financial professional's primary obligation concerns the financial data integrity. In the US there is a frequently used phrase 'ledgers and liability' that clearly shows this emphasis. Indeed, even in the US, which has been the prime country as regards due diligence developments, due diligence is not yet a focus that the educational community has recognised. As a separate discipline, due diligence is not taught in law schools. Within the business education community, due diligence is only covered within the accounting world, typically integrated as an audit topic.

The practice of due diligence has evolved over decades by tradition in the US. Practitioners comment that not only is it customary to never question whether the foundations of traditional due diligence are sound, but it is a widespread practice to look no further into a prospective acquisition or merger beyond the mere basics.

In the US business community, when a transaction is underway and as the target or the candidate becomes more impatient as the due diligence process gets longer, the reluctance to push for important information also rises. Subsequently the due diligence process is curtailed. As the time lengthens and the costs rise, it becomes easier to justify minimising or even ignoring anything that is more difficult. The reality is that many, if not most, firms suffer from mild to extreme reluctance to consult more qualified outside expertise. Finally, all too often, the deal is done before any real due diligence, traditional or otherwise, takes place.

The application of due diligence 1.5

Different meanings of due diligence can apply in different situations:

- a buyer or seller of a company or business or assets – the investigation of the assets and liabilities of a company or business for the purposes of buying or selling its assets;
- a lender providing finance – the assessment of the viability of the project and the status of the borrower; this will often involve the banker's lawyers checking the due diligence undertaken by the borrower's lawyers where the borrower is acquiring another company;

- a potential joint venture partner – the investigation of the assets being transferred to the joint venture vehicle and potential joint venture partners;
- the state enterprise for the purposes of privatisation;
- a company listing on a recognised Stock Exchange – the process of verifying when preparing the listing prospectus; and entering into a contract – an analysis of the ability of the other party, or parties, to the contract to perform; or
- a state body undergoing privatisation – the investigation of its assets and liabilities.

Essentially in all such situations there will be:

- the transfer of assets from one party to another or the creation of obligations;
- the existence of risks that may affect the future value of such assets or obligations; and
- the need to apportion the risks between the parties.

Therefore, due diligence is applicable in varying extents or degrees whenever there are asset swap transactions and the creation of obligations. In view of the many circumstances that may arise it is important to note that the same level of due diligence cannot be applied uniformly. Even where the due diligence process is required for a corporate sale or business the documentation should be applied selectively. It is self evident that the detail, scope and intensity of the process will have to be adapted according to the size, value and significance of the transaction, as well as having regard to the human resources available to each of the parties to the transaction. For instance, in a deal that involves a company purchase run by the chief executive of the purchaser, the finance director of the vendor's group of companies and one partner from each of the purchaser's and vendor's lawyers and accountants the organisation of dedicated due diligence teams and data centres will not be required.

Legal due diligence

The purpose 1.6

The purpose of legal due diligence by a purchaser (buyer) or party entering a joint venture is to ensure that:

- the assets have the value that the vendor (seller) has given them;
- the vendor has good title to those assets free from encumbrances, including intellectual property and – in particular – the key assets that are being acquired;

- there are no risks, liabilities or commitments that reduce the value or use of the assets, for example another party having the right to use them;

- there are no other existing or potential liabilities that may adversely affect the object of the due diligence (the target or candidate).

As a priority, therefore, the purpose of the legal due diligence relates to the verification of the legal affairs and good standing of the target, which, in turn, impact on or verify the consideration being given.

Topics usually covered by traditional due diligence	
Topic	**Examples**
Assets	Primarily, assets are considered tangible property, such as buildings, computers, furniture, etc. However, other important assets include people, contractors, business ideas, and product relevance in the marketplace.
Contracts	Contracts for work to be done and commitments by others to do work for the company. The contract can be with individuals or companies. Keep in mind that it is not just the contract terms but whether the terms are in fact enforceable. A lot of employment contracts have appropriate terms, but if the individual has a serious accident and is incapacitated, none of the work related terms may be enforceable.
Customers	Customers for products and services are important elements – who they are and where they are. When reviewing this topic, consider whether there is a secondary market for the resale of products such as through Amazon or eBay. Customer support may start to come from locations not anticipated.
Employee agreements	This requires appropriate legal support to make sure that the agreement is not so restrictive that the employee could easily break the agreement for it being unfair, etc. These agreements may also require consistency which is a process that due diligence can support.

Topics usually covered by traditional due diligence	
Topic	**Examples**
Employee benefits	This is not just about health insurance. Due diligence requires the comparison of planned benefits with the benefits that are actually received.
Environmental issues	These can form a significant part of any due diligence activity. Environmental impact statements have to be considered a never-ending part of business operations as well as the business planning. Regulators from government agencies as well as non-governmental organised groups can delay or prevent a specific development project (see also **CHAPTER 16** for further discussion of environmental issues).
Facilities, plant and equipment	Classically, this item is included within the asset category. It is separated here to indicate the requirements for a continuing due diligence for the potential retirement or sale of any old facility that is no longer effectively supporting the enterprise business. Examples of this can be old buildings. In the US recently many municipalities have torn down old sports stadiums to construct new ones with 21st century features like adequate bathrooms and enough executive suites.
Financial condition	Traditionally this is the province of the accountant. It has expanded to recognise the confluence of cash availability, debt limitations and restrictions, the industry's economic climate, the country's economic climate and the global economy. All of these components can be monitored on a continuing basis as part of the overall financial review of the business.

Topics usually covered by traditional due diligence	
Topic	**Examples**
Foreign operations and activities	Globalisation is the major element of 21st century business. Outsourcing, multiple worldwide locations, different business and governmental regulations, currency conversions, transportation issues, employees and, cultural differences all add up to substantial impact on company operations.
Legal factors	Legal issues from country to country, state to state, municipality to municipality all have to be considered, and monitored.
Product issues	Product life cycles need to focus on old products, products about to be launched and products in the development pipeline. Moreover, due diligence includes the need to monitor competitor's products. It is a growing issue considering the expanding global economy.
Supplier issues	Companies are segmenting the manufacture and delivery of products and services. Some companies want to control all aspects of manufacture. As noted, trends in today's economic climate show that more companies are outsourcing parts of all of the development cycle. Due diligence needs to include the viability of the supplier's ability to deliver on time, on budget and within the established quality parameters. If the supplier declares bankruptcy there may be significant issues impacting the completion of company products as well as the financial impact of not receiving value for payments already made.

Topics usually covered by traditional due diligence	
Topic	**Examples**
Tax issues	Tax increases, tax decreases and taxing authorities all need to be monitored. Due diligence needs to include the potential liability of taxes. On the other side of this coin is the potential for the impact of economic loss on the tax liability. In some cases, the liability may be greatly reduced and/or turn to a cash refund. In this event, due diligence needs to make sure that this is an accurate calculation and then consider what to do with the returned funds.

These topics are not the only categories of information that an appropriate due diligence exercise will want to assess and analyse. However they do demonstrate the range of issues that should be addressed in order to fulfil the purpose of due diligence. When establishing due diligence activities, it is essential that the assessments have to expand beyond the most basic levels. This is true wherever the exercise is taking place (for an example of a simple due diligence exercise in India see **APPENDIX 1**).

From a 21st century perspective, the traditional due diligence methods account for ten to 25 per cent of a complete due diligence process, especially in light of the fact that approximately two thirds of all mergers and acquisitions fail completely, or fail to deliver the value expected. Further comment on the level of failure is made below (see also **CHAPTERS 2** and **8**). For the present purpose it should be noted that not all failures are due to the lack of data gathering and appropriate due diligence procedures. Quite often, it is the process of due diligence that reveals appropriate information that this merger is not a good deal for one or both parties. For example the cultural due diligence exercise can assist toward more successful outcomes (see **CHAPTER 8**).

With enormous emphasis placed on the short term by most firms, especially those in the public markets, a quick increase in price or earnings for the stock market is in many cases the sole reason for a merger and acquisition transaction. Easy revenue increases or cost reductions often make management the 'hero', albeit almost always at the cost of the stakeholders in the long term. Having regard to the purpose of due diligence, in most places traditional business views are based on an historical perspective. Examples of this type of view are:

- what were the level of earnings?;
- what did the CEO do?;

- how were sales made in the last quarter?; and
- more analysis of completed transactions.

Following this model, traditional due diligence looks at the past, not the future, to assess what a company can accomplish from today onwards. This analytical method is often performed too quickly and with too narrow a focus on completed transactions that may not be an appropriate predictor of future behaviour. Clearly there are now more modern tools and resources available that can support a more thorough analytical process.

The process 1.7

In most jurisdictions the typical traditional due diligence process usually looks at the target or candidate company, its financial performance, products, market and employees – usually in that order of priority. The typical traditional process starts with a legal questionnaire and disclosure documents attested by the candidate, and is coupled with a review, compilation, or audit of financial records. Generally, a regulatory agency records search is performed. Most often various public records are searched. Often, research is added in areas such as the industry niche(s) of the candidate, and sometimes the media. Also, additional research is sometimes added by contacting various industry and government organisations. In the course of this process the use of warranties and indemnities has developed to facilitate the transaction. Although, as is discussed further below, the vendor may provide warranties that provide assurances on the above; the purchaser should check them in any event. The verification approach reduces the potential for conflict because problems can then be identified at an early stage. It can happen, for example, that the vendor is unaware of the problems or issues that emerge through the due diligence process. It is important to note that warranties and indemnities given by vendors often are qualified in certain respects as follows:

- generally they are subject to time limits and last for only a few years, by operation of law and contract;
- usually they are limited by amount, including a maximum liability under the warranties and an aggregate level of claims;
- the compensation can be incomplete either because of the inadmissibility of the claim or the restrictive method of calculating the damages, as well as the fact that sometimes it is difficult to evaluate the compensation accurately as in the case of loss of brand reputation (which is discussed in detail in **CHAPTER 7**);
- normally there is a de minimis limit in respect of individual claims;
- any claim will only be successful if it is conducted in accordance with the requirements for the conduct of claims;
- claims can be disputed, thereby they can be expensive and drain resources;

- even in the absence of a defence it can be costly and time consuming – as well distracting – to claim; and
- the terms of the warranties and indemnities can be difficult to enforce.

The repercussions of entering litigation, as well as the available alternatives, are set out in **CHAPTER 5** when considering some organisational issues. In the light of the above, in many developed economies the due diligence process requires the vendor to disclose all relevant information. Such disclosure enables the purchaser to evaluate the target or candidate properly and to negotiate from the perspective of a level playing field. Therefore the process of legal due diligence can be used to provide information on the legal affairs of the target. Such information can, of course, assist in the decision whether or not to purchase and can check:

- the assumptions made by the purchaser or agreed with the vendor, for instance the rate of cancellation of contracts by customers;
- the valuation of the target;
- the operational capability of the target; and
- the identification of any adverse factors.

Once the purchaser has such information the process enables options that include:

- proceeding with the transaction as agreed;
- cancelling or 'walking away' from the transaction;
- negotiating specific indemnities; or
- changing the terms of the transaction.

Another important benefit of the legal due diligence process is that the information reveals how the target has been managed. This is very relevant to the overall discussion in the handbook that relates to corporate governance and the broader understanding of due diligence as an ongoing internal tool. It can take account of the background to the target and candidate and its objectives, including its chosen structure in the form of a company, partnership or owner manager operation. While many due diligence exercises involve very large transactions there are also many smaller deals that attract the due diligence process. The chosen vehicle and the alternatives available are considered in **CHAPTER 4**. While some of the issues in the process are more suited to the larger transaction, others are equally applicable whatever its size. For example, late or inaccurate returns to the authorities, such as the Inland Revenue and corporate registries, will reflect upon the management of the business. They may also indicate financial difficulties, such as in financial statements lodged late with the corporate registries. Moreover, as suggested, once the purchase is completed the due diligence information can provide an invaluable tool in the ongoing management of the target. Some of the management aspects of a business are considered more fully throughout the handbook.

Areas of legal due diligence 1.8

Key topics that are usually covered in the course of the due diligence process to fulfil its purpose have been indicated above. Moreover **CHAPTER 2** sets out the relevant areas in the context of corporate finance. Many competent due diligence professionals have begun to use a more comprehensive due diligence process. They have established a broad range of topics that have to be monitored and reviewed on a continuous basis that can also be useful in the overall context of corporate governance. As can also be seen form the discussion in **CHAPTER 9,** 21st century technology is the tool that enables this kind of continuous or ongoing due diligence. These areas include, but are not limited to, the following list:

- accounting issues;
- behaviour;
- business organisation streamlining;
- business planning;
- business processes streamlining;
- change;
- competitive analysis;
- culture;
- customer defections;
- distribution channels;
- employee retention;
- environmental issues;
- executive retention;
- global business operations;
- global partnering;
- human resources;
- information systems integration and compatibility;
- intellectual capital;
- intellectual property;
- internal auditing;
- interviewing customers;
- interviewing former employees;
- joint venture partner and other alliance reviews;
- legal contract reviews;
- litigation or claims reviews;

- logistics costs;
- manufacturing;
- operations;
- post-deal planning and integration;
- potential market growth;
- potential revenue growth;
- problem identification;
- product distribution;
- product portfolio expansion;
- quality assurance;
- quality control;
- research and development expansion;
- revenue losses;
- risk management;
- security matters;
- strategic planning;
- strategic questions;
- suppliers;
- supply chain;
- tax issues;
- technology – internal/external;
- technology – Internet//Intranet/Extranet;
- technology planning;
- vendors; and
- warehousing.

It has been commented that the traditional due diligence process is largely driven from a legal perspective. The legal practitioner typically delivers numerous documents with the goal of being as legally reliable as possible. The legal professional specifically builds a careful legal paper trail, and gathers as much detail as possible regarding the legal condition of the candidate firm. With the benefit of technology, the point should once again be made that 21st century due diligence should enable an improved approach through ongoing information gathering that keeps the business abreast of its status in terms of corporate governance.

In the UK, the typical legal due diligence exercise will cover some or all of the areas listed. As has been noted in this handbook the selected checklists can only exemplify some of the general concerns, and provide the basic concerns that should be amended to reflect the individual circumstances of the transaction. Therefore, the list below does not purport to be conclusive and will require tailoring according to such factors as:

- whether the transaction is a share purchase or asset purchase;
- the target's industrial sector;
- the geographic location of its activities; and
- the size of the transaction.

The principal selected headings of traditional due diligence are:

- corporate structure;
- company secretarial;
- corporate acquisitions and disposals;
- compliance programmes;
- trading activities;
- competition law;
- personnel;
- health and safety liabilities;
- pension schemes;
- land and buildings;
- environmental management;
- plant, equipment and other fixed assets;
- computer software;
- intellectual property;
- investments;
- lending to third parties;
- banking facilities/borrowing from third parties/financial grants;
- guarantees/indemnities/letters of credit;
- product liability;
- investigations, litigation, disputes;
- insurance;
- taxation;
- non-compliance with agreements/change of control;
- voidable transactions/reconstructions;
- impending legislative changes;

- compliance with special industry sector legislation; and
- the effect of the Euro on contracts, including payment arrangements.

Transaction documentation

Types of documents **1.9**

In a transaction, such as the sale of a business, the materials and main documents will typically include some of the following documentation:

(*a*) pre-exchange – including:

- due diligence preliminary enquiries, further enquiries and the seller's responses to them;
- heads of agreement;
- due diligence terms of reference – engagement letter – accountants;
- instructions for accountants' short and long form report;
- due diligence terms of reference – engagement letter – lawyers;
- environmental audit – engagement letter;
- exclusivity agreement;
- confidentiality agreement;
- data room letter;
- due diligence rules of engagement;
- funding comfort letter;
- environmental comfort letter;
- due diligence reports;
- report on title; and
- legal opinion of foreign lawyers.

(*b*) exchange of contracts – including:

- disclosure letter;
- sale and purchase agreement;
- guarantees;
- tax indemnity;
- service agreements;
- intellectual property assignments and licences;
- real property transfers;
- assignment of contracts; and
- directors' meeting minutes.

(*c*) The closing or completion of the transaction – including:

- shareholders' resolutions and circular;
- announcements;
- letters to customers;
- stamp duty and company registry forms;
- governmental consents;
- release of charges and guarantees; and
- closing agenda.

Ongoing action for due diligence teams regarding transaction documentation 1.10

Throughout the transaction, the legal due diligence team leader for both the seller and the buyer needs to peruse any transaction documentation which might affect the due diligence exercise. Each draft of the documentation should be checked for any amendments made affecting the due diligence. One document that the legal due diligence teams on both sides are usually responsible for is the disclosure letter. As will be seen from the discussion below and the appendices to this chapter, following the initial meeting with the client, the buyer or seller and its lawyers will have a number of tasks to undertake, including:

- selecting and instructing other advisers;
- the preparation of legal due diligence enquiries;
- agreeing terms of engagement and reference with them and any in-house due diligence teams;
- planning the campaign;
- agreeing various preliminary documents with the other side, including:
  - heads of agreement;
  - a confidentiality agreement;
  - a lock-out agreement; and
  - rules of engagement;
- project and data management; and
- starting the data acquisition/disclosure process.

Typical transaction procedures 1.11

This chapter has already commented on the usual due diligence process. However, the due diligence process may vary according to the following situations:

- sale by auction; or
- sale by treaty.

In the case of both a sale by auction and a sale by treaty, a number of preliminary steps may be taken:

(*a*) the seller will obtain a valuation of the target and possibly instruct a corporate financier to find a buyer;

(*b*) the seller will:

  (i) instruct its corporate financier (if any), accountants and lawyers to do some analysis of the target to identify any major issues to be addressed (such as subsidiaries to be split off);

  (ii) analyse the taxation ramifications of the sale for the seller; and

  (iii) begin to prepare or 'groom' the target for sale;

(*c*) the seller may prepare an information memorandum regarding the target for potential interested parties;

(*d*) the seller will require any interested parties to execute a confidentiality agreement before issuing the information memorandum to them;

(*e*) the buyer may undertake some basic due diligence into markets, political risk, compatibility of organisational cultures;

(*f*) the buyer may also involve its accountants in some preliminary analysis of the seller's financial accounts.

Sale by auction or tender **1.12**

It is evident that the seller can improve the terms of its sale by creating a competitive environment where a number of bidders are given access to the due diligence data and make bids for the target. In many cases the sale is not strictly an auction in that the seller is not obliged to accept the highest bid. The typical sequence of events in a sale by auction is as follows:

- with the issue of the information memorandum, the seller may request that bidders respond by a specified date with an indication of the price to be offered and the assets desired;
- the seller is likely to issue a data room agreement to each bidder;
- each bidder will be allotted a certain amount of access to the data room and possibly to target personnel for additional information;
- the bidders may submit requisitions for further information which the seller may respond to;
- the buyer's in-house due diligence team, if any, prepares their due diligence reports;

- the buyer's accountants prepare their draft due diligence report. Often, this is sent to the seller for comment on any inaccuracies;
- the buyer's lawyers may provide a due diligence report, which is not usually provided to the seller;
- the bidders submit their bids;
- the seller will select one or more bidders with whom to continue negotiations unless a bidder persuades the seller to enter into an exclusivity agreement whereby the seller agrees not to negotiate with any other party for a period of time or not to conclude a sale with any other party;
- the buyer may negotiate basic heads of terms with the seller;
- the buyer may be permitted additional time to undertake due diligence;
- the negotiations concerning the warranties, which have been ongoing for some time, are finalised close to exchange of the sale and purchase agreement; and
- the seller's lawyers produce a draft disclosure letter, exempting facts and documents from the warranties.

It should be appreciated that this is a basic structure for the process of conducting a sale by auction or tender. An example of a sales process letter is also found in **APPENDIX 3**. All of these should, of course, be amended to reflect the individual circumstances of the transaction. Similar comment applies to the discussion of sale by treaty below in **1.13**.

Sale by treaty **1.13**

The typical sequence of events in a sale by treaty is as follows:

- the buyer may negotiate basic heads of terms with the seller;
- the buyer may insist that the seller enters into an exclusivity agreement whereby the seller agrees not to negotiate with any other party for a period of time or not to conclude a sale with any other party;
- the seller nominates either a member of its own staff or of the target to handle the due diligence enquiries from the buyer and its professional teams;
- the buyer sends its in-house due diligence team to the target's offices where it is usually given a room or a data room is set up by the buyer (where the seller wishes to keep the buyer away from its offices). Alternatively, the data may all be sent to the buyer to analyse at their own offices;
- the buyer instructs its accountants to commence due diligence. They will also usually be based at the target's offices;
- the buyer and seller instruct their lawyers;

- in addition to the commencement of preparation of documentation, the buyer's lawyers forward a set of preliminary enquiries to the seller's lawyers. Normally, the seller will warrant the accuracy and completeness of the written responses;
- the buyer's lawyers are rarely based at the target's offices;
- at the same time as the due diligence teams commence work, the parties and their lawyers start to negotiate the various agreements. These activities will continue in parallel during the due diligence exercise, with the agreements and their terms being amended to reflect the results of the due diligence exercise;
- the seller's lawyers pass the preliminary enquiries to the seller's nominee for handling enquiries, who will arrange for the collation of requested documentation and for answers to the buyer's questions;
- the seller's lawyers will vet, filter and qualify the representative's responses and, where appropriate, restate them in their own terminology before sending them to the buyer's lawyers;
- the documents the seller has collected will usually be indexed and sorted into separate sets of folders known as the 'disclosure bundle'. The written responses will refer to the appropriate documents by their index number in the disclosure bundle;
- the seller's lawyers prepare further enquiries from time to time, based on the answers to the preliminary enquiries and earlier further enquiries and results of its independent information collection activities;
- omissions in the disclosure bundle and any gaps in the written responses by the seller's lawyers are made good during the negotiations with additional written responses and deliveries of documents;
- the buyer's in-house due diligence team, if any, prepares their due diligence reports;
- the buyer's accountants prepared their draft due diligence report. Often, this is sent to the seller for comment on any inaccuracies;
- the buyer's lawyers may provide a due diligence report, which is not usually provided to the seller;
- the negotiations concerning the warranties, which have been ongoing for some time, are finalised close to exchange of the sale and purchase agreement; and
- the seller's lawyers produce a draft disclosure letter, exempting facts and documents from the warranties.

The formal and informal due diligence processes **1.14**

When considering the typical process concerning non-data room due diligence exercises, it can be seen that there are two due diligence processes underway. These are:

- the formal process, involving the buyer's lawyers; and
- the informal process being undertaken by the buyer, the accountants, the merchant bank (if any) and any other advisers.

Often, the facts and data that the seller warrants the accuracy and completeness of are those supplied under the formal process, that is the facts and data that are:

- provided in the answers to the preliminary enquiries and further enquiries posed by the seller's lawyers; and
- stated in the warranties.

The formal process tends to exclude most data that does not refer to the legal affairs of the target or candidate, thereby excluding much significant information obtained by the buyer, its accountants and other advisers.

Disclosure – the traditional approach 1.15

The traditional approach to handling the legal affairs of an acquisition target has often been to place primary reliance on the warranties provided by the seller. Some analysis of documentation and data on the target's legal affairs is made, but this tends to be largely confined to the documentation and data disclosed by the seller against the warranties that they have given. An example would be a warranty with which there is no outstanding litigation and a disclosure made against this in the disclosure letter, which outlines existing litigation and attaches relevant papers.

Disclosure is carried out by the seller in response to:

- the preliminary enquiries raised by the buyer's lawyers;
- the obligations on the seller to provide details of the warranted items contained in the warranties in the sale and purchase agreement; and
- the benefit offered to the seller by disclosing information known as the 'excepted items' against the warranties in order to dilute the warranties.

The disclosure process often does not commence until the seller and buyer have entered into an agreement, in principle, or in the case of a controlled auction, until a non-binding bid is made. It should be noted that there is a certain amount of overlap between disclosure and due diligence which causes some confusion when referring to the respective exercises.

As a result, due diligence investigations into the legal affairs of targets or candidates are undertaken to varying degrees. Some buyers' lawyers do not undertake extensive investigations in advance of receiving the sellers' disclosures. This can mean that the extent of the buyer's knowledge of the legal affairs will depend on the extent of the 'warranted items' negotiated into the warranties. A problem with this approach is that it is always possible that the seller will negotiate down on the warranties and remove the reference to warranted items, so that very little information is actually provided to the buyer. Another disadvantage of relying on the disclosure exercise for information is that the seller will often only turn its mind to the process of collecting the disclosure documents once the negotiation of the warranties is complete, immediately prior to exchange. This will leave little time for quality data to be collected, let alone adequate time for the buyer's due diligence team to review and report on the data.

Due diligence and the seller's duty to disclose **1.16**

The obligations on the seller to disclose information are discussed here. As noted, this discussion principally relates to the English transaction.

The classical common law position is that there is no duty on the parties to disclose material facts to the other party. The historic approach of English contract law has been to allow each party to look after their own interests. This was the 'apotheosis of nineteenth-century individualism' and a result of the application of the doctrine of 'laissez faire'. Accordingly, the maxims *caveat emptor* (let the buyer beware) and *caveat venditor* (let the seller beware) are the starting point for consideration of this issue. Over the years the traditional common law position has been gradually eroded and a variety of exceptions to the general non-disclosure rule evolved. These include:

- fiduciary relationships, including *uberrimae fides* (of the utmost good faith);
- the duty of disclosure on sellers in real property transactions; and
- the obligation on the seller and the buyer not to make misrepresentations.

Fiduciary relationships **1.17**

These are relationships between persons that involve trust and confidence. Typical examples are those between solicitor and client, trustee and beneficiaries. In all such relationships there is a duty of disclosure of all material facts. This also applies to relationships that are *uberrimae fides*, that is of the utmost good faith, such as between insurer and insured. A failure to comply with the duty of good faith in an insurance situation will lead to the insurer being able to treat the contract of insurance as voidable. The obligation of utmost good faith can also be imposed contractually by a buyer and seller in a corporate transaction. The parties can expressly contract to create a duty of utmost good faith and thereby place a duty of disclosure of all material facts.

Real property transactions **1.18**

In the UK, particular rules have been developed by the courts in relation to transactions regarding land and buildings that is real property. This real property framework has been based on the fact that, traditionally, the nature and state of the seller's title could only be determined by the buyer questioning the seller. However this duty has been weakened to some extent by the regime for registering land which was introduced by the *Law of Property Act 1925*. As a result the seller is under an obligation to disclose to the buyer any latent defects in the title to the property which will not be removed before the completion of the transaction, or closing. According to English law a defect may be described as latent if:

- it cannot be discovered by the exercise of reasonable care on the inspection of the property; or
- it amounts to a deficiency in the seller's documentary title which might affect his or her ownership of the property or his or her right to deal with it.

It has been suggested that there is no reason why the common law duty on the seller to disclose any latent defects in title should not apply in an asset purchase. This is because the property conveyance is occurring in much the same way as in any other real property transaction. However, the question is less clear and therefore more arguable in the case of a share acquisition as the transfer of the real property occurs as a corollary of the share purchase.

Matters subject to the duty of disclosure 1.19

The seller is under an obligation to disclose such information as that referred to here following, including as was decided by the Court of Appeal as long ago as 1899 in the case *Re Brewer and Hankin's Contract (1899) 80 LT 127* where they were outside his knowledge:

- type and tenure of title;
- leases and tenancies;
- matters registered at the Central Land Charges Department;
- local land charges;
- restrictive covenants;
- easements; and
- exceptions and reservations.

Due to the heavy burden of common law disclosure on the seller, the parties may agree to vary the duty. This is done as a matter of course in English property transactions.

Remedies 1.20

The remedies for breach of the duty of disclosure by the seller in the case of real property transactions fall into three categories flowing from the type of breach.

1. Non-disclosure

   Where the non-disclosure relates to a substantial defect, the buyer will be entitled to rescind the contract before the completion of the transaction, or closing. They may alternatively seek a reduction in the purchase price.

2. Misdescription

   The failure to disclose a matter in relation to the type or tenure of the seller's title or a misleading physical description of the property in the particulars of sale leads to liability in misdescription rather than non-disclosure. The remedies for misdescription are similar to those for non-disclosure.

3. Misrepresentation

   This is a vast area of discussion that has led to legislation and extensive case law that have been the subject of textbooks. Therefore it should be appreciated that the discussion of the law of misrepresentation that is set out below is only intended to cover the position in the broadest sense in order to provide some guidance to personnel involved in the due diligence exercise.

Misrepresentation – an outline **1.21**

There are two general circumstances in which the seller might make a misrepresentation to the buyer:

- by way of an incorrect statement (often a warranty) contained in the sale and purchase agreement or a statement made in the disclosure letter (including the disclosure bundle); and
- by way of an incorrect statement outside the formal due diligence or transactional process, such as a written statement in documentation not contained in the disclosure bundle or an oral statement.

The seller should monitor and vet all data disclosed to the buyer throughout the transaction, both at the time it is made and as at the date of exchange, to ensure it is still accurate. The personnel of the seller should also take care to avoid making statements outside the sale and purchase agreement or disclosure letter/disclosure bundle, whether orally or in writing, which might later be used by the buyer to rescind the sale and purchase agreement or claim damages from the seller. Conversely, the personnel acting for the buyer should take steps to accurately record any statements which the seller makes to them, which they rely upon in entering into the agreement.

Generally, agreements relating to the sale and purchase of a business or a joint venture will specifically list those statements that the buyer is relying upon in entering into the agreement. These warranties are usually negotiated and worded with some care. There is usually an 'entire agreement' provision in the agreement, whereby the buyer acknowledges that it has not entered into the agreement in reliance on any representations other than those contained in the agreement or the disclosure letter. However, such a provision will only be effective if the seller can prove that it satisfies the requirement of reasonableness as stated in *section 11(1)* of the *Unfair Contract Terms Act 1977. Section 11(1)* provides the requirement of reasonableness. This requires that the term should have been a fair and reasonable one to be included, having regard to the circumstances which were, or ought reasonably to have been, known to or in the contemplation of the parties when the contract was made.

In a commercial transaction, particularly one where the parties are seasoned business people, the buyer may have difficulty in persuading a court that it was not reasonable to include an entire agreement clause in an agreement which contained an extensive list of warranties. The buyer may also have difficulty in persuading the court to find that he or she relied on the representation in entering into the agreement, when he or she had the opportunity to include this in the warranties but did not do so.

Nevertheless, the courts have held entire agreement clauses in commercial contracts to be unfair. In *Goff v Gauthier (1991) (62 P & CR 388)*, which involved a contract for sale of land, the seller orally represented to the buyer that he would withdraw from negotiations with the buyer unless contracts were exchanged immediately. The court found that the seller had no such intention and his statement was a misrepresentation that induced the buyer to enter into the contract. The buyer was permitted to rescind the contract notwithstanding the entire agreement clause which was held to be unreasonable within the meaning of *section 11(1)*.

Accordingly, circumstances can arise where a misrepresentation might be actionable by a buyer in a business acquisition, avoiding an entire agreement clause. This might particularly apply where the misrepresentation was outside the usual ambit of the warranties, as in the *Goff* case, or where the warranties were brief. Entire agreement clauses may also be held to be unreasonable if they purport to exclude claims based on misrepresentation even where there was fraud. For instance in *Thomas Witter Ltd v TBP Industries Ltd [1996] 2 All ER 573* the court held the clause to be void on the grounds of unreasonableness because it purported to exclude liability for misrepresentation. Therefore, the seller should draft the provision so as to preserve the buyer's rights in the case of fraud.

The courts have also shown that they will construe entire agreement clauses narrowly. Accordingly, sellers are well advised to seek an acknowledgement from the buyer that it does not enter into the agreement in reliance on any other representations.

There are many cases in which statements will be made during the pre-contractual negotiations and will be at risk of change during the course of those negotiations. For example *With v O'Flanagan [1936] Ch 575, [1936] 1 All ER 727*, demonstrated this concern. A dental practice was to be sold and accurate supporting financial information was provided. The negotiation took five months to conclude during which time the seller fell ill and income was lost. These losses were not notified to the buyer who purchased the business in good faith based on the original figures. It was held that, as the buyer had continued to place reliance on the original figures believing them to be true, those figures being uncorrected amounted to a misrepresentation.

Another matter relates to silence. It has long been understood that, by itself, silence will not amount to a misrepresentation (see *Keats v Lord Cadogan (1851) 10 CB 591*). However, if a party is deliberately silent regarding certain information which, if known, would distort the information provided, then there may be an actionable misrepresentation. There is also no reason why the parties to the agreement cannot place themselves under an obligation of disclosure, as long as that obligation was carefully drafted, so that the detail of the disclosure required was unambiguous. In addition, if a party makes a disclosure upon a particular matter which details certain facts but omits others, this may be regarded as a misrepresentation. The test is whether, by only proving certain facts, they would mislead the party placing reliance upon them in the absence of having knowledge of the omitted set of facts.

One of the key sources of information that a buyer will rely on will be the audited financial statements of the target. The question therefore arises whether the buyer has any recourse against the seller's accountants if those accounts prove to be incorrect. Under English law in 1990 the matter was considered and decided in *Caparo Industries plc v Dickman [1990] 1 All ER 568.* The House of Lords unanimously rejected the concept that auditors owed an unlimited duty of care in the preparation of accounts which extended to anyone who used those accounts. Accordingly, the buyer should identify those facts and statements presented in the financial accounts on which it is relying and seek warranties on them from the seller.

There may be other circumstances where the target's accountants may be liable to the buyer and each case should be considered on its facts. The buyer may actually experience some difficulty claiming against the target's auditors even they had prepared draft accounts at the request of the buyer and make certain statements regarding the level of profits (see *McNaughton (James) Papers Group Ltd v Hicks Anderson & Co [1991] 1 All ER 134*). Accordingly, a buyer should ensure that it takes appropriate warranties from the seller to increase its protection and to avoid any defence of contributory negligence by the accountants for not having done so.

Where the buyer relies on a prospectus previously issued by the target, the buyer may have a remedy under *section 150* or *section 166* of the *Financial Services Act 1986.* These sections provide protection for those acquiring shares. It should be noted, however, that the House of Lords' decision in the old reported case of *Peek v Gurney [1861–73] All ER 116, LR 6 HL 377,* supported in *Al-Nakib Investments (Jersey) Ltd v Longcroft [1990] 3 All ER 321,* held that the purchaser of shares must be an original allotee to succeed in a action for misstatement in a prospectus.

Remedies for misrepresentation 1.22

Where the buyer wishes to make a claim that it has been induced to enter into the contract in reliance on a misrepresentation, its remedy will depend on whether the misrepresentation was contained in the warranties or otherwise. The sale and purchase agreement will usually provide for remedies in the event of breach of warranty, which will generally be damages or, rather exceptionally, rescission (setting aside a voidable contract). Alternatively, the buyer may be able to claim rescission or damages on the basis of misrepresentation. Misrepresentations made outside the formal documentation would also be made under this heading.

Criminal sanctions 1.23

A key concern for the seller and its advisers must be to ensure that they do not become exposed to criminal sanctions when dealing with the buyer. This may be as a result of regulation making it an offence to make misleading statements in a sale of shares knowing them to be misleading or reckless as to their truth, for the purpose of inducing any person to enter into an agreement to acquire securities. This will affect the sale of shares but not the sale of assets. In addition, two or more persons may be guilty of the common law offence of conspiracy to defraud if fraudulent statements are made with the intention to deceive. This would apply to both sales of shares and sales of assets.

Contract subject to due diligence 1.24

In the vast majority of commercial transactions to which due diligence is applied, the due diligence exercise is carried out prior to signing and exchange of contracts. However there may be cases, especially where the seller is desperate to dispose of the target or candidate company or business, where contracts are signed and exchanged subject to the buyer being satisfied with the results of a subsequent due diligence exercise. The due diligence is then carried out within specific time limits. Generally such an arrangement is unsatisfactory from the perspective of the seller because it gives the unscrupulous buyer an open-ended opportunity to withdraw from the contract. Moreover, the obligations on the buyer to act in good faith and de minimis provisions are capable of being disregarded. It is often argued that such a deal structure amounts in reality to the grant by the seller to the buyer of an exclusive option to purchase the target during a specific period of time. Therefore, however desperate the seller may be, the disadvantages may well outweigh any advantages.

Legal due diligence – the limitations 1.25

The key to a successful due diligence exercise is the data. It is important to bear in mind that all investigations undergone in the due diligence process depend very much on the quantity and quality of the data supplied by and on behalf of the vendor. The purchaser has to rely on such information in making any decision regarding the target. Therefore the purchaser is very much exposed to potential non-disclosure and misrepresentation by the vendor. This is considered further above (see for example **1.15** and **1.21**) in the context of English law.

Having regard to such limitation in the due diligence process it is important for the purchaser to support the due diligence investigation by obtaining warranties from the vendor on those matters that are uneconomic or impossible for the purchaser to check. It is of course important for the vendor to warrant the completeness and accuracy of the date supplied to the purchaser.

Financial professionals concentrate primarily on comprehending and delivering historical financial reports, and often a valuation of the candidate – traditional financial review focusing on what was, rather than what lies ahead. Over the past decade there has been significant growth in understanding the need to focus on all aspects of future financial issues. This is not only for the repayment of outstanding debt, such as a mortgage; rather it is for the consideration of all current and future financial obligations. This includes knowing what obligations are falling due and when, synchronised with the company's capability to make payment as scheduled. Thus due diligence is both an internal and an external looking effort.

The legal or the financial viewpoints are not focused on the actual validation of the acquisition or the merger deal. Traditional due diligence is essentially an audit of legal and financial aspects of deals. Legal and financial reviewers, lawyers and accountants work within the same context and structure as due diligence is an essential part of any decision process.

Financial due diligence **1.26**

Financial due diligence entails:

- identification of the commercial rationale for the transaction or deal in terms of growth, technology or synergy savings;
- the differentiation between facts, assumptions and projections; and
- the financial facts, assumptions and projections.

The underlying process regarding the facts involves accountancy policies with a degree of central control. The accuracy is affected by the access to auditors, late adjustments and the management accounts and financial accounts. Qualitative issues relate to the budgeting style, performance pressure and internal audit.

As regards the financial audit, the basis for assumptions will often be:

- seller knowledge;
- buyer knowledge;

- target information; and
- speculation.

The basis for projections are the financial facts and assumptions, as well as commercial facts and assumptions. The gap between due diligence conclusions and contract has to be bridged by accounting warranties, completion accounts and commercial warranties. Due diligence conclusions are often a mix of facts, assumptions and projections. It may be said that value judgements with commercial and sustainable views, together with real advice, goes beyond such conclusions in order to realise the due diligence objective and achieve a successful outcome.

Risk and insurance due diligence 1.27

Traditionally insurance brokers or advisers have focused on the target's existing insurance arrangements as the relevant element of risk financing. However as this meant a rather limited exercise, the discipline of risk and insurance due diligence evolved. The more modern approach considers as a priority the target's business, enabling a detailed risk profile to be developed. A clear risk profile is achieved by thorough identification, analysis and evaluation of the risks and exposures of the business, including:

- an assessment of hazard and risk by industry sector and territory;
- an analysis of past, present and future activities and product range;
- an analysis of the loss experience, specifically examining frequency, severity and trends and may include the preparation of a claims survey if sufficient data is available; and
- a review of potential liabilities and relevant outstanding litigation.

An examination and analysis of the target's current organisational and operational structures are conducted in relation to key issues, including:

- the overall risk management philosophy;
- environmental management;
- health and safety programmes;
- crisis management or disaster recovery planning;
- product safety procedures; and
- asset protection procedures and methodologies.

The risk financing and risk retention decisions are also reviewed in order to analyse the target's risk financing arrangements. Current and past risk transfer arrangements are audited. Where relevant, an analysis of the involvement of a captive insurance company is also carried out. The main headings of this review are:

- risk financing methodology;
- critical analysis of insurance policies;
- determination of uninsured or underinsured risks;
- examination of self insured risks; and
- analysis of insurers' financial security.

Through the preparation of a risk profile the buyer can understand the candidate's past and present activities in order to assess its potential exposure to unexpected and unbudgeted risks. The profile will cover any 'forgotten' past activities such as:

- any potential liabilities arising from historic joint ventures and other strategic alliances;
- the past trading activities or product range of dormant companies;
- latent disease liabilities such as asbestos or industrial deafness that could be the subject to future employers liability insurance claims that are unforeseen or unplanned;
- any discontinued product lines that could create a future liability; or
- past exposure of employees or third parties to a variety of substances or situations known to cause harm or damage.

It is important to determine the extent of the target's exposure in comparison with the buyer's current activities or product lines. The priorities are to:

- ensure that all activities or product lines are identified;
- review the additional overseas exposures, including information on overseas subsidiaries and legislation or regulation of the respective jurisdictions;
- identify any dependency on a supplier, customer, assembler or manufacturer, particularly in the era of outsourcing; and
- check whether or not the target undertakes design work or provides advice to third parties for a fee as this could mean a substantial increase in any exposure to negligence actions.

One critical area that the buyer must review is the outstanding litigation against the target. As a matter of course a schedule of outstanding litigation should be analysed in conjunction with the target's insurance arrangements. The problem of litigation and the threat to the successful business is discussed in more detail in **CHAPTER 5**.

The risk and insurance due diligence exercise enables the buyer to understand the level of exposure, as well as the adequacy of available cover to deal with the exposed risks. Indeed the key benefits to the buyer are:

- a clearer understanding of the target's business and its risks;
- the identifying of potential 'deal breakers' that mean the need to review the transaction as a whole;
- the prioritising of problem risk issues and potential solutions;
- a recognition of likely cost implications that can enable a keener purchase price to be negotiated; and
- the focusing of necessary contractual warranties and indemnities.

Beneficiaries of due diligence 1.28

It should be noted that with the ever-increasing pressure from regulators, security exchanges, and stakeholders there are a growing number of beneficiaries of the due diligence process. When the parties are establishing the methodologies for the due diligence tasks it is important that the user of this information is considered. For example in many places, if a government regulator, there are specific forms and formats for data to be presented. It will be very frustrating and more expensive to have to recast the information multiple times just to conform to the regulator's penchant for specificity.

Shareholders, investors, and stakeholders can be satisfied with accurate and timely information but generally more concerned with the overview or bottom line. In fact, most would prefer simpler rather than complex information. They may be making decisions about company compliance with a specific regulation, but they are also concerned with understanding the company's ability to survive and prosper.

It is important to ensure that employees are not forgotten in this process. Clerks, middle managers, management and all other related individuals who receive compensation from the company enjoy hearing about the company. Due diligence can include preparation of reports, without violating the rules of privacy and government regulations.

Transactional and operational concerns – integration value post-merger 1.29

In this era of heightened corporate governance it is important to assess the real value of a transaction once the deal has been done, having regard to the growing number of stakeholders. Despite the many processes that may be in place and the detailed checklists that have evolved in commercial circles, the question is often raised: how hard do firms try to gain value from acquisitions? For instance, one assertion is that the majority of acquisitions and mergers do not deliver shareholder value. Work by the international consultant firm

McKinsey in 1998 concluded that around 60 per cent of mergers fail in financial terms. An earlier study in 1987 highlighted that 58 per cent of acquisitions were later divested due to under-performance. Other commentators have noted that 70 per cent do not deliver intended value. Whatever is the actual figure today there is no doubt that it will also be high. Key questions that have been asked are:

- why is effective acquisition integration, that is the method whereby the paper strategy of combination begins to realise value, so elusive?
- is it that target firms are chosen wrongly in the first place? Or
- is it that doing the deal is seen as the conclusion to the merger rather than the beginning of the process of integration?

For in-house and advisory lawyers, the issue of integration is an important area. It is clear that in actually doing the deal itself, the momentum, pressure to close and limited time to assess all information can result in decisions that will seriously impact the ability to integrate in the longer term. This will range from operational detail through to more organisational concerns such as the loss of tacit, codified corporate knowledge that exists in the heads of key personnel that have left the business. A further example is in information technology (IT) where often only the most basic inventory of physical hardware is undertaken, when programming skills and long-term single-source service deals may in fact be the critical factors to consider.

In all these areas, acquirers need advice and assurance that they will have the ability to retain and modify key assets (in the broadest sense of the word) post-merger. To this extent, a key input from legal advisers will be to anticipate the effect on integration that contractual financial arrangements, remuneration practice and the like will have.

In the 1970s and 1980s, acquisitions tended to be typified by the conglomerate approach – acquired companies were viewed as assets to be financially engineered to improve profits, and often a diverse set of companies was brought under one corporation to achieve a diversified portfolio. As such, acquired targets were often only integrated to the extent of being handed a heavyweight financial reporting handbook. Today, such an approach actually attracts a 'conglomerate discount' from analysts evaluating the share price of firms – shareholders can create a diversified portfolio of companies themselves.

In the 1990s, the concept of related diversification by acquisition has been touted as a more effective route to wealth creation. That is, for effective acquisition, acquirers should bring a synergy to target firms that will create more value than the previous entities when they were not combined. Such synergy may be brand image, manufacturing know-how, physical asset overlap, research and development skills or just pure economies of scale. Such issues are considered further in **CHAPTERS 7** and **8** in particular.

In reality, accessing true synergy in acquisitions means that:

- the acquirer is fully aware of the depth and breadth of its core skills;
- it is similarly aware of the target's skills; and
- there is always a clear and focused set of acquisition objectives for integration post-merger that will create new wealth from an optimised combination of assets whether intellectual, physical or informational.

Therefore the business school mantra of achieving synergy which is intellectually satisfying, leads rather quickly to a potentially vast and complex web of practical detail.

Competitive advantage through successful acquisitions 1.30

In the more straightforward days of conglomerate acquisition, financial control was often the only overlap that had to be managed. However, when a 'synergistic' acquisition takes place, marketing, manufacturing and IT functions may also have to be combined. Thus the risk and complexity of the acquisition is much higher.

Yet, as is clear from the chapters that follow, this risk and complexity, however daunting, is a vital area of corporate life that has to be managed effectively. There are a number of reasons for this. At the strategic level, it is likely that the ability to achieve successful acquisition will become a competitive advantage as corporations look more and more to this method of accessing growth and differentiation.

Tactics and price 1.31

At the tactical level, the actual price paid for a target firm will reflect the proposed synergies to be achieved – for example one way of looking at the price of any target is:

- the fair market value of the firm;
- add or subtract any applicable premium or discounts; and
- add the perceived operational synergies created by the merger.

If such synergies are not to be a major factor, then the process is merely asset capture (simply put, the purchase of an ongoing business' revenue stream) and the price paid should reflect this. If synergies are to be a major factor in price determination, then the realisation of post-merger synergies should be a priority for business managers. What is the reason for paying a premium for intended synergy and then not integrating properly to get the benefits paid for up front? In a sense this is a double blow for shareholders and other parties involved. A premium is paid for the target firm above true market value. If integration does not then follow, the price paid has been too high and the intended synergies have not materialised, investors lose both ways. Organisational and strategic research has led to the conclusion that unsuccessful or piecemeal integration often leads to merger and acquisition failure.

Integration options **1.32**

It has, therefore, also been asked whether the assessments of the above mentioned 1987 study (see **1.29**) are correct. Do firms find that acquisitions become too difficult, and that asset capture remains the easiest and most worthwhile route – irrespective of whether or not this affects the overall success of the acquisition? Some answers are available from a study of all UK acquisitions in the period 1991–1994. The study[1] conducted by the Warwick Business School used a framework which usefully highlighted the different type of integrations undertaken in real life by acquiring corporations, as follows:

- absorption – full integration of the acquisition by the buyer;
- preservation – where the acquisition is not integrated at all and held at arm's length;
- symbiotic – where there is mutual dependence (but not integration) between the two companies; and
- holding – which indicates that the acquired company is held to be (possibly) traded at a later date.

It has been argued that the simplest way to view these four alternatives is to note that they are separated along two dimensions:

- the amount of resource transferred; and
- the degree to which the acquired company is left independent or not.

Data has shown that over three quarters of the companies studied had not attempted to integrate resources significantly. It was concluded that this was due to the inherently difficult nature of integration (and hence risk) and, in terms of acquiring good performing assets, the instinct was not to 'contaminate' the good work for risk of harming performance. The stark conclusion is that in the UK during the 1990s most acquisitions were of a 'purchasing asset' variety involving little or no transfer of resources between buyer and target.

The best option **1.33**

Although evidence has revealed that only in a minority of cases does extensive integration appears to occur, a number of key changes do happen in all acquisition types. The most common changes are ones which are symbolically important, signalling progress to staff and the City. Therefore management changes, financial reporting changes and communication changes are all quick to take place. Changes that are relatively easy to accomplish and have high impact, such as senior staff movement, will be pursued first. Longer term, if at all, the more complex areas will be tackled such as site rationalisations and IT systems. Therefore, given the focus on share price and City evaluation that drives much business strategy, not least because of the share options link between senior management and company stock price, acquisition integration tends to focus on early indications of success which is a less risky and preferable approach to a longer term involved integration which may take time to signal success.

In the long run, however, the early win approach may be detrimental to value. If the acquired target is not brought in to the wider corporate structure and is left to continue much as before, it is unlikely that the acquirer will have accessed that premium he paid for in the deal. This may well be part of the reason for the number of divestments that have to be undertaken (see also **CHAPTER 8**).

The fact that the above study (see **1.32**) was conducted in the UK means that the stock-market impact on managerial action will inevitably be a major factor, as it will be in the US – that is countries with highly active and liquid stock-markets. It may, however, be that in European or Asian countries with more conservative stock markets that the focus on short-term delivery is less intrusive – this may facilitate more measured analysis and implementation of integration, and a comparative analysis would be useful. True value-creation from acquisitions therefore appears to be only pursued in a minority of actual instances – a clear opportunity for genuine wealth-creation therefore remains.

Priorities 1.34

Textbooks and journals on how to integrate successfully are now commonplace in management literature. They tend to cite sensible advice such as:

- communicate thoroughly with all employees;
- set clear objectives and initiate appropriate organisational change;
- have clear milestones;
- ensure normal business is not hampered; and
- act quickly to avoid losing momentum and enthusiasm.

Both the Warwick study (see **1.32**) and this integration literature tend to be descriptive and have not attempted to draw out contingent circumstances of acquisitions. That is, it may well be inappropriate (although technically preferable) to try all elements of integration depending on what the acquirer's objectives actually are. Indeed, there is evidence to suggest that post-merger synergy may not be a primary objective at all – competitive reaction, revenue capture for growth or other strategic issues may actually dominate, and require that resource is only expended on specific key areas. Acquirers should, however, stand back and realise what it is they are actually stating by following this thinking: 'I am deliberately purchasing assets to which I will not add value, and from whom I will probably not gain value in the long run'.

Whatever the case, recent merger activity in the UK suggests that the integration issue is receiving more attention. Interviewed about the merger of Price Waterhouse and Coopers & Lybrand, the integration manager (a good asset) noted a simple but insightful point about the finer details:

> 'The two firms bill their clients and do their time sheets differently. It will probably take two years to sort all that out, and its going to take a huge investment.'

This is a realistic assessment of the kind of effort required and highlights the reality that even with best intentions it is often very difficult in the short term to make significant progress in certain areas – a realistic and flexible evaluation of these areas could ensure expectation about progress is managed.

For example, integrating the culture of two combining firms (often given high profile in merging episodes both in publications on the subject and by management) is very likely not to occur much beneath the surface in the time-scale of integration – irrespective of what best practice advises. In many instances, even after several years, a parent culture tends to linger with many individuals. It might be better to realise that the cultures will continue to have a certain flavour and that this is either acceptable or, if it is not, a replacement of key staff may have to occur. In any event, a fundamental review as to how important this is actually going to be should be undertaken and, if it seems manageable but resource-consuming, this fact should be reflected in the purchase price. Cultural issues are worthy of a separate debate and are considered to an extent in **CHAPTER 8**.

It is clear that extensive post-merger integration, if attempted, will often be a difficult and time-consuming task. It therefore would seem to make sense that as part of a negotiated target price this investment in resource and time be used to counter inflated premiums for the target firm. This addresses, in part, the problem of shareholder value return and in fact may be a salutary action to undertake in terms of assessing how viable the wider acquisition or merger will actually be.

In-house lawyers are often key players in determining the strategy and implementation of acquisitions and so it is important that they can provide guidance to their colleagues on the importance of effective integration. Furthermore, when advising on a transaction both in-house and private practice lawyers need to be aware of the buyer's purpose for making the acquisition and integration will be a factor in this. This will influence the due diligence, the transactional documentation and the negotiations. For instance, the purchase of a service sector business, such as an insurance brokerage, will involve a different set of legal analysis and transaction if the personnel are not being acquired.

The facts are, however, that often mergers and acquisitions provide such a strong momentum of their own that it tends to sweep aside all but the most obvious of post-merger integration considerations. In order to realise the objectives of good corporate governance and to achieve realistic purchase prices, and fully access post-merger value, these aspects should be prioritised before rather than after the event.

Other business issues 1.35

At this stage it is useful to bear in mind the due diligence exercise in the context of other business issues and the general objective to understand the:

- risks and rewards of due diligence;
- repercussions of the analysis; and
- interaction with risk management and corporate governance.

Practitioners emphasise the importance of understanding how due diligence must not be blinded by only looking within the company. The external environment in which the company operates, hires personnel, deals with suppliers etc is a significant component. In this context the term 'environment' is used to group together the elements that support and impact the company.

To recap, the mission of any due diligence exercise starts with verification of the company goals. The goals have to be prepared and be able to be articulated. A statement such as: 'we want to compete in the bicycle market' is much too vague and does not provide any support for people who are trying to implement a business plan. Clarity and specificity is much better than vague assumptions and generalities if the business goals are to be understood and achieved. Another important reason for clarity is to enable the due diligence team to be able to recognise when the business is heading off course and when the company is moving ahead – the objective is not to find reasons and justifications to keep going. While there are always exceptions to every procedure, the more business operations are guided by consistent principles, the easier it is to identify the exceptions and determine whether this time such an exception is justified or not. With enough exceptions, a company's operations handbook needs to be adjusted to embrace this exception as it is now a standard procedure, not an exception.

There are several key objectives for company operations to support a due diligence environment. Ongoing due diligence demonstrates:

- the company business capabilities;
- the stability of the company, the industry and the overall economy;
- the revenue and expense flows for business operations;
- company support methods;
- risks – business, personnel, economy, environment;
- a clear strengths and weaknesses assessment of the company; and
- specific business continuity issues.

Ongoing due diligence helps to create the framework that enables companies to operate effectively. While each company is different, the objectives are typically the same, to:

- sustain profitability;
- reduce risk to the company;
- achieve business goals and objectives; and
- improve quality of life.

Drivers for ongoing due diligence **1.36**

Some consideration has been given above to the drivers for the due diligence process mainly in the context of transactions. Mention has also been made of the growing importance of ongoing internal due diligence that supports the corporate culture and good corporate governance positively. In view of the concerns raised by the high failure integration rate it is useful to close this chapter with some of the highlights of the business issues, thereby setting the scene for the chapters that follow.

External drivers **1.37**

External drivers include:

- regulatory issues;
- company standards;
- corporate governance issues and trends;
- investor/lender/stakeholder confidence;
- consumer confidence;
- consumer satisfaction; and
- government compliance issues.

This list continues to expand – what is essential to understand is how the environment outside the organisation needs to be included within the elements that due diligence reviews. The due diligence team needs access to where this information can be gathered on an ongoing basis. Very often trade associations are a good source for this information and the data collected can be used to compare the company operations with other companies, industries and countries.

Internal drivers **1.38**

These include:

- employee satisfaction;

- management satisfaction;
- supplier relationships;
- operational procedures;
- operational implementation; and
- multi-office relationships.

While this list is more limited than the external list, this does not mean that any of this information has limits and boundaries that are set in stone. As regards internal due diligence, the team needs to be attentive to the results of interactions among employees and management. Proper implementation requires overall co-ordination and consistency by the people who are empowered to perform the various company tasks. For example, if a bookkeeper is allowed to change the data of a sales entry unilaterally without any transaction trail, then chaos could result. There could also be a problem that could lead to fraud and/or the commission of criminal acts. Exposure to such risks is not what business wants and supports the appropriate use of due diligent activities.

Practical issues 1.39

In the context of this discussion, practical examples of combining internal and external activities are joint deals, transactions, joint ventures and other relationships. For this work due diligence is actually two-sided. Each company within the relationship will have due diligence to perform:

- company one will want to investigate and examine company two and vice versa; and
- company two will need to assess company one.

Mergers are especially the subject of a due diligence exercise by each company, their lawyers, their accountants, government regulators (if public companies), insurance advisers and so on.

Bearing in mind the importance of the quality of data, the key is to have each due diligence team determine the level of exposure based on what can or can not be answered. Many deals or negotiations never get past due diligence because there is not enough documentation about the company's operations. The deal can be filled with risk as there cannot be a total investigation just as there is never a complete investigation of the medical, emotional, financial history for each party to a marriage. There has to be a balance that is part of the risk reward formula for all due diligence activities. For example, in the case of a £50m deal, not being able to verify a £1,000 transaction may not be worth the thousands that it takes to validate the transaction.

This is where the experience and capability of the due diligence team is essential.

First, they have to possess training and expertise to be able to recognise the important and the unimportant. Second, they need to have the appropriate tools necessary to perform their tasks. The tools can include, but not be limited to:

- internet research capability;
- legal data bases – especially for lawyers;
- tax data bases – especially for accountants and lawyers;
- industry perspective and data;
- access to company personnel;
- access to all relevant regulations – securities, government, environment, etc; and
- appropriate computer and resource tools that support this work.

As due diligence continues to expand, companies will rely on the information gathering that can sustain the enterprise, reduce the risk of business activity and reward the various stakeholders.

Other due diligence drivers 1.40

There are other drivers for the due diligence process that exist and should be mentioned by way of a summary. They are described below.

Macro and micro issues 1.41

The company operations has to be able feed appropriate data into the due diligence mixing bowl. Macro issues include the larger pictures, especially those issues that the company has absolutely no control over. For example, the global economy, global politics and global terrorism are examples of very real issues that have to be monitored with appropriate plans in place and ready to be implemented, the moment that a due diligence alert has been sounded.

If country one, which is a buyer of your goods and services, has a major weather related disaster, a number of events may be triggered. All sales to that country may be suspended, employees may not be reachable or worse, an inventory already in place may be destroyed. As is discussed further in **CHAPTER 5,** due diligence efforts can include monitoring possible disasters, and if predictable, taking precautions in the days leading up to the event. Another part of due diligence is understanding the potential of such risks occurring and establishing procedures way in advance of any possible catastrophe that has a full set of operational procedures of what to do. The quality control needed by senior management is to verify that such procedures are in place and employees have been alerted and trained.

The interaction with risk management and corporate governance 1.42

Risk is the dealing with the unknown. It requires making decisions without all of the facts that are absolutely required to determine the outcome. Horse racing as a sport provides the gambler with a chance to exercise risk decision making. He can establish all he can about the horse, speed of the track, weather conditions, etc and then make a bet on who he or she thinks will win, place or show. In business, of course, responsible players have to reduce the gambling aspect and lower the risk of being wrong.

Due diligence methods can also monitor situations that can be a level of risk. Airlines know that bad weather can be a risk to any plane. Consequently, it is essential for all aviation personnel be provided with all data about weather between where they are and where they are going. In this way the risk of being hit by lightning, or worse, can be reduced to a manageable procedure. Airline pilots quite frequently request permission from ground controllers to change their altitude to avoid endangering the plane, passengers and people on the ground. The pilots and ground controllers are trained to be diligent about the risk that can lie ahead. In this example, the passengers are merely freight and along for the ride. Further discussion of exposure to such types of risks and their management is found in **CHAPTER 5**.

Transactional and operational assessments 1.43

Transaction auditing is a well-known component of the auditor's world. In this activity, the auditor, internal or external, selects some number of transactions to determine their accuracy with the company's entire processing cycle. For due diligence, it is also essential to be able to identify specific operational transactions – financial or business. For example, if a patent application requires a series of steps then it is essential that each step be documented so that the work can be proven as being completed.

Operational assessments refer to how the company conducts its business. Planning or zoning requirements vary by country and region. International transportation of goods and services has different regulations and requirements than national business activities. The due diligence team is required to be able to access all that encompasses the business – before, during and after a specific business transaction.

Transactional issues 1.44

Globalisation is also a potential mine field filled with places that require guidance and careful management. In some respects the due diligence exercise starts with the knowledge of the company's desire to do business within another country. The due diligence team takes over to establish the legal, financial and government regulations that need to be understood. Then procedures can be implemented to make sure that company actions fall within the required guidelines. In this case, it is an absolute requirement to perform due diligence prior to commencing any business activity. It would certainly be expensive, in terms of both time and money, to perform due diligence after the fact and reference should be made to the comments in the chapters that cover the international dimension of due diligence and corporate governance, particularly **CHAPTERS 11** to **15**.

References

1. Angwin, R and Wensley, R. The Acquisition Challenge, *Realising the Potential of Your Purchase.* Warwick Business School Paper, Vol 1, Number 4, 1997.

Appendix I

Overview of main issues for a due diligence exercise

Submitted by Kesar Dass B and Associates, Corporate Lawyers, Delhi

> *'Almost half of all acquisitions fail or do not achieve expectations. Improved advanced planning, target evaluation and due diligence can improve chances.'*

The underlying purpose of due diligence is to observe, analyse and examine all the key areas and, as well as the minute details involved in any business, the company, immovable property etc that is under consideration. In the present summary we have classified the requirements of due diligence under three headings.

1. Acquisitions: the act of contracting or assuming/acquiring possession of something; 'the acquisition of wealth'; 'the acquisition of one company by another'.
2. Joint ventures: a partnership or conglomerate, formed often to share risk or expertise.
3. Purchase of immovable property, for example land and building etc.

After the initial stages of buying a business/immovable property or entering a joint venture are complied with, such as that of search of a seller and interaction with the prospective seller to formulate a strategy and conclusion of a transaction, then follows the stage of due diligence. This is the last stage in the buying process. This is the time when the buyer will have access to all of the company's/owner's books, records and files. The buyer will have a pre-determined due diligence period in which to investigate the information that it has been given so far to ensure that it is true and accurate. Due diligence is the way to discover everything before the actual purchase that the seller knows. Once the deal is closed there is little or nothing that can be done about it.

The basic requirements for acquisition of a business/joint venture are listed below.

1. Corporate books and records:
   - original certificate of incorporation of the company and all amendments thereto;
   - memorandum and articles;
   - closing record books for any material corporate transactions (eg reorganisation into holding company structure, joint ventures, etc);
   - other relevant legal documents governing the organisation and management of the company;
   - minutes of annual general meetings and other board meetings;
   - shareholder list and other stock records, any shareholder agreements or any agreements relating to shareholders; and
   - annual reports and other quarterly and special interim reports since most recent annual report.
2. Financial information:
   - consolidated financial statements, monthly income statements for most recent twelve months, internal financial (profit and loss, capital expenditures, etc) projections and all supporting information, most recent business plan, list of any off-balance sheet liabilities not appearing in most recent financial statements, auditors' reports ('management letters'), management responses and summary of accounting policies to the extent not disclosed in financial statements;
   - tax materials and documents and records; and
   - debt obligations.
3. Employee materials:
   - employment agreements (including, but not limited to, contracts with management personnel or entities affiliated with management personnel) and all other agreements with employees in any regard;
   - any labour disputes either pending or disposed of;
   - organisational information including detailed organisation chart, list of all directors and officers, biographies of senior management and any outside directors, schedule showing number of employees for each year and interim periods and list and description of current operations of each key business unit.

4. Contingent liabilities:

   litigation:

   - list of all pending, disposed of or threatened litigation, appeals, arbitration, administrative or other proceedings involving the company, any subsidiary or any joint venture involving the company or any subsidiary, or any officer or director (including parties, remedies sought and nature of action);
   - list and description of all pending, threatened government or other investigations involving the company, any subsidiary or any officer or director;
   - pleadings and other material documents in material litigation, arbitration and investigations and other proceedings;
   - decrees, judgments etc, under which there are continuing or contingent obligations;
   - letters from lawyers to auditors concerning litigation and other legal proceedings;

   regulatory compliance:

   - description of any violations of governmental laws or regulations;
   - material reports to governmental agencies;
   - reports, notices or other correspondence concerning any known or alleged violation of central or state laws and regulations;
   - agreements or commitments with governmental entities or other persons relating to clean-up obligations or other environmental liabilities;
   - copies of correspondence between central or state government agencies and the company; and
   - list of all governmental filings and consents required for a purchase of the stock of the company.

5. Contracts, agreements and other arrangements.
6. Intellectual property rights owned by the company, complete information and documentation.
7. Plant, real property and equipment.
8. Insurance information and documentation and up to date records.
9. Sales/marketing policies/strategies/records.

The basic requirements for acquisition of an immovable property are listed below:

- files and other records with respect to the immovable property in possession of the land owner;
- files and other records with respect to the immovable property in possession of the revenue authorities;

- files, records and other documents with the Register of Companies.
- contingent liabilities:

  litigation:
  - list of all pending, disposed of or threatened litigation, appeals, arbitration, administrative or other proceedings involving the owner of the immovable property;
  - list and description of all pending or threatened government or other investigations involving the owner of the immovable property;
  - pleadings and other material documents in material litigation, arbitration and investigations and other proceedings;
  - decrees, judgments etc, under which there are continuing or contingent obligations;
  - letters from lawyers to the owner or any other legal proceedings;

  regulatory compliance:
  - description of any violations of governmental laws or regulations;
  - material reports to governmental agencies;
  - reports, notices or other correspondence concerning any known or alleged violation of central or state laws and regulations;
  - agreements or commitments with governmental entities or other persons relating to clean-up obligations or other environmental liabilities; and
  - copies of correspondence between central or state government agencies and the owner.
- contracts, agreements, sale deeds, lease deeds, rent agreements and any other arrangements pertaining to the said immovable property.

Appendix 2

Initial steps in transactional due diligence in the UK

The initial steps taken in transactional due diligence in the UK are:

(*a*) outline of general action;

(*b*) typically a number of preliminary steps before the negotiations commence in earnest, in an asset transfer transaction such as a share purchase, asset purchase or joint venture, these include:

- the seller takes a strategic decision to sell the company or business or to enter into a joint venture. This may have been instigated by the seller or initiated by an approach from a merchant bank or purchaser or potential joint venture partner; or
- the buyer takes a strategic decision to buy the company or business or to enter the joint venture. Again, this may have been instigated by the seller or initiated by an approach from a merchant bank or purchaser or potential joint venture partner;

(*c*) in larger transactions, the seller may undertake its own due diligence, particularly where a controlled bid is being employed;

(*d*) discussion takes place between the buyer and seller and an agreement in principle is reached;

(*e*) lawyers, accountants and other advisers are instructed;

(*f*) the first meeting between the seller or buyer and its advisers;

(*g*) the first meeting between the seller or buyer and its advisers is worthy of specific mention as it encompasses an overview of the issues affecting the transaction as a whole. These will include the following:

Overview of transactional issues	
Issue	**Explanation**
Identity and residence of the buyer	Identity is important from the consideration of the strength of the covenant of the party. Residence is relevant from the point of view of enforceability of any agreements and the possibility of tax sheltering any profits arising from the transaction.
What are the key assets the buyer wishes to obtain in making the purchase?	Some typical key assets are: • management or technical expertise; • personnel; • client or customer base; • suppliers; • brand name(s); • technology, including intellectual property such as know how; • production capacity; • distribution network; • land and buildings; • regulatory approvals and licences.

Security	Where the agreements are likely to provide for obligations which continue after closing, then the covenant of the obligated party may require to be backed up or reinforced by some form of guarantee or other security.
Consideration	Some input from the lawyers on general valuation considerations may be appreciated at the initial meeting.
Timescale	It will be helpful if the lawyers have an influence on determining a realistic timetable for the deal.
International aspects	International aspects of the transaction will need to be addressed at an early stage and the question of engaging foreign professional advisers should be tabled.
The structure of the transaction	● parties to the transaction; ● asset or share purchase? ● key assets being acquired? ● valuation considerations; ● assets hive down prior to sale of shares? ● key liabilities to be minimised; ● tax considerations; ● general approach regarding indemnities; ● security appropriate for indemnities? ● restrictive covenants? ● staggered closing appropriate?
Funding	How is the transaction being funded? Consider the issues arising from this.
Stock exchange requirements	Are any of the parties governed by stock exchange rules? If so, what will this mean?
Other regulatory issues	Are there any other regulatory issues affecting the transaction or the parties individually, for example competition law, foreign exchange requirements?
Consents and conditions	Are there other consents and conditions to be obtained or satisfied, eg shareholder approval, key customers or suppliers?

Confidentiality	Are confidentiality agreements or undertakings in place: • from the buyer; • from the buyer's shareholders; • from the buyer's employees? Is there any confidential information, such as the customer list or secret processes, that the seller wishes to restrict?
Insurance	Will existing insurance arrangements pass?
Lock-out	Is an exclusivity arrangement negotiable for a period of, say, six months?
Heads of agreement	Any entered into? If so, consider the terms. If none, should they be prepared?
Funding comfort letter	Sometimes, whether or not a merchant bank is involved in the transaction, the buyer will be required to supply to the seller a letter confirming it has access to sufficient funds to enable it to close the transaction and, expressly or by implication, satisfy the purchase consideration.
Due diligence overall issues	Discuss the overall strategy. Consider due diligence checklist with buyer and its professional advisers.
Instruction of advisers	It is to the benefit of client and professional adviser to record appointments in connection with specific transactions in letters of engagement, to reduce the possibility of any subsequent disagreement or dispute.
Rules of engagement between buyer and seller	The parties will wish to set out guidelines for carrying out the due diligence exercise and these are normally set out in rules of engagement either in heads of agreement between the parties or a separate letter.
Data room letter	If a data room is to be used, then rules of engagement may be set out in a data room letter
Other pre-closing documents	Refer to the list of transaction documents

Civil sanctions for non-disclosure	The seller and target personnel should be reminded that misrepresentations, whether innocent or otherwise, can have serious civil repercussions.

Issues to be considered	
Issue	**Explanation**
Specialised industry sectors issues	Such as the Chemical Sector, Food Sector or Automotobile Sector regulations etc
Employment matters	• application of transfer of employment regulations; • new terms of employment; • restrictive covenants; • pension provisions.
Insurances	In major transactions it is not uncommon to employ insurance experts to assess risk and risk cover and review insurances generally.
Intellectual property and information technology transfers	In the case of a business acquisition, intellectual property assets such as copyrights, registered designs, trademarks and patents will require to be transferred by the seller to the buyer under simple forms of assignment. In the case of share acquisitions, a review of the title to the assets will be required.
Real property transfers	In the case of a transfer of a business any real property assets will be transferred from the seller to the buyer under normal conveyancing procedures. In the case of leasehold property the landlord's licence or consent to the assignment of the lease will normally be required. In the case of share acquisitions, a review of the title to the assets will be required.
Position between exchange and closing	• consents; • closing accounts; • searches; • documents to be agreed; • all charges released.

Basic differences between a share acquisition and an asset purchase	
Share acquisition	**Asset purchase**
Involves sale of share capital in the target by its shareholders to the buyer	Involves the target itself selling assets to the buyer.
Liabilities of the target continue to affect the target after sale, unless agreed otherwise	More limited liabilities and commitments pass with the transfer of the target's assets.
The target's rights are not affected after sale, unless there are 'change of control' provisions in effect	Title to the assets and rights have to be transferred by the target to the buyer. Where those rights involve a contract with a third party, their consent may be required to the assignment or the agreement novated with them being a party, unless there is no burden attaching to the right and consent is not required by the agreement.

The due diligence exercise for a share acquisition is therefore complicated by a greater concern of the buyer to identify the liabilities attaching to the target. The potential for unknown liabilities raises a level of uncertainty less present in asset purchases, making asset purchases generally more desirable. However, asset purchases are often impracticable in larger businesses because of the volume of assets and rights which require assignment and novation. This problem of assigning rights has been overcome by certain legislation in England which allows the rights and liabilities of businesses to be transferred by operation of law and without the need to obtain the individual parties' consent. Such legislation applies for instance to asset transfers by building societies and insurance companies.

Appendix 3

Specimen auction or tender process letter

This process could alternatively be contained in an agreement or in the information memorandum itself.

Dear Mr [ ]

SALE OF [name business] ('target')

We attach an information memorandum relating to the above company. This information memorandum has been or will be sent to qualified interested parties, all of whom are bound by similar confidentiality agreements in order to assist them in deciding whether they wish to enter into negotiations to acquire the whole of the issued share capital of the target or part thereof, some or all of the target's assets or a combination of the above.

This information memorandum is provided to you in commercial confidence and on the terms of the confidentiality agreement dated [ ].

Would you please forward to us your proposal in writing for the acquisition of target to our offices by [ ] on [ ], covering the following:

- the purchase price in [state currency] or formula for determining the same, that you are prepared to offer for 100 per cent of the issued shares of target, subject to contract. This would be on the basis that [target's bank and inter-company borrowings have been discharged];
- your intentions regarding any reorganisation or redundancies of target personnel;
- confirmation that your bid is made as principal on your own account;
- details of your financing arrangements for the transaction and a letter from your financial adviser or bank that you have the necessary finance;
- any specific due diligence issues that you wish to address;
- details of any regulatory or other consents required (including internal approvals), and the timing therefor;
- an estimate of your share of the [ ] market in [state territory]. Details of any competition clearances or notifications required. If none, a statement as to the reason why none are required;
- confirmation that there are no other conditions attached to your bid;
- a statement of your ultimate ownership, together with copies of your annual reports for (two or three years).

We do not envisage providing more information prior to indicative bids being submitted. However, if you do have queries they should be raised directly with [name] who can be contacted by telephone on [#] or email on [#]. (Where contact point is at target – if [he/she] is not immediately available you should leave your name, the company you represent and your number so that [he/she] can return the call. Under no circumstances should you try to discuss this sale or leave more substantive messages with any other member of staff at the target.

Process

Following receipt of indicative offers, the vendor intends to select a small number of bidders to proceed to the final stage of the process, involving:

- a meeting with the target management and a visit to the target's premises;
- access to a data room maintained at [state location];
- copies of due diligence reports prepared by the vendor's advisers (these will be arranged through us);
- a draft of the sale documentation will be distributed and bidders will be asked to submit their final offer for the target together with a copy of the documentation with any suggested amendments. Such offers will be binding on the bidders; or
- following receipt of such offers, the vendor will select a bidder with a view to finalising the transaction.

The vendor shall be under no obligation to accept the highest bid offered or any bid at all. The vendor reserves the right at any time and without notice or and without assigning any reason therefore to vary or discontinue the sale process or to sell the target to any person.

Yours sincerely,

[ ]

For and on behalf of [ ].

- [illegible] the large [illegible] [illegible] pre [illegible]
- [illegible] data [illegible] [illegible]
- [illegible] of the [illegible] reports [illegible]
- [illegible] will [illegible] and [illegible] for the [illegible]
- [illegible] the [illegible] [illegible]

[illegible]

Chapter 2
Due Diligence In Corporate Finance

Introduction 2.1

The expression 'corporate finance' generally refers to the mechanisms and processes by which businesses raise capital or enhance capital values for operations and growth. Capital can be raised in a variety of ways, such as by the issue of shares or debentures, or the provision of loans or banking facilities. Capital values can also be enhanced or protected when businesses are merged, acquired or restructured. The mechanisms and processes can either be private or public in nature, depending on the objectives of the corporate finance exercise.

The mechanism or processes by which corporate finance is provided, or the corporate finance transaction executed, as well as the nature of the transaction contemplated, will determine the level or type of due diligence required or available.

For example, the level of due diligence required for a secured borrowing facility will be significantly less than that required for an equity investment. The lender will simply be looking to ensure:

(*a*) his or her loan will be repaid together with interest over the term and that there is sufficient net cash flow cover to ensure repayment; and

(*b*) the value of the security is sufficient to cover the loan to value given.

A highly liquid publicly traded stock will have a higher loan to value than an illiquid stock and certain types of real estate, such as development or commercial properties, will have lower loan to values than residential properties. On the other hand, an equity investor, such as a venture capitalist, will be concerned to ensure that the profitability and positioning of the company over the given investment time horizon will be sufficient:

(*a*) to yield the required internal rate of return on the investment; and

(*b*) for that return to be crystallised by an exit event, such as initial public offering of floatation, or a trade sale.

The focus will therefore be more on the ability of the management to deliver the returns based on the forecasts or projections for the business.

In some cases, the degree of due diligence available or required will be limited or restricted by the nature of the business or transaction contemplated. The offering circular or document by a publicly traded company which is subject to on-going disclosure requirements to its shareholders, governing regulator or exchange will differ from that for a private company where there has been no such public disclosure. The due diligence available on a recommended bid for a company will also differ dramatically from that for a hostile bid where the bid company simply is not given access to the relevant internal financial and management information, and therefore has to subject the bid to a number of conditions, which it will have power to amend or vary subject to what is discovered or obtained by way of due diligence in the bid process.

In each corporate finance situation, however varying or differing, invariably the same fundamental issues are being addressed. What is the investment return available or the level of finance affordable by the business, and what risks are there which would result in the expected return not being achieved or the financing not being repaid? The due diligence process will define the return or pricing of the investment or financing, the amount of the investment or financing to be made available and the structure of investment or financing.

Invariably too, the due diligence process is forcing an alignment of interests, of risk and reward, as well as expectation. There are competing interests. What a business or company may perceive as a fair return for an investment may differ substantially from what the investor or financier may require. Alignment of interests will not just address the issue of price, but also management commitment, focus and vision and the ability to deliver the return based on the expectation created. There is also the issue of evaluating external factors, which will affect the return and reward, factors that are either systematic, such as interest rates, inflation and political events that are common to the business and industry as whole, or events that are specific to the business or industry itself and which can be compared to similar businesses. Risk can thus be defined and isolated and then appraised relative to the expected return. A process generally known as risk adjusted return.

The weight given to the critical financial analysis tools used by lenders and equity investors alike will vary. A lender will be evaluating financial risk ratios such as interest coverage and interest coverage adjusted for cash flow and debt to equity ratios, whereas an equity investor will be more concerned with profitability ratios looking at return on capital and return on equity. Both will be concerned, however, with historical business information such as matters affecting the essential ability of the business to pay its debts as they fall due and to turn its inventory or assets over in a sufficient period of days to create cash flow and remain solvent and profitable. Of course emphasis on all or any of these will vary from business to business and many of these core financial ratios were dispensed with during the technology boom of the late 1990s when forecasts and projections were solely used as determinants of value and investment.

Often, for a corporate finance transaction there will be a mixture of equity or debt or variations of this, and this will give rise to an appropriate review as to what the mixture should be and how it should be priced. The weighted average cost of capital calculation would be a familiar calculation used to determine the correct balance of each and the risks arising.

In coming to a conclusion as whether or not to invest or finance or to undertake the transaction and in evaluating the business or transaction as a whole, various factors will need to be considered. The critical due diligence process would be the assessment of these factors and undertaken by a team of professionals of varying disciplines to examine and investigate different aspects of the business or transaction.

The due diligence process undertaken will vary considerably from one business or transaction to another and the initial requests for information or the basis of investigation and research will have to be tailored to the business or transaction concerned. Whilst there are broad categories of information that are normally reviewed and requested, it would be a serious misconception to think that one size fits all. The due diligence process will be an important part of the evaluation and structuring process and decisive of the ultimate success or failure of the investment or financing proposed for the business or transaction.

Depending on the business or transaction being proposed or considered, from a high level approach, the due diligence process will then reach down to a more detailed and distilled consideration of the various issues affecting the value of the business – often after combing through a myriad of legal, technical and commercial matters. These considerations will then form the basis of a report or reports on which ultimate investing or credit decisions are taken.

A caveat, therefore, has to be made that, in discussing due diligence in corporate finance, it is very much an introduction and specialised professional advice will often be sought on the various matters to be reviewed and analysed. This chapter therefore looks at some typical due diligence topics arising in corporate finance. It is certainly not intended to be a comprehensive checklist, rather a list of examples of due diligence exercises that might be undertaken and the type of issues addressed.

Due diligence process **2.2**

The due diligence team would consist of various professional advisers, normally legal, financial, technical, environmental, insurance and actuarial experts working in tandem with each other. These advisers should be brought on board as soon as practicable to give them sufficient time to cover comprehensively the issues involved. The due diligence process is often driven by the lending institution or investment bank who will co-ordinate the due diligence team.

The team that undertakes the due diligence exercise needs to be given clear instructions as to the objective of the exercise and its parameters. Understanding what the business or transaction entails and what the exercise is intended to achieve will help the team streamline the exercise, focus on the relevant issues and make it more time and cost efficient. If a company plans to acquire a target with the intention of developing and selling properties, starting or developing, for example a hotel with a casino, it has to convey these ideas and strategies to the due diligence team. Whether these strategies are realisable and what complexities they entail are the answers that the investing company would wish to find out from the due diligence exercise. For example, before embarking on a proposed transaction involving the purchase, by loan financing, of a hotel operation in India, a first tenet would be to determine whether or not profits could be remitted from India to repay the loan and whether the lender could perfect security in India over the assets being acquired. It is often very surprising how advanced transactions may proceed before critical corporate finance issues are discovered. High level information overviews are therefore advisable from a very early stage.

To highlight this point, an overview list of requested information with respect to the purchase of a regulated financial services business is set out in the appendix to this chapter. At the core of this acquisition would be the regulated status of the business and its compliance with regulatory provisions. From the information disclosed, the basis on which the transaction is financed and structured, documented and completed can then be negotiated and finalised.

The responses given to the detailed questionnaire would normally be expected to form part of 'disclosures' to warranties and representations to be made by or in respect of the business or transaction. Once disclosures are made, for instance, as regards pending litigation, breaches of overdraft facilities or arrangements with creditors, the liabilities would be quantified, the price adjusted or the purchase price deferred and the disclosures warranted as being complete and accurate in themselves, so that the extent of the liability is correctly provided for.

An important part of the inception of the due diligence process is the exchange of confidential undertakings. These structure the environment in which often price sensitive and valuable information relating to the business or transaction is passed by the business to lenders or investors securely without risk of leakage into the public domain and thereby potentially damaging the value and reputation of the business. Generally, the confidentiality undertakings will expressly refer to what is to be disclosed and how it is to be identified as confidential, usually by reference to the initial or subsequent due diligence lists or questionnaires, to whom it may be disclosed, for what purpose and how it is to be returned or dealt with if the transaction does not proceed.

The undertakings may also extend to non-compete clauses and lock-up periods during which the negotiating parties agree to deal exclusively with each other for a certain time period with respect to the business or transaction concerned. This enhances confidence among the parties and creates a safe environment in which disclosure can take place. Creating the right environment is paramount as no less than full disclosure will determine the right analysis of the appropriate risk and reward. The due diligence team and its advisers will normally be ring-fenced and each participant required to give its undertaking to be bound by the rules of confidentiality on which engagement takes place.

Important issues to be borne in mind are:

- who is keeping the master disclosures lists;
- who is co-ordinating disclosures; and
- who is responsible for evaluating and analysing the disclosures given in the context of the transaction as whole.

In a typical transaction, one would expect to see a combination of reports being assembled for review from the due diligence team. For example, financial reports including reviews of management accounts and forecasts, asset valuations, working capital and historical annual accounts, a legal report on regulatory and compliance issues and title to assets and the business, tax reports on the tax implications for the parties, actuarial reports on pension funds, environmental reports and other specific valuations and project implementation reports – depending on the business or transaction and the nature of the corporate finance activity.

Anti-money laundering due diligence 2.3

A chapter on corporate finance due diligence would not be correctly balanced if it were not also prefaced by a section on anti-money laundering due diligence. This is dealt with as a subject in much greater detail in the chapter on anti-money laundering due diligence (**CHAPTER 3**)

Money laundering is now an essential prerequisite and integral part of the due diligence exercise for any corporate finance transaction in light of the now very extensive, stringent and punitive money laundering regulations that have been adopted by most countries. The scope of the regulations can be very wide. In the UK, for example, *Money Laundering Regulations 2003* (*SI 2003/3075*) now cover any transfer, conversion, removal, concealment or disguise of funds which constitute the proceeds of crime. This means that in looking at a business or transaction as a whole it will need to be borne in mind that questionable accounting practices or possible tax evasion discovered as part of the due diligence exercise, and which is or could be a criminal offence, may then lead to independent money laundering disclosures obligations for the due diligence team. First steps in any corporate finance transaction would be undertaking appropriate due diligence on the business owners or management, or ensuring that the proposed funds to be invested or lent to the business as part of the corporate finance transaction are clean. The exercise would then extend to evaluating the proper sources of funds coming into and out of a business' operations. Where businesses have significant cash revenues, or operate on a cash based system, this could give rise to a significant number of problems and potential disclosures. Moreover, the professional members of the due diligence team will usually have independent disclosure obligations and may be required to make disclosures to the regulatory authorities without reference to the client or other members of the due diligence team if they have concerns or suspicions. It could be a money laundering offence if an adviser who is regulated fails to detect money laundering when they should have detected or suspected it.

The terms of any due diligence engagement should be very clear on this issue and certainly disclosure will be an exception carved out of any confidentiality agreements.

Financial due diligence 2.4

Financial investigations would typically be carried out with the help of a firm of accountants or an investment bank. The due diligence intends to verify whether all books of accounts and other financial records are up to date and have been accurately maintained. It also seeks to search out the current trading position and prospects of the business, which are not evident from the historical or filed accounts. The fact that the books are audited provides some comfort but usually this information will be significantly out of date. Businesses are dynamic and every day something may change that will affect the overall financial position of the company prior to conclusion of the transaction. For example, on a lending transaction, a lender will frequently require extensive drawdown (drawing of loan at agreed intervals) conditions under which very up-to-date information is provided on the company's affairs prior to drawdown, for example up-to-date debtors and creditors lists, cash balances and reassurance that no material changes or events have taken place.

There is much that can be gleaned from disclosed financial information and the accompanying financial ratios. However, there is still much that can lurk behind the figures, irrespective of the interpretation of the entries and calculations of financial ratios. The information identified from detailed investigation will be highly relevant in the ongoing negotiations for the cost of finance and its structure. At a minimum, accounts should be audited as required by best accounting practices and should be in conformity with the local law. An important part of historical analysis is to look at any disclaimers or qualifications on the audit reports or a frequent change of auditors. Equally, at a minimum, management accounts should adopt accounting policies and practices that are consistent with the audited accounts. A more vital review is that the accounting systems used to record information are robust, postings are accurate and properly posted and the basis on which postings are made are consistent and reliable.

Revenues need to be carefully reviewed. Contracts can be entered into and invoices raised even though there may be no underlying delivery or agreement as to delivery. There also needs to be a careful review that the true costs of revenues are shown – and not hidden so as to inflate earnings and profitability.

Capital commitments have to be carefully considered for onerous or unusual payment provisions, for example, advance payments that do not necessarily oblige delivery or performance. Financial projections based on the projection of capital expenditures and sources of capital for the payment of such expenditure have to be matched against the reality of true cost, delivery and implementation. A capital financing proposed may merely be used to prop up working capital requirements and be insufficient to fund the capital needs of the business on which future growth hinges. Often management understate the true working capital needs in order to enhance earnings and returns.

The inventory and the work in progress should be valued without including profit but taking into account anticipated losses. It is important to examine the inventory physically to ensure that the stock and the raw materials are not obsolete or redundant and the value is accurately reflected in the books. On site stock checks on the closing of a transaction would not be uncommon.

Contingent, disputed and other liabilities, including claims under contracts, should be specifically reviewed, as well as any default or cross defaults occurring under existing borrowing facilities as a result of the financing and the consequences of those defaults considered in detail.

There should be a clear analysis of those accounting arrangements which are by definition subjective, for instance, provision for bad or doubtful debts, treatment of capital or revenue leases, off-balance sheet risks and liabilities, the use of favourable inventory methodology, treatment of capital expenses and recurrent expenses. For example, one of the key frauds leading to the collapse of the US telecommunications group Worldcom in 2002 was the capitalisation of costs that were clearly recurrent leading to a massive overstatement of earnings.

In the US, it is also worth noting that the *Sarbanes-Oxley Act of 2002* (*SOX*) was enacted in the aftermath of the US corporate accounting scandals necessitating revision of corporate governance standards and increasing the disclosures requirement to protect the interest of the investors. The statute establishes an oversight board to oversee the audit of the public companies, which are subject to the securities laws. *Sections 302* and *906* require the Chief Executive Officer (CEO) and Chief Finance Officer (CFO) to certify that the information contained in the periodic report required to be filed with the US Securities and Exchange Commission (SEC) fairly represents, in all material respects, the financial conditions of the company and the adequacy of the internal controls. Any false certification will attract criminal liability. For further details on the SOX see **CHAPTER 11**.

Oversight of critical financial issues in the due diligence exercise could result in a company investing in the US having a huge task of organising the target company's books post transaction to be able to make accurate quarterly and annual reports to the SEC.

Tax **2.5**

Tax implications will vary depending on the country in which the business operates as well as the country of the lender or investor. Tax is a very important part of the due diligence process and the tax consequences of a transaction or investment could have significant pricing implications, for example, the withdrawal of reliefs or crystallisation of charges previously held and the understatement or overstatement of deferred tax charges and tax assets as shown in the books of accounts. Historic tax computations will need to be reviewed and the impact of the transaction considered. This will often lead to the manner in which the transaction is structured in order to maximise tax efficiencies, for example, whether as a share acquisition or asset purchase or a financing by way of debt or equity, deferred consideration or instalments.

The tax review would extend to the whole gambit of applicable taxes, for instance, income or capital gains taxes, estate or inheritance taxes, valued added tax or sales or service tax and customs and excise duties and charges. In each case, an analysis would be necessary of the likely impact. For example, loans may attract withholding taxes on interest payable, asset purchases may attract a value added tax and an overseas equity investment may attract a punitive exit charge as a result of foreign exchange controls. As part of any documentation following such review, it would be usual to see tax warranties and representations dealing with the issues of concern as well as full tax indemnities being given in respect of any liabilities or contingent liabilities likely to arise. For instance, the sale of shares may or may not attract capital gains tax. However, in case of the acquisition of a company by a purchase of shares, capital gains tax may be attracted when the underlying asset is then sold. Stamp duty or transaction duty on acquisitions by the purchase of assets can in some countries be as high as eight times the stamp duty on the purchase of shares. Value Added Tax (VAT) is generally not payable on the sale of shares.

Explicit tax benefits may flow from the transaction for both parties. For instance, in some instances reliefs and tax deductions may be available on exiting the investment or financing. On other occasions, tax may be payable at a reduced rate as a result of group relief or applicable treaty provisions.

Commercial due diligence 2.6

The focus of this aspect of due diligence is on the true business potential and impact of the transaction. The investor will concern itself with the market conditions, the competition, the brand value if any, the image of the products, the integration and interaction of the employees at different levels, as well as the cost of negotiations with any existing lenders or investors and the consequences. It will be an extensive appraisal of the business plan in the context of these issues. The heart of the investigation will focus on assessing the true risk of the return.

Prior borrowing arrangements 2.7

The prior borrowings, including debentures, overdrafts and loans made available to the business, and the terms of the loans, should be carefully reviewed. Any new borrowings made available should not exceed the permitted existing facilities and the total borrowings and the new debt should be within the borrowing limitation prescribed by the constitution of the business. If not, the constitution will need to be amended before extending the additional facilities to the company.

The existing facilities should be in continuity and in force. The terms of the existing facilities should be vetted to ensure that default provisions of the facilities and the implications of amendments are well understood. Additionally, covenants must be taken to the effect that the business has not breached any of the provisions of the existing facilities and relevant disclosures should be obtained if it has.

Where new debt is to be made available in addition to existing debt, it may be necessary to subordinate the existing debt or ensure the prioritisation of the existing debt as part of the transaction. If the existing or new facilities are to be subordinated, subordination arrangements will need to be entered into. Avoiding the triggering of default or cross default provisions under existing facilities as a result of the new investment arrangements or transaction will be a primary concern. Change of control is often an important event of default which is triggered by a corporate finance transaction as well as the breaching of important debt-to-equity ratios as well as cash flow and interest coverage covenants.

Commercial contracts and other agreements or arrangements 2.8

The business would have entered into other various commercial contracts that would typically include its suppliers, distributors, marketing agents and export agents. These contracts may be key to business operations and sales. It therefore becomes important to ensure that these contracts are valid and existing, ascertaining the obligations under each of them, and to ensure that the business is not in default nor would be in default as a result of the transaction. In particular, outstanding obligations would need to be considered, such as monies owed or payable.

If the entire business is being acquired, contracts may need to be either assigned or novated (substituted with the agreement of all parties). The consent of the contracting party will be required if liabilities are to be novated, or the assignment otherwise requires consent. This could be quite a cumbersome and time consuming process.

Termination provisions of the contracts may also be unfavourable, may have lengthy notice periods and may be activated upon a change of control.

Special care will also have to be taken if the business has issued any guarantee, surety or indemnity in favour of any third party for the obligations of a group company or otherwise. Powers of attorneys, option agreements, indemnities, agreements granting pre-emption rights, comfort letters, credit extensions or credit grants or any offer or tender or such other document which could be converted into an obligation for the business as a result of the transaction should be carefully examined. If there are contracts or arrangements, the benefit under which could be assigned to a group company or to a third party, suitable covenants should be taken to prevent the business from doing so if this were to be against the investor's or lender's best interests. All the contracts in which the directors are interested or are not of arm's length may also have to be renegotiated.

Licences and intellectual property rights 2.9

Many corporate finance transactions take place and place value on the brands or licences of the business. The licensing arrangements and the protection afforded to the intellectual property rights should be investigated in detail to ensure value is preserved. For example, a licence may stipulate termination if there is a change in control of the business, or a company may have exclusive or non-exclusive rights to the technology or may have rights exclusive to a particular area or country. The licences should be valid and in force.

If the intention is to acquire the assignment of the license, one will need to ensure that the licence is not personal to the party in whose favour it is given in which case it may not be capable of being assigned or sub-licensed.

Intellectual property rights (IPRs) capable of registration can be ascertained from the records maintained by the Patents and Trademark Office in the relevant country of use. It is important to bear in mind that registrable IPRs may be territorial and not global. Again, what can be registered and what cannot may differ from country to country. According to the UK Patent Office, computer software may be capable of being patented if it results in a 'technical effect', generally interpreted as resulting in an improvement in technology and it has to be essentially in the technology sector. Whereas in the US, as acknowledged by the UK Patent Office, the approach taken is more liberal and the software that would be allowed to be patented would not necessarily qualify for registration in the UK. It is important to establish the source from where the IPR emanates. Warranties and representations should be taken particularly where the IPRs are not registrable. Some jurisdictions may require registration of ownership – license or sub-license of the IPRs may have an additional requirement of registering it within a particular time period.

Care must also be taken that the technology or information employed by the business does not amount to infringement of third party IPRs and that the business is authorised to use such technology or confidential information. Employees dealing with confidential information should be bound by confidentiality agreements.

IT 2.10

The business should have an IT system in place, and have trained support staff to ensure proper handling, operation and monitoring of the business' hardware and software. The system should be geared to handle disaster recovery of the date ensuring that the hardware or the software can be replaced or substituted without material disruption. However, it is prudent to have insurance to cover data loss risk. The company should have the benefit of a comprehensive support and maintenance agreement with appropriate confidentiality clauses.

Real estate and business assets **2.11**

The business may hold real estate as an owner or as a lessee. If title is registered, it should be relatively straightforward to check the title of the business and any encumbrances on the land as all these are registered with the land registry and will be a matter of public record. For unregistered titles, there will have to be a title investigation to ascertain the quality of the title held.

Equitable charges are not required to be recorded as the title deeds are in possession of the chargee, which should serve as a notice to a potential purchaser or charge holder. If the transaction results in the transfer of the property the stamp duty implications may be significant. Many transactions are therefore structured so as to avoid the stamp duty charge. Stamp duty laws of the relevant jurisdiction will have to be considered before choosing a structure.

If the property is mortgaged or otherwise encumbered, depending on what the intention of the transaction is, the provisions of mortgage should:

(*a*) allow transfer of the mortgage to the new owner; and

(*b*) not trigger an event of default, or acceleration.

However, if the intention is to seek a first charge or a subsequent charge on the property the existing mortgage will have to be amended with the consent of the existing lender.

Planning permissions should permit necessary extensions or alterations if that is essential to the transaction. If the business is a tenant, all necessary protections under the tenancy statute must have been procured. If tenancies and licences cannot be registered with the land registry, the relevant agreements will determine the right of the business to the property. The tenure of the tenancy or the licence would in that case assume importance. The property should have all the essential utilities and the outgoings should not be prohibitively expensive.

For a transfer of business assets, plant and machinery and the stock in trade may have stamp duty implications. In most jurisdictions plant and machinery capable of delivery is exempt from stamp duty.

Employees, consultants and directors **2.12**

Normally, applicable regulations will protect the rights of the employees when a business, whole or part, is transferred to another. The employees are automatically transferred on the same terms and conditions. Transfers that do not involve the transfer of business will not be attracted by the provisions of such regulations. Therefore, the transfer of shares will not be affected by the regulations as the employer entity stays the same. Under a business transfer, the new employer assumes all the rights and obligations arising from the employment contracts, including all the collective agreements made on behalf of the employees, but not the benefits relating to occupational pension schemes, which would have to be transferred separately.

Redundancy of any employee as a result of the transfer of business may be deemed unfair dismissal unless the reason for the dismissal is an economic, technical or organisational reason entailing changes in the workforce. Regulations may apply irrespective of the size of the undertaking. Consultation with trade or employee union representatives will also need to be considered as part of any transaction which involves a change of ownership.

It is also important to check the restrictive covenants in the employment contracts. The restrictive covenants would be covenants relating to non-competition, non-disclosure and confidentiality. It would therefore be essential to confirm if the employment contracts have a clause relating to the assignment of the covenants and whether such assignment would be valid in the state where the contract is executed.

Service agreements with directors would likewise need to be reviewed as well as consultancy arrangements.

Pension arrangements 2.13

An actuarial expert will advise the company on various financial aspects of the pension arrangements such as:

(*a*) debts of the business to the pension scheme;

(*b*) contributions to be made to the pension scheme; and

(*c*) projected increases and losses to the pension scheme.

In one case, it was discovered that the target's subsidiary had a pension plan since 1976, which it had not accounted for, and the liability under the US Generally Accepted Accounting Principles (GAAP) treatment amounted to $US 20m. Employee benefit liabilities in the other subsidiaries, which were originally estimated to be $US 25m, turned out to be nearly $US 200m[1].

In the UK, a company is obliged to provide comparable pension arrangements to the employees. In the case of a business transfer, it will also have to ensure that the employees who were members of the pension scheme are transferred to the new pension scheme and that they are properly advised and given complete information on the new pension scheme (Transfer *of Undertakings* (*Protection of Employment*) *Regulations 1981* (*SI 1981/1794*).

Insurance **2.14**

The insurance cover should be as comprehensive as possible and cover all the assets and risks of an insurable nature – the key to which would be ascertaining the exposures and the risk involved. This in turn will determine the types and the amounts of the insurance covers. Insurance policies generally do not cover terrorism risk, which would require payment of an additional premium. The loss payee provisions and the assignment of the policies are a key in securing a debt and hence it is crucial the laws of the local jurisdiction do not prejudice the investor or lender.

Environment **2.15**

In an age of awareness of increased risk to eco-systems due to global warming and environmental pollutions, the regulatory authorities have endeavoured to keep the environmental pollution in check by forcing businesses to comply with the environmental laws. If the business requires authorisation to conduct any of its business activities, such authorisations should be valid and existing.

Warranties and indemnities **2.16**

While due diligence reveals information on the operations of the business, it does not necessarily address the concerns raised by the due diligence exercise. Following a due diligence exercise, a lender or investor may then seek to get extensive warranties and representations from the business owners and management which can be relied on and form part of the risk return analysis.

The warranties may take the form of indemnities which are undertakings that reimbursement or the making good of a loss will be made if a specified liability should occur. Indemnities are invariably taken in case of tax liabilities on sale of shares, litigation claims, environmental risks, doubtful claims, third party liabilities and any other matter, which has the likelihood of it resulting into a liability.

As due diligence is an integral part of an acquisition or investment or borrowing, it should be approached with a clear purpose. The purpose will depend on the strategy of the investor behind the investment, or simply the perceived lending risks. The emphasis should be on ascertaining the true risk and reward equation.

A KPMG survey on global deals[2] reveals that the priority accorded to due diligence has increased significantly with 35 per cent of the respondents citing it as the most important pre-deal activity.

Identification of key issues, clear guidelines and accurate analysis and interpretation of the findings is what the due diligence exercise achieves. It is an invaluable tool in a deal if employed properly.

References

1. Warner, E. *Acquisitions Monthly*. pp 33–36. October 2003 supplement.

2. Pratt, W and Issitt, M. *Acquisitions Monthly*. pp 12–15. October 2003 supplement.

Appendix

Example of initial due diligence list for the purchase of a regulated financial services business

Definitions	
Company	ABC and each of its subsidiaries
	Directors and compliance officer and other personnel having supervisory responsibilities for regulatory purposes
Regulator	Financial Services Authority (UK)
Accounts Date	31 December
Valuation Date	

1. Individuals/owners:

   - due diligence on each of the individuals (eg passport, two utility bills/bank statements showing proof of address and professional reference letter); and

   - confirmation of regulatory good standing and any material disclosures made to Regulator in respect of Individuals.

2. General information on the company (includes each of its subsidiaries), including financial information, tax, business arrangements and corporate governance:

   - full accounts of the company for the five years prior to valuation date;

- management accounts of the company for the period from the last accounting date to the valuation date;
- management forecasts produced prior to the valuation date;
- the proposed use of any cash balances in the balance sheet;
- any significant additional bank borrowings anticipated;
- capital commitments;
- anticipated capital expenditure, not committed;
- memorandum and articles of association;
- shareholders' agreement;
- estimates of property and investment valuations;
- details of property leases (particularly to estimate liabilities such as dilapidation), service property commitments, and capital or operating lease obligations in respect of assets and equipment;
- details of any assets not shown on the balance sheet, such as intellectual property rights;
- details of key directors and employees contracts including any consultancy contracts or arrangements as well as all service, benefit and pension or bonus arrangements proposed or made;
- tax computations for the same periods as accounts as well as confirmation that no outstanding tax obligations or liabilities exist;
- full disclosure of material non-recurring items;
- copies of minutes for Board, remuneration committee, audit committee and risk and credit committee;
- details of all pending or threatened litigation or dispute matters;
- details of any surplus assets;
- details of all computer software and hardware, licences and dealing platforms and data feeds and any plans to update or replace these;
- full details of aged:
  - debtors; and
  - creditors outstanding as at the valuation date and all book debts written off or provided for;
- full details of all professional indemnity and other insurance cover and any claims made;
- details of all material trading exposures and positions at valuation date;
- details of all bank borrowings and banking facilities;
- full inventory, asset and investment lists as at the valuation date;

- full details of any dividends declared or proposed subsequent to valuation date;
- full details of shareholder resolutions recently passed or made since valuation date;
- details of all joint venture and other business associations or agreements
- group structure chart and group organisation charts;
- compliance or operating manuals;
- full copy of customer documentation; and
- full details of all agency and commission arrangements.

3. Regulatory position and standing:
   - details of all licences, permissions and consents issued by regulator and confirmation that all remain good standing and have not been revoked, qualified or modified;
   - details of all material correspondence with regulator, site visit reports, external audit and management reports and steps taken to resolve and satisfy regulator on any issues arising;
   - details of any customer complaints and notifications to regulator and steps taken towards resolution;
   - evidence of compliance in last twelve months of all capital adequacy and/or solvency or liquidity requirements imposed by regulator or under governing regulation and law.
4. The company's current trading position (including each subsidiary) and prospects:
   - breakdown of revenue generation prior to valuation date and after accounts date:
     - fees; and
     - commissions;
   - revenue forecasts prepared prior to valuation date for next six months;
   - major clients and fee income or commissions representing greater than five per cent of revenue;
   - major contracts being performed, eg funds under management or being managed;
   - major future contracts anticipated at the valuation date;
   - capital expenditure either incurred or anticipated that would have an effect on revenue;

- sales brochures, giving an understanding of the product, for existing or new products;
- details of any new products or services likely to come on stream in the near future;
- spending on technology etc, for instance for the previous three years;
- details of any changes of the business in recent years, such as the closure of a division;
- is there any new legislation or expected legislation likely to affect the business?
- details of proposed hiring and recruitment of new employees or directors; and
- details of establishment of new divisions or operating subsidiaries.

Chapter 3
Money Laundering

Introduction 3.1

The primary objective in 'laundering' money is to conceal the source of the funds acquired illicitly and convert the funds so that they appear to have been obtained by legal means. This generally involves mixing of illicit funds with the clean funds to avoid detection by the authorities concerned. The stages widely recognised in money laundering are placement of money in the banking system, layering by creating complex structures (usually entailing a chain of bank accounts to throw the authorities off the paper trail) and integration by investing in legitimate business.

It is difficult to estimate the amount of money being laundered across the world. According to the International Monetary Fund (IMF) the aggregate size of money laundering in the world is between two and five per cent of the world's gross domestic product. Using 1996 statistics, the Financial Action Task Force (FATF) (see **3.3**) calculated the money laundered to be in the range of $US 590bn and $US 1.5 trillion.

History 3.2

The term money laundering is believed to have originated from the laundromats owned by the mafia. The funds earned from gambling, prostitution, extortion etc were mixed with the cash earning of the laundromat. These laundromats thus offered an opportunity to wash or launder the tainted funds into clean funds.

In the US a criminal investigation into money laundering was first established in 1919. Tax evasion was rampant during those times as it was not customary for a bank to ask about the source of funds before accepting deposits. The *Currency and Foreign Transactions Reporting Act*, better known as the *Bank Secrecy Act of 1970* (*BSA*) was passed which required banks to establish a paper trail. After the enactment of the *BSA* other statutes relating to currency transactions and financial accounts were enacted in the US. Money laundering was formally made an offence in US when the *Money Laundering Control Act of 1986* was passed.

The Financial Action Task Force 3.3

The Financial Action Task Force (FATF) was established in 1989 by the G–7 Summit in Paris to address the increasing drug problem and to counter organised crime and money laundering. FATF is an inter-governmental body which forms policies and procedures to tackle money laundering at national as well as international level. It has 31 member countries from the initial 16. FATF's original Forty Recommendations made in 1990 were updated in 1996 to reflect 'evolving money laundering typologies'. These 1996 Recommendations were endorsed by 136 countries. In October 2001, FATF included an additional eight recommendations on terrorist financing. The FATF Forty and Eight Special Recommendations have been recognised by the International Monetary Fund and the World Bank as the international standards for combating money laundering and financing of terrorism.

Other international bodies 3.4

Various international organisations are involved in formulating policies and raising awareness of money laundering mechanisms and means of fighting them. A brief description of some of these organisations is given below:

The Basel Committee 3.5

The Basel Committee was formed in 1974 and is made up of representatives from Belgium, Canada, France, Germany, Italy, Japan, Luxembourg, the Netherlands, Spain, Sweden, Switzerland, the United Kingdom and the United States. The Basel Committee prescribes broad supervisory standards and guidelines and recommends best practices for internationally active banks.

Though the Committee does not have any formal supervisory authority and the standards it sets have a persuasive effect. The Committee provides a platform for the countries to formulate their banking policies and have in fact been adopted by various countries. Its aim is to implement a common supervisory standard among the banks to ensure that no bank escapes supervision.

In October 2001 and 2003, the Committee issued the know-your-customer (KYC) programme which incorporated:

- a customer acceptance policy;
- customer identification;
- on-going monitoring of higher risk accounts; and
- risk management.

This Committee emphasised the need to have an appropriate legal framework to facilitate cross border sharing of information. In June 2003, the Committee participated in the joint issuance with the International Organization of Securities Commissions (IOSCO) and International Association of Insurance Supervisors (IAIS) on the assessment of their anti-money laundering/ combating financing of terrorism approach.

International Organization of Securities Commissions **3.6**

International Organization of Securities Commissions (IOSCO) is an international body formed to promote co-operation among securities regulators around the world. It has set standards for enforcement and exchange of information among the securities regulators. IOSCO was formed in 1983 and has 181 members.

European Union **3.7**

The European Parliament and the Council of the European Union (EU) aspires to strengthen the money laundering checks throughout Europe and have issued directives on prevention of the use of the financial system for the purpose of money laundering.

United Nations **3.8**

The United Nations' Global Programme Against Money Laundering is instrumental in influencing the money laundering policies of its member states. The United Nations (UN) co-ordinates and participates in initiatives taken by other international organisations in combating money laundering.

International Money Laundering Information Network **3.9**

The International Money Laundering Information Network (IMoLIN) is a web-based platform where agencies involved in combating money laundering can share and exchange information. IMoLin was created in 1996 under an agreement among various anti money laundering agencies. IMoLin can be accessed at www.imolin.org.

Money laundering offences **3.10**

FATF recommends that all countries should ensure that a money laundering offence applies to all predicate offences. A predicate offence is described as reference to:

(*a*) all serious offences; or

(*b*) to a threshold linked to a category of serious offence; or (*c*) offences punishable by a maximum penalty of more than one year's imprisonment; or

(*d*) a combination of these approaches.

The principal money laundering offences that most jurisdictions recognise are:

- conducting or attempting to conduct a financial transaction which involves proceeds from a criminal activity;
- transferring, transporting, converting or removing (or facilitating in any of these arrangements) a monetary instrument or funds or property of any kind where a benefit can be derived from a criminal conduct;
- concealing or disguising the nature, location, the source, the ownership, or the control of proceeds of a criminal activity;
- failure by a person to disclose a money laundering offence to the authorities where that person knows or suspects or has reasonable grounds for knowing or suspecting or where a person should have known or suspected even if there is no actual knowledge or suspicion;
- acquiring, using or possessing funds or any property representing proceeds of crime;
- where a money laundering investigation is being carried out, it is an offence to disclose such investigation to the person against whom it is being or intended to be carried out; and
- any transaction with the funds derived from tax evasion or making a false statement on a return will amount to a money laundering offence as the source of the funds then becomes illegitimate.

Indicators of money laundering offences 3.11

The Financial Intelligence Units (FIUs) of various countries had their first meeting in Brussels in June 1995. The FIUs analyse and process the information that may involve money laundering offences. This information is then passed on to appropriate governmental authorities for further action. The network of FIUs is known as the Egmont Group, a name derived from the venue in Brussels where the first meeting of the FIUs was held. The Egmont Group consists of 84 member countries. The Egmont Group recently compiled a list of 100 cases handled by FIUs around the world. The most common indicators of money laundering offences as discovered by the Egmont Group from these cases are:

- large scale cash transactions;
- unusual underlying business action (cross-border travel to undertake simple transaction);
- unrealistic business turnover;
- large and/or rapid transfer of funds;

- unrealistic wealth compared to client's profile;
- unusually high rates of return for a low risk business activity;
- unrealistic explanation given by customer for account activity;
- unwarranted high security risk – personal transfer of valuable asset;
- new customer attempting large transactions with no supporting rationale;
- unusually complex method of purchasing financial products;
- deposits at a variety of branches and times for no logical reason (possible evidence of 'smurfing', that is, making a number of small deposits under the threshold);
- identifying re-emergence of known financial criminal in financial services sector;
- lack of knowledge by individual atypical to trade practitioners;
- attempts to avoid identifying final beneficiaries of accounts;
- unnecessarily complex fund structure;
- possible client relationship to previous crimes;
- identification of false documentation;
- suspicious activity of client associates;
- deliberate concealment of fund ownership;
- change of account behaviour without explanation;
- multiple amounts paid into personal account without explanation;
- transfer of assets at well below (or above) market rates;
- over-complex fund movements;
- multiple transactions below threshold;
- illogical business activity – sending multiple cheques for cashing at a higher charge;
- multiple use of money transmission services; and
- unconnected parties channelling funds to a single account.

FATF's aim is to ensure that all the countries and territories have regulations in place to detect and penalise money laundering offences. FATF has compiled a list of twenty-five criteria to identify the detrimental rules and policies of the countries which limit or prevent international measures to combat money laundering offences. In its annual review of June 2003 FATF identified seven non-cooperative countries or territories (Cook Islands, Guatemala, Indonesia, Myanmar, Nauru, Nigeria and Philippines). Financial institutions should be cautious of transactions relating to any person or companies from these non-cooperative countries or territories.

Measures to prevent money laundering 3.12

FATF has recommended measures to be taken by the financial institutions, designated non-financial businesses and professionals to prevent money laundering. Apart from the indicators that are listed above, measures should be taken to procure more information about the customer and the transactions that are being carried out for such customer. Financial institutions are persons or entities whose activities include:

- lending;
- financial leasing;
- transfer of money or value;
- trading in:
  - foreign exchange;
  - derivatives;
  - securities;
  - commodities futures; and
  - other derivative transactions;
- life insurance and other investment related insurance.

The measures to prevent money laundering are described below.

1. Before opening accounts, the financial institutions should undertake customer due diligence to ensure that no anonymous or fictitious accounts are opened. This is done by verifying the identity of the customer. Documents like passport, a photocard driving licence and at least two recent utility bills should be obtained for the purpose of ascertaining the identity of natural persons. In case of other persons, proof of incorporation or similar proof to establish legal status, and the identity of directors or trustees or partners and the proof of authority from which they derive power to bind such person. If the information on a customer is obtained from a third person, it should be ensured that the source is independent and reliable and is regulated and supervised and has anti money laundering controls in place.
2. The ultimate beneficiary should be known and his or her identity should be ascertained. This is important to understand the ownership and control of the customer.
3. Ongoing due diligence should be conducted of the transactions undertaken at the behest of the customer and of the customer to ensure that information submitted by the customer is consistent.
4. It is imperative that the source of the funds is legitimate and checks should be made and evidence procured to make certain that that is the case.

5. Enhanced due diligence and monitoring for politically exposed persons as these prominent public figures may abuse their positions.

6. While dealing with a cross-border correspondent bank or other firm, additional due diligence measures should be adopted which include:

   (*a*) understanding the business of the correspondent institution;

   (*b*) determining that it has money laundering checks and regulatory supervision in place; and

   (*c*) ascertaining if any money laundering investigation was carried out against it.

7. Records on the customers and the transactions should be maintained for a minimum of five years. This is to provide a paper trail to the authorities in case they need to take legal action against the customer for any criminal activity.

8. The transactions should be examined in detail especially those which are complex for no apparent reason. Notes on findings and any other information relating thereto should also be maintained.

9. Customer due diligence and record-keeping requirements are also applicable to designated non-financial businesses and professions in the following situations:

   (*a*) casinos – when transactions are equal to or above the applicable designated threshold;

   (*b*) estate agents – transactions involving buying and selling of real estate;

   (*c*) dealers in precious metals and dealers in precious stones – cash transactions with equal to or above the applicable designated threshold;

   (*d*) lawyers, notaries, other independent legal professionals and accountants –transactions concerning the following activities:

   - buying and selling real estate;
   - managing client money, securities or other assets;
   - management of bank, savings or securities accounts;
   - organisation of contributions for the creation, operation or management of companies; and
   - creation, operation or management of legal persons or arrangements, and buying and selling of business entities;

   (*e*) trust and company service providers.

   The designated threshold in the case of:

   (*a*) casinos, including internet casinos, shall be €3000; and (*b*) dealers in precious metals and dealers in precious stones when engaged in any cash transaction shall be €15,000.

10. Any suspicion that funds are proceeds of a criminal activity should be reported to the FIU concerned.
11. Institutions should have internal controls to ensure adequate compliance management, employee training and policies which do not restrict employee disclosure of suspicious activities.

Most jurisdictions, which have money laundering regulations in place, require financial institutions, designated non-financials and professionals to have a money laundering officer. The employees are required to report suspicious activities to the money laundering officer who will then take it up with the FIU. Employees are required to receive appropriate training on anti-money laundering procedures, internal reporting procedures on suspicious transactions and reporting lines. Employers may be prosecuted if they fail to give money laundering training to their employees.

Reporting requirements and exceptions 3.13

A person is required to report suspicion of money laundering to the FIU or to the designated money laundering officer. A person reporting after the commission of any money laundering offence should have good reasons for not reporting before the commission of offence and the disclosure should have been made as soon as it was practicable. Where the disclosure should have been made but was not made, the person failing to do so should have reasonable grounds for not making such disclosure.

Under the UK law, to prove a money laundering offence it is not necessary that the benefit received should in fact be proceeds of crime. It is enough to have reasonable grounds to believe or suspect that such benefit consists of proceeds of crime (*R v Montilla [2004] 1 WLR 624*). However statutes now make it amply clear that it is an offence not to disclose a money laundering offence even if the person did not know but should have known about it. It is evident that the courts view money laundering as a serious offence and will interpret it strictly and it will be up to the accused to prove his innocence. The burden of proof required is civil and the courts may make far reaching assumptions to require a defendant to provide information for the court to rely on evidence even if not called at trial. (*R v Levin [2004] EWCA Crim 408, [2004] 07 LS Gaz R 35*)

It is an offence to disclose to a person who is reported in the suspicious activity report or against whom any money laundering investigation is being carried out or intended to be carried out which commonly referred to as 'tipping off'. However if the person perceived to be tipping off did not know or suspect that his or her disclosure is likely to prejudice any ongoing or intended investigation, he or she shall not be liable for committing a money laundering offence. Another exception to tipping off would be privileged communication.

There is lack of clarity on whether:

(*a*) a lawyer is liable to non-disclosure of a suspected offence if his or her knowledge of that information is part of the privileged communication; and

(*b*) the lawyer's duty to inform the client and give best advice amounts to tipping off.

These issues were raised in an English case of *P v P [2003] EWHC 2260 Fam, [2003] All ER (D) 141 (Oct).* pursuant to which the Law Society of England and Wales formulated a guidance note for the lawyers. This guidance note clarifies that a professional legal adviser does commit an offence of tipping off if he or she makes a disclosure to his or her client in privileged circumstances, that is, in connection with the giving of legal advice to the client, or to any person in connection with legal proceedings or contemplated legal proceedings. It was held in *P v P* that the duty to keep one's client informed was paramount and therefore it was inappropriate to withhold any information concerning the client.

According to the guidance note, there is no absolute duty imposed upon the legal adviser to inform the clients that he or she has made or intends to make a report to the FIU. If the legal adviser confers with their client to make a report to the FIU and if the client disagrees, then the legal adviser should withdraw from the case and consider carefully in accordance with the Law Society guidelines, while making a report to the FIU. A legal adviser does not commit an offence when in the course of giving legal advice to the client or acting in connection of actual or contemplated legal proceedings, they inform the client that they have made or intend to make a report to the FIU.

Cost effectiveness **3.14**

Financial institutions and designated non-financial professionals and businesses are likely to see a fair amount of time, man power and money being expended on implementing money laundering regulations. The US government's position on money laundering regulations is that it is losing billions of dollars on tax evasion and money laundered which escapes the net of the authorities. The UK's Financial Services Authority (FSA) appointed PricewaterhouseCoopers LLP (PwC) to conduct a cost benefit analysis of implementing a requirement for financial institutions to undertake a specific anti money laundering 'Current Customer Review' ('CCR'). The current customers were those who became customers before the institutions were required to comply with the money laundering regulations. PwC recommended two approaches:

(*a*) review of all the institutions by a fixed completion date; and

(*b*) issuing general guidance and making the institutions largely responsible for compliance.

It was found that approach (*a*) would cost £174m and approach (*b*) would cost £92m. Approximately 90 per cent of these costs would be borne by the institutions to implement adequate compliance and the balance would be assumed by the customers, National Criminal Intelligence Service (NCIS) (the FIU in UK) and other law enforcement agencies. The FSA concluded that it would not impose a requirement on the institution to carry out a review on the pre-money laundering regulations customers. The decision of the FSA was based on:

(*a*) the existing FSA regulations on money laundering that oblige the institutions to maintain adequate systems and controls to check money laundering;

(*b*) the findings which did not convince the FSA that the benefit of imposing CCR on the institutions was proportionate to the cost that would be incurred on the exercise;

(*c*) the fact that most institutions (33 per cent for approach (*a*) and 46 per cent for approach (b)) that were interviewed did not consider taking any further action if CCR was introduced as they already had undertaken a similar review or had adequate systems and controls in place;

(*d*) an increase in the suspicious activity report which would burden the NCIS when it was already under so much pressure; and

(*e*) the fact that the CCR would be just one aspect of money laundering controls.

The other areas include training, new customer due diligence, record keeping and monitoring. It was thought best to let the institutions decide how they want to manage internal controls and procedures to deal with their money laundering obligations in the most efficient manner.

The FSA, however, emphasised that their decision does not undermine the importance of customer identification and forms an elementary part of the anti money laundering exercise. The institutions are not complaining as they realise that the money laundering regulations are here to stay, especially in the wake of 9/11 which revealed close nexus of money laundering to terrorism financing. Governments have shown serious determination to tackle this at grassroots and any lapse on the part of the financial institution and professional bodies and businesses in complying with their obligations could cost them dear.

Data Protection Act 1998 3.15

The UK *Data Protection Act 1998* (*DPA 1998*) provides that an individual is entitled to be informed by the data controller whether his or her data is being processed, the description and purpose of the data that is being processed, description of the recipient whom the data is being disclosed, and entitled to be communicated in an intelligible form as to what constitutes his personal data and the source of that data. An exception is carved out to enable the data controller to withhold the information which would likely prevent detection of a crime or prosecution of the offender. The UK government published a paper in April 2002 exploring the interrelation between the tipping off provision under the statutes dealing with money laundering and *DPA 1998*. The government clarified that that the paper has no legal status.

At the first instance it would seem that where a suspicious transaction report (STR) is made the exception under the *DPA 1998* would apply and the data controller is not obliged to disclose the information on STR to the individual concerned. However, the paper warns that no assumption should be made that the exception under the *DPA* automatically applies to STRs. A case by case approach should be taken to determine the applicability of the exception to STRs. When in doubt it is best to consult the NCIS.

Conclusion **3.16**

Tackling money laundering requires a co-ordinated effort at a global level. All countries have to make a concerted effort to implement a regulatory framework to allow transparency and adequate supervision. Countering money laundering requires co-operation from all the countries and the institutions involved in financial transactions as well as the professional advisers who structure the deals for the money launderers. Many countries have already implemented or are in the process of implementing regulations extending to professional advisers which impose a duty on them to disclose any suspicious money laundering activity. Under the guise of secrecy and confidentiality, many financial institutions prevent detection of money laundering crimes.

Countries should amend such secrecy laws and endeavour to give mutual legal assistance. Authorities should be able to promptly act on requests for freezing the accounts or confiscating the criminal property. Regulations to combat money laundering have evolved in the last few years, particularly with the countries realising what consequences it can have on its economy and on its people. International agencies like FATF, Basel Committee, the United Nations and the European Union have issued directives and money laundering guidelines which offer a basis for countries to formulate their money laundering regulations. Many countries have been proactive in forming specialised task forces and groups to exclusively deal with and monitor the money laundering activities. These actions have increased the detection of money laundering related crimes. However, awareness at macro and micro level is required to understand the seriousness of the consequences of money laundering.

Chapter 4

Drivers for Due Diligence and Corporate Governance

Introduction 4.1

There is no doubt that the business environment requires that due diligence should be understood as a core feature of doing business in today's world. Whereas the impact of due diligence was originally felt by larger organisations, nowadays drivers such as supply chain pressures, reputation issues, regulatory and voluntary frameworks have meant that due diligence is a matter of concern to most organisations. This is especially true bearing in mind the interaction with risk management and corporate governance. As a result, the business environment demands due diligence as an ongoing tool to deal with both internal and external requirements.

It has been seen in **CHAPTER 1** that internal pressures come about through the deals, transactions, joint relationships and operational issues of the business. In addition, it may be noted that the key external drivers are regulatory and reporting standards that affect stakeholder and insurer confidence. The issues and concerns that were once only the domain of large businesses have crept into that of small businesses and small and medium-sized enterprises (SMEs) as they deal with the implications of today's business environment and scrutiny of bureaucracy, regulation, customers and non-governmental organisations (NGOs) as well as the media. New business relationships are forcing improved standards on organisations of all sizes and indeed relationships can impact on the reputation of all those involved.

While a jurisdiction's company law governs companies incorporated in that jurisdiction, a jurisdiction's securities laws and regulations govern companies, investors and intermediaries involved in the buying or selling of securities in that jurisdiction. Jurisdictions may have private and 'over-the-counter' markets for equity securities, but the most important equity markets are stock exchanges. These also apply to international organisations. For instance, in the oil sectors, most of the major listed companies have their primary listings on a stock exchange in the US, and two-thirds of those with primary listings on other exchanges have secondary listings on a US stock exchange. Therefore changes in US requirements have major implications for how business is done generally. For this reason the analysis of the disclosure requirements of the US Act *Sarbanes-Oxley Act of 2002 (SOX),* set out in **CHAPTER 11,** is quite detailed.

The US regulatory framework has been considered to be a major driver for changes in Europe and Asia (see comparative developments of corporate governance in **CHAPTERS 11** to **15**). This reflects the importance of the US equity market within the global capital markets. It makes the US securities laws and regulations generally a critical source of corporate governance law for major companies in these industries, and an important influence over the laws of other jurisdictions seeking to attract global equity capital for their domestic companies. It should also be noted that the UK is the other main source of the global capital markets alongside the US and has influence as a result on the trends in corporate governance (as is clear from discussions in later chapters: see **CHAPTER 10**, **12** and **14** in particular).

One other driver is the community perspective on corporate governance, particularly with the international human rights framework governing corporations. Increasing attention is being paid to the use of proxy votes of shareholders. For example, public pension funds, religious organisation funds and labour union investment funds use the shareholder proposal process to pressure companies regarding environmental and labour-related issues. The wide distribution of company proxy statements make them a powerful foundation for a media campaign to direct attention at the activities of multinational corporations. If there is a serious prospect that a proposal could garner a majority vote, and hence represent, at least formally, the will of the shareholders, proponents of resolutions are often in a position to seek company concessions in exchange for withdrawing the proposals. Proponents of shareholder resolutions generally make some effort to associate their proposals with increasing shareholder value, but they are often investing in the companies simply in order to qualify as shareholders in order to make their proposals.

One example relates to Baker Hughes. Baker Hughes is engaged in the oilfield and process industry segments. It also manufactures and sells other products and provides services to industries that are not related to the oilfield or continuous process industries. A recent proxy statement included a proposal to implement the MacBride Principles in Northern Ireland. The MacBride Principles, consisting of nine fair employment, affirmative action principles, are a corporate code of conduct for US companies doing business in Northern Ireland. The Principles do not call for quotas, reverse discrimination, divestment (the withdrawal of US companies from Northern Ireland) or disinvestment (the withdrawal of funds now invested in firms with operations in Northern Ireland). They are intended to encourage non-discriminatory US investment in Northern Ireland.

Another example is ChevronTexaco's resolution requesting reporting on renewable energy, and Exxon Mobil's six proposals dealing with social responsibilities issues. When activist shareholders began attempting to use this pressure technique with BP, it began including instructions for 'members' requisitioned resolutions' on its investor centre website to clarify the differences between UK law and US law, particularly the UK prohibition against shareholder resolutions which simply express opinion. Nevertheless it is clear that developments in one jurisdiction influence trends and decisions elsewhere, especially with the powerful forces of the media and technology at work.

As will be seen below, the business benefits achieved by businesses that follow proper due diligence procedures and good corporate governance can be demonstrated in bottom line performance and stakeholder confidence. It is increasingly recognised that those businesses that manage their business ethically and having regard to these concepts are the better managed businesses overall. They are also closely aligned to the other well-known goals of sustainable development and sustainability (see below and **CHAPTER 16**).With this in mind an organisation will need to establish whether its management programme is consistent with its best practice policy and has regard to relevant developments in risk management, as well as current practice in corporate governance (see **CHAPTER 10**).

As regards listed companies, the primary exercise by shareholders of their right to participate in the governance of their corporation is in connection with election of the board of directors. Specific information is required in connection with the election of directors. This includes for each person nominated for election by the shareholders:

- their age;
- business experience over the last five years;
- directorships of other listed companies;
- any position they have had with the company;
- any understandings with others related to their selection as a nominee;
- legal proceedings where the director's interests are adverse to the company's;
- any bankruptcies of companies they were executives or partners of;
- various types of legal proceedings they have been involved in; and
- indebtedness, financial transactions or other relationships with the company.

In view of the independent review of the role of non-executive directors (NEDs), the Higgs Review, (see also **CHAPTERS 7 AND 10**) the appointment of NEDs is becoming more sensitive as the level of responsibility and integrity is raised. Moreover, whereas the review was aimed at large businesses, advisers have noted that, as with other regulations and standards relevant to ongoing due diligence and good corporate governance, there are impacts also on small businesses. Bearing in mind the increased risks and the increased vigilance of shareholders, as well as other stakeholders, a proactive approach can assist business performance, whatever the size of the organisation.

Business benefits through risk/reward balance 4.2

Risk is part of the entrepreneurial culture needed by any organisation that wishes to develop, expand and improve. An effective risk management structure allows an organisation to understand the risks in any initiative and to take informed decisions on whether and how the risks should be managed. Risk management is now a practical feature of business, just as due diligence should be recognised as another proactive tool that is of vital assistance in the management of risk. Nowadays, no matter where an organisation operates and whatever its size, risk management is becoming an increasingly essential part of today's business environment and is of interest to:

- chief executive officers (CEOs), financial directors, other executive and non-executive directors;
- company secretaries;
- heads of internal audit, risk management or other assurance functions; and
- all managers with an interest in risk management.

Risk should be regarded as an opportunity to improve not only the management of the particular risk or uncertainty in a specific project but also the business as a whole (see also **CHAPTER 10** As is recognised in the UK's risk community, projects involve an element of risk and the success of any project depends on managing this risk. Controlling risk requires an understanding of the dynamics of change and a healthy respect for the unexpected.

The discipline of risk management extends to, for example, environmental management and provides a range of tools and techniques to help organisations, and the individuals within them, to succeed in a complex world. Although risk management will never be a substitute for prudent judgment, it can sharpen innate wisdom and improve decision making to help organisations both survive and thrive. This is very important in the UK in the light of the Turnbull Report – a report that has helped companies formalise risk management (which is referred to in **CHAPTERS 7** and **10.**The date for compliance with Turnbull was 23 December 2000. Turnbull has required that any risk management programme should focus on the way to achieve full disclosure to ensure that processes are embedded in corporate policy at Board level for listed companies. Any additional practical advice on how to create an embedded and ongoing risk management process should be valuable. Benchmarking with other companies is a beneficial tool to help in this endeavour (see also **CHAPTER 6** regarding commercial due diligence). A process to conduct a benchmarking study that embraces the concepts of risk management in a selected sector having regard to the size of the business is clearly useful in today's business environment. The overall objective will be to determine how participating companies manage these risks.

Benchmarking – an illustration 4.3

By way of example, a utility that has inherited a great deal of contaminated land as part of its asset base would prioritise its environmental management programme. In view of the fact that the management of contaminated land is an increasingly important concern in the UK and elsewhere, this is considered in more depth in **CHAPTER 16** For the purpose of the current example, in order to objectively assess environmental programmes relative to contaminated land, an organisation might find it helpful to engage experts to carry out the following four step benchmarking study.

1. Map the various processes and options available in a typical contaminated land site management programme.
2. Define the values and objectives of 'best practice' in the context of any environmental policy for the management of contaminated land.
3. Design a benchmarking study that will provide a consistent framework to evaluate the peer industries relative to their management of contaminated land.
4. Carry out the study, compare the results to the programme and report the findings.

In mapping the processes the experts would look at both the business and technical aspects of decisions and actions that would be involved with the management of a contaminated site. Particular emphasis would be placed on how these decisions and actions could impact the value chain for an industry. The value chain in this context would not only be the bottom line profitability of the company, but also include other factors that impact value like reputation, public heath and welfare, risk of regulatory enforcement and risk management. This is particularly important having regard to both ongoing corporate policy and specific projects. This would then enable the organisation to define the basis of best practice. Often 'best practice' really means 'effective practice'. It is a question of where the company wants to fit (that is the best, top ten per cent, above the median) and with what group it is to be compared.

For example, in its policy statement prior to the merger with Transco (the gas transporter), National Grid stated:

> 'Following the principle of integrating environmental considerations into all aspects of the group's activities we will:
>
> - as a minimum, meet the statutory requirements of environmental regulations;
> - seek to keep known adverse effects on the environment to a reasonably practicable minimum; and
> - seek to achieve as high an overall level of environmental protection as is reasonably practicable.'

It further went on to compare itself using the FTSE 100 Environmental index for the Utility Sector and Business in the Environment's Index (BiE) where issues of corporate governance, sustainability and ethical investment are significant. This would appear to be an appropriate starting point for the basis of the design of the benchmarking study, that is, a comparison as to how other utilities (or possibly non-utilities) in the UK are dealing with contaminated sites. If there are more effective methods known in the EU or US spheres these could be reviewed.

However, one needs to keep in mind that National Grid is functioning within the UK regulatory context. National Grid improved its placing in the FTSE 100 table from number 44 last year to 24 in the latest index. They also came tenth out of 20 companies in the utilities sector, which has the highest average score for all the groups.

In such an instance one would review the existing policies and extract from that meeting a better understanding of values and objectives that would be expected to fulfil the intent of the policies relative to the management of contaminated sites.

The design of the benchmarking study would build onto the previous task. A series of questions would be developed that would extract from peer companies how they manage the values and objectives that have been identified as important to National Grid. The environmental policies, organisational, structure, financial management and corporate objectives for contaminated site management would be included in the benchmarking study. It would be important to identify a list of several peer companies to conduct the survey. The companies could be all UK based or include companies based in the US or other countries. Also, the companies could include all utilities or may include other related industries. The final benchmarking targets would be dependent upon the definition of the values and objectives determined important by National Grid.

With gas site remediation projects in mind, the benchmarking programme would focus on the ways the participating companies:

- manage risk in the project environment;
- implement a risk analysis framework;
- define qualitative and quantitative risk analysis;
- discover the role of assurance providers and how they can add value;
- analyse the different tools and techniques available for project risk management; and
- create an effective risk response plan.

Once the benchmarking has been completed the data would be reduced, evaluated and reported to the organisation. The key deliverable for the final benchmarking study would be a report that summarises:

- how the companies manage the key values and objectives previously identified;
- how the business relates to the peer practices; and

- a list of other effective practices that could further enhance its management programme.

Several of these issues are also relevant to the discussion in **CHAPTER 16** regarding environmental due diligence, as well as in the consideration of risk management and the SERM case study referred to in **4.24**. Moreover the approach to risk management has relevance to the culture of the company or business, as is noted in **CHAPTER 8.**

Regulatory background and drivers 4.4

For the purposes of this handbook, the question of drivers is a broad subject that embraces a review of the economic development, value and community perspectives on corporate governance. Those perspectives drive the formal and informal frameworks of responsibility and accountability applicable to the people and institutions that control, manage and influence organisations. It is not intended to go into the details of the formal and informal frameworks of laws, regulations and norms such as the company laws that govern shareholder participation in governance as well as the international human rights framework governing less formal responsibilities of companies – these can be found in specialist texts. Moreover, the key frameworks that pertain to corporate governance are referred to in **CHAPTER 10**.

Instead it is intended to raise awareness of the increasing scope of what has come to be due diligence and corporate governance in today's business environment with just a brief mention of the main areas of relevant legislation, that is company law and securities law.

In the UK some of the main legislative instruments are:

- the *Companies Act 1985*;
- the *Business Names Act 1985*;
- the *Companies Consolidation (Consequential Provisions) Act 1985*;
- the *Insolvency Act 1986*;
- the *Company Directors Disqualification Act 1986*;
- the *Financial Services Act 1986*;
- the *Companies Act 1989*; and
- the *Financial Services and Markets Act 2000.*

As regards the EU, there has been considerable activity over the years in the areas of company law, as well as more recent initiatives to deal with corporate mismanagement concerns. The European Commission has been working on the harmonisation of the rules relating to company law and corporate governance, as well as to accounting and auditing. This is considered essential for the proper functioning of the internal market.

While company law defines the issues that shareholders may or must decide upon at annual general meetings, as indicated, for listed companies the applicable securities laws and regulations governing proxy solicitation have a significant practical impact on the extent to which the vast majority of shareholders, who do not attend the meetings, actually participate in the decision making at those meetings. The term proxy solicitation has evolved from the key proxy solicitation rules – in that context they are:

- the need for proxies to include the ability to vote against, as well as for, management's proposals;
- the information which must accompany proxy solicitations; and
- the facilitation of shareholder proxy solicitation and proposals.

Generally, company law governs the authority and rights of the different organs of the corporation and their responsibilities toward each other. It applies to all companies under the law's jurisdiction, whether closed, privately held or listed, and is primarily designed to make corporations effective entities for doing business.

By way of example, in March 1998, the British government set out, in the paper 'Modern Company Law for a Competitive Economy', its intentions of taking a broad approach to review the framework of company law. A consultation document was subsequently issued in March 2000, entitled 'Modern Company Law For a Competitive Economy: Developing the Framework' by the Company Law Review Steering Group (available from www.dti.gov.uk/cld/modcolaw.htm). The draft statutory statement of directors' duties comprised the following:

- compliance and loyalty;
- independence of judgement;
- conflict of interest;
- fairness; and
- care, skill and diligence.

As regards the latter, a director must exercise the care, skill and diligence which would be reasonably expected of a director in his position and any additional knowledge, skill and experience which he has.

The Steering Group presented its Final Report to the Secretary of State on 26 July 2001. The government published its response to the Company Law Review's major recommendations in a White Paper: Modernising Company Law (Cm 5553) published on 16 July 2002 (see also below in the discussion of SRI in para **4.32**).

Securities laws and regulations govern the rights and obligations of corporations that issue securities vis-à-vis investors in those securities. They also deal with the rights and obligations of buyers, sellers and intermediaries involved in the trading of those securities. They are designed to protect investors and encourage the proper direction of capital markets. In the context of corporate governance, these laws generally deal with disclosure of information material to a decision to buy, sell or hold shares in a company. Business must also take account of stock exchange listing rules, which impose governance standards on the companies that list on the exchanges, and cross border issues, as they relate to shareholder participation.

It should also be noted that the dividing line between what is the proper subject for company laws, or securities laws and regulations, or listing rules differs from one legal regime to another, such as the listing requirements for the London Stock Exchange or the Alternative Investment Market or the New York Stock Exchange, and they do not apply symmetrically to corporations. Global companies incorporated in one country often list their securities on the exchange of another country, and have investors in yet other countries, and the legal framework applicable to their corporate governance can be complex. The most important aspects of these areas of the law – in relation to both the company and securities frameworks, as they affect the corporate governance of listed companies, can also impact on smaller organisations. Therefore they too must be aware of:

- laws and regulations governing the board of directors and management;
- accountability of the board and management; and
- the corporate governance structures and processes that companies have in place to carry out good corporate governance in practice.

Moreover in today's fast moving pace of business development it is important that the growing business should be aware of the potential trends and challenges that will require ongoing attention and decision-making.

The choice of vehicle, structure and liability in business 4.5

Part of the above discussion (for example in **4.1** and **4.4**) assumes that the reader is a larger and more advanced business entity that has been in operation for some time it is as well to recognise also the core decisions that must be made in the business environment in order to establish a vehicle that suits the choice of operations. Moreover, as is seen in other chapters, such as **CHAPTER 9** regarding e-commerce, it is important to recognise the legal status of any party that the business deals with. A brief analysis of the advantages and disadvantages of the available business vehicles is therefore relevant here. This decision can lead to one of the most important management decisions made. The discussion focuses on the circumstances in the UK. It should be noted that other practical issues, such as the appointment of a finance director at an increasingly early stage and the appointment of non executive directors (NEDs) should also be considered as early as is practicable. While the choices and decisions made may not remove risk or liability (see **4.6**) they can mitigate repercussions.

Sole trader, partnership or limited liability company? 4.6

It is easier to set up a business organisation in Britain than in most industrial nations. Whether or not the business prospers is another matter, but it is unusual for bureaucracy to be the cause of failure. Getting the structure in place can be remarkably simple. This chapter considers briefly which business structure is most appropriate for the contemplated or existing business, with particular emphasis on how to limit or reduce personal liability for business mistakes or business insolvency.

Firstly, it is appropriate to consider the sort of circumstances which might lead to personal liability on the part of the trader and place his or her personal assets at risk. It probably goes without saying that dishonest or reckless conduct by the trader in the course of his or her business, whether or not as a sole trader, partner in a firm or company director may well lead to a personal liability for losses suffered by others as a result. Being honest and conscientious, however, is no guarantee that personal liability will not arise and even experienced or skilled business people are likely at some time to make an error of judgment. The one-man financial adviser who unwittingly gives what turns out to be bad advice, the restaurateur whose suppliers supply sound looking but contaminated food that is served to a customer and the international auditing practice who failed to uncover a fraud buried deep in the accounts may all be at risk.

Being careful is advisable, but it is not enough. How risk can be reduced will depend to some extent on how the business entity is organised. The more common alternatives are considered briefly below.

The sole trader 4.7

This is the simplest form of business organisation. With a few exceptions, the trader will be free to set up and run his or her business without prior registration, licence or formality. Like everyone else, the trader will have to keep business records so that he or she may properly account to the Inland Revenue. Depending on annual turnover, the trader may have to register for VAT purposes (at risk of being personally liable for uncollected VAT if he or she fails to do so). For some businesses (for example where public health maybe involved), it may be necessary to first obtain formal registration or a permit to trade. In the case of most professions, it will be necessary to comply with professional rules and restrictions. For the majority of businesses, however, no such obstacles or restrictions apply.

The business may be carried on in the name of the trader or the trader may use a business name. If the business name is not offensive and does not suggest that the business is, for example, a limited company, or connected with government, the trader may generally adopt whatever name he or she pleases without the need to obtain permission or register it.

If the sole trader incurs a liability in the course of his or her business his or her personal assets will be at risk. When that happens, there is no differentiation between the assets employed in the business and personal assets outside the business. All the sole trader can really do is:

- trade honestly and carefully;
- choose and supervise employees, agents and contractors carefully (he or she may be liable for their mistakes);
- if appropriate in the business, trade only on the terms of his or her own written trading terms and conditions limiting the extent of responsibility. This option is not generally available to the professions;
- insure against loss and liability; and
- operate good financial and credit controls

Paying lawyers to draft a tailor-made set of trading terms and conditions is likely to be expensive, but should be cost effective in the long term. Sometimes membership of a trade association enables the member to use the association's own standard trading terms and conditions designed to protect its members. One off liability insurance policies can be expensive. Sometimes a better insurance deal is also available through membership of a trade association.

The sole trader may consider forming a limited liability company (see **4.9**) through which to operate his or her business. This can be done and still be a sole trader in reality. He or she should then be careful to make it clear to those with whom he or she trades that he or she is not the person with whom they do business (particularly if his or her clients have become used to dealing with him or her as a sole trader). He or she is merely the agent (as a director) of the company.

These days many companies, including major companies, engage consultants to work on a regular basis but without the consultants becoming employees of the company. Sometimes, to avoid it being said that the consultant has in reality (and in the eyes of the Inland Revenue) become an employee or part-time employee of the company, the company insists that each individual consultant must render his or her services as a limited company. This is an example of how a sole trader, otherwise content to trade as such, may decide to conduct his or her business through a company even though, when one looks behind the company, there is no-one there except that former sole trader.

The partnership **4.8**

When two or more traders decide to combine operations and share assets they may form a partnership which lawyers, but not everyone else, know as a firm. Alternatively, they may, for example, like doctors in a group practice or barristers in chambers, work together only in the sense of sharing accommodation and overheads whilst continuing to trade or practice as individuals. In the latter case, they remain sole traders. A partnership can be easily formed and it can quickly and easily be dissolved, but the consequences of dissolution can be far-reaching. The distinguishing feature of a partnership is that each partner is personally liable for the whole debts of the partnership. Again, that means personal assets, not just partnership or business assets. This applies even when the client had never previously had any business dealings with the partner who has become liable for his or her partner's loss or error of judgment. Perhaps then the first rule of risk management in a partnership is to choose one's partner (or partners) carefully.

A partnership can be created at will. Whilst it is advisable to have a partnership deed spelling out the structure of the partnership and, in particular, the personal liabilities of the partners towards each other in the event of partnership insolvency, the partnership can come into being without any formality whatsoever. Each partner places his or her personal assets at risk on entering into a partnership. In these circumstances one wonders why this form of business entity should ever be adopted. It has the advantage of great simplicity and that is perhaps its primary attraction. However, some professions (for example solicitors) have for many years been obliged to conduct their business either as sole traders or partners. The theory behind not allowing some professions to limit liability by incorporation is that it keeps the practitioners 'on their toes' and by preventing outsiders from participating in the profits it keeps the practice in the hands of the professionals. With compulsory insurance to compensate for losses, this obstacle to incorporation is increasingly being regarded as outmoded.

Partnerships can be very useful for occasional, as opposed to full time, business or, for example, when friends or members of a family pool their resources to purchase a property to let for investment purposes.

Generally speaking, professional rules, regulations and restrictions aside, most or many partnerships could probably operate satisfactorily as limited liability companies. Some traders, however, prefer to remain in partnership rather than become directors of their own limited company and then be treated by the Inland Revenue as employees and taxed accordingly. Another attraction may be that partnerships are not (unlike limited companies) required to file, for public record, partnership accounts. A self-employed partner in a business may conclude that the personal tax advantages he or she may enjoy outweigh the peace of mind of being a director and/or shareholder of a limited liability company. Perhaps we will see more rapid development of the limited liability partnership (see **4.9**) that lies somewhere between a partnership and a limited liability company.

Personal liability, or the risk of it, can be reduced in the same ways as the sole trader, added to this must be the caution to choose one's partners carefully. Partners should also recognise the need to diplomatically 'police' each other's activities.

What often happens is that the sole trader progresses to a partnership. The partners then conclude that it is cheaper and therefore preferable to run the business as a limited liability company than to pay high liability insurance premiums.

The limited liability company **4.9**

Unlike setting up as a sole practitioner or as a partner in a firm, establishing a limited liability company requires a certain formality. It has become a common business vehicle in the UK. In essence the nature of a limited liability company is that:

- it is a legal 'entity' and is regarded in law as existing entirely separately to its shareholders and directors. It is the company that carries on the business not its directors, officers or owners;
- it is owned by its shareholders. The business of the company is under the control of its directors, who may or may not be its shareholders, and who are appointed by the shareholders;
- the personal liability of the owners of the company (the shareholders) to those with whom the company comes into contact is, with few exceptions, limited to what is owed by the shareholders for shares in the company which they have taken up but for which they have not yet paid; and
- although the shareholders appoint the directors, they have very limited control over the day to day running of the company (subject to the comments above bullets).

If the company incurs liabilities which it cannot meet from its own resources, the circumstances in which the shareholders or directors will be called upon to make up any shortfall are very limited, provided the loss has not been brought about by the dishonest activity of the shareholders or directors. There are exceptions to this rule. For example, directors may be personally liable for unpaid income tax and national insurance contributions which have been collected from employees but not paid over to the authorities. The tendency over the years has been to increase the circumstances in which directors' personal liabilities may be involved by increasing legislation (see examples in **4.4**) intended to ensure good corporate management. Compared to a partnership, the directors and shareholders are much less exposed. Directors' liability insurance is generally available.

A limited company is a creature born of statute. It owes its existence to due observation of rituals required by the *Companies Act 1985* (*CA 1985*). Its formation and constitution is a matter of public record. Companies are required to keep and file up to date information concerning the shareholders, directors and location of registered office, etc. Companies are also required to file accounts each year showing turnover, profit and loss, etc. All this information becomes a matter of public record and can be obtained by any member of the public making a search at Companies House or via the Internet where information can be downloaded or faxed to the applicant on payment of the appropriate fee.

The structure and operation of the company will be determined by its Memorandum and Articles of Association (its charter), by the *CA 1985* and by the great body of law that has come into being over the years since the concept arose of a business organisation existing independently of its members.

Despite the due solemnity which necessarily accompanies the formation of a company, a new company can be purchased for a relatively cheap price. Company formation agents will sell you a 'ready made' company 'off the shelf', that is a company which has already been incorporated with temporary shareholders and directors and is available for instant transfer to its new owners. They will arrange for the temporary directors and shareholders to withdraw so that the new directors and shareholders can take over. Someone can walk in, buy the company over the counter and start trading immediately. They can also arrange to change the company's name, which will usually take a few days. Alternatively, if not in a desperate hurry, the formation agent will make a 'bespoke' company according to requirements and with the name of choice, if that name is available. Solicitors and accountants also use the services of company formation agents. A company can be bought elsewhere that has already traded and may already have an on-going business.

Private companies (identified by the word 'Limited' or abbreviation 'Ltd' or its Welsh equivalent 'Cyfyngedig' or abbreviation 'Cyf' at the end of its name) are not allowed to offer their shares to the public. If the business of the company succeeds and outside capital is required, it may be appropriate to transfer the business of the company to a public limited company (whose name will end in 'Public Limited Company' or 'PLC).

Limited companies also have their less obvious uses. Leaseholders in a block of flats who decide to exercise their collective right to acquire the freehold may arrange for the freehold to be held by a limited company in which each leaseholder becomes a shareholder. Ownership of the freehold by a limited liability company provides a democratic structure with all the protection that is provided by company law (see **4.4** above). Charities or clubs may also be incorporated into a company, although a variant form is usually used – the company limited by guarantee. Such a company has no share capital and therefore no shareholders. It has a board of directors (or trustees) who operate the company and whose personal liability is limited to a predetermined, and perhaps nominal, sum which is the extent of contribution that can be called up in the event of the company being unable to meet its liabilities. For further discussion of Charities and charitable companies see **CHAPTER 17**.

The company can limit its own liability and the risk of liability in the same way as a partnership, but the owners' and directors' personal liability for the debts of the company are already limited. Shareholders should take care who they appoint as directors, and the directors should police each other.

Summary **4.10**

The below list provides a summary of choice of vehicle, structure and liability in business.

- Sole traders remain personally liable for the debts of their business. Turnover, profit and loss are not a matter of public record.
- Partners in a partnership (firm) are each personally liable for the debts of the partnership. Turnover, profit and loss are not a matter of public record. The partners may or may not be treated as employees, according to their status within the partnership.
- Partnerships are an easy way to get started and may eventually lead to a transfer of the business to a limited company.
- Limited liability companies, including public limited companies, provide limited liability for their shareholders, directors and officers. Ownership, turnover, profit and loss are all a matter of public record as is the identity of the board of directors.
- Big business, requiring the employment of outside capital, tends to be conducted through public limited companies. Private limited companies are ideal for small businesses, including the sole trader, as well as major family owned businesses.
- A well tailored liability insurance policy will take care of liabilities that arise when good and honest business practices and trusted colleagues are not enough to prevent losses arising. It should be noted, nevertheless, that many of the trends in the business environment are creating major ramifications for today's insurance industry.

The Limited Liability Partnership **4.11**

The *Limited Liability Partnerships Act 2000* (*LLPA 2000*), which came into force on 6 April 2001, created the business entity called the Limited Liability Partnership (LLP). The Act is much more accessible than many other legislative instruments. It is unusually short, consisting of 19 sections and a schedule. These cover the core characteristics of the LLP, including:

- the requirements for incorporation;
- membership issues (including, becoming and ceasing to be a member);
- the LLP agreement;
- taxation;
- regulation and definitions;
- commencement; and
- application.

The implementing regulations are the *Limited Liability Partnerships Regulations 2001* (*SI 2001/1090*).

Main characteristics of the LLP 4.12

The LLP is a body corporate, that is an entity with a separate legal personality from its members and it has unlimited capacity. The LLP is the subject of the duties and liabilities of the business. On the winding-up or dissolution of the LLP, the liability of its members to contribute will be limited to whatever has been agreed in this respect. Every member of an LLP stands as an agent of the LLP, subject to some exceptions. The LLP protects its own members from personal liability for both their own acts and those of their fellow members.

In order to incorporate an LLP there must be two or more persons associated for carrying on a lawful business with a view to profit, and they must subscribe their names to an incorporation document. The two persons may be natural persons, bodies incorporate, trustees or partnerships and they may reside anywhere in the world. There must be at least two designated members or provision that every member is to be a designated member (see **4.13**).The incorporation document is delivered or submitted in paper form to the Registrar of Companies. The approved form of incorporation document is form LLP2, which sets out a clear format for disclosing the information prescribed by the *LLPA 2000* at the time of incorporation. The LLP2 must provide:

- the name of the LLP which must end with 'Limited Liability Partnership', 'llp' or 'LLP;
- the legal domicile of the LLP, such as England and Wales or Scotland;
- the address of the registered office of the LLP;
- the name and address of each member of the LLP; and
- the details of the designated members.

A well drafted written LLP private agreement should also be formulated which, unlike the memorandum and articles of a company, are not public documents.

Designated members 4.13

The designated members have the responsibility to ensure that the LLP should meet its disclosure obligations under the *LLPA 2000*. As it is a vehicle with limited liability, the LLP is also subject to the compliance regime or requirements of the *Companies Act 1985*. Therefore, like a private limited company, the LLP must file an annual return and statutory accounts with the registrar of companies at Companies House. In addition an LLP is similarly subject to the provisions of the *Insolvency Act* 1986 (*IA 1986*) and the *Company Directors Disqualification Act 1986* (*CCDA 1986*). Accordingly, members of an

LLP are subject to potential personal liability under the provisions of the *IA 1986* relating to 'wrongful trading'. They are also subject to disqualification under the *CCDA 1986* in the same way as company directors.

Comparison with UK private limited company 4.14

The LLP differs from the UK private limited company in three important ways.

1. An LLP can establish the decision-making and profit distribution arrangements more or less as the members wish. Whereas certain other provisions of the *CA 1985* apply, the LLP is not subject to the strict rules concerning share capital, the management of companies, as well as the meetings and resolutions which govern companies.
2. Unlike a company there is no distinction built into the structure of an LLP between the roles of proprietors and management. In effect this means that the decision-making and profit distribution arrangements are a matter for private agreement. These should be set out in a written LLP agreement that is a private agreement and not a matter of public record. It is important to understand that in the absence of a properly drafted agreement the default provisions that are contained in the *Limited Liability Partnerships Regulations 2001* apply and lead to unforeseen and unwelcome repercussions.
3. Unlike a company, which is a separate fiscal entity, a UK LLP has, what has become known as, fiscal transparency (see **4.15**).

Fiscal transparency 4.15

Essentially, *section 10* of the *LLPA 2000* provides that a trade, profession or business carried on by an LLP shall be treated as though carried on in partnership by its members. This means that the members of an LLP are taxed under the self-assessment rules as if they were partners in a partnership. Therefore, like a partnership under UK law, the LLP is fiscally transparent. There is no taxation at the entity level. Profits and losses flow through the members themselves, subject to the important caveat that such transparency is lost if the LLP goes into insolvent liquidation or is wound up for avoidance of UK taxation. This fiscal transparency is very important in the context of tax planning. It means that, in view of the fact that the LLP is not required by law to have UK resident members (whether individuals or companies), the LLP has significant application for cross-border trading arrangements.

It should be noted that in the case of a UK LLP with non-UK resident members, UK taxation is only chargeable on profits derived by such a member if either the profits are derived from a trade subsisting in the UK or if the profits otherwise have a UK source. There is no single test to establish whether or not a trade is exercised in the UK.

Investment activities 4.16

One question that is pertinent is whether a UK LLP can function as an investment vehicle. This could happen, for instance, in the case of shares held in a non-UK company with a view to receiving dividend income or realising capital gains. The tax transparent nature of the LLP is based upon the LLP carrying on business with a view to profit. The issue is whether the conduct of a pure investment holding function is the carrying on of a business. This can also be considered on another occasion, along with consideration of the LLP, as a potential tax planning vehicle.

Comparison with limited partnerships 4.17

The limited partnership is a long established alternative structure as compared with the LLP. It was created by the *Limited Partnership Act 1907*. Like an LLP, a limited partnership is fiscally transparent. However the characteristics can vary to a small or large degree depending on a particular case and the needs of the founders. As with LLPs the limited partnership can be subject to English or Scottish law. It should be noted that under English law the limited partnership has no separate legal identity whereas under Scottish law it does.

Important distinctions arise relating to the principal place of business and the filing of accounts. In an age of increasing disclosure requirements the transparency of the LLP may prove to be a useful advantage.

There is some evident value in understanding the nature of the LLP as an alternative to other more traditional vehicles. Its existence may provide a commercial and tax efficient opportunity for small businesses, depending upon the objectives of the members. Having regard to today's highly sensitive regulatory and business requirements relating to corporate governance, the LLP may enable a viable choice for small business representatives to consider subject to, of course, any professional advice taken and their individual needs and circumstances.

Governance and the balance of interests 4.18

Regardless of the choice of vehicle and whether or not the organisation operates in the profit or not for profit sector (see also **CHAPTER 17**) one of the main drivers for due diligence and good corporate governance is the trend toward increased regulation and operating standards. Indeed, it is interesting to note that as the business world shrinks and the speed and amount of information grows, transparency is the key note of the day. This is true of all businesses, whatever their size or place of operation. Moreover, the number of interested parties is on the increase as the use of technology enables access to information in a much more cost effective manner. As this handbook emphasises it is advisable for small businesses also to take appropriate steps in risk management (see also **4.45**). They must also bear the impacts of such matters as:

- data protection laws;
- product liability legislation;
- health and safety regulations;
- environmental requirements; and
- comprehensive EU developments (increasingly important with the enlargement programme meaning a union of 25 countries since May 2004).

Materialism has largely left few unaffected and the normal human nature is to want as much as it can now – from an increasingly early age. Meanwhile, those who are not benefiting from the overload of consumer items will try increasingly hard to penetrate business practices and prevent unfairness, as well as illegal practices. It is therefore unlikely that the major scandals have failed to impact developments in most places. In some respects this can lead to a real clash of interests between the developed and developing worlds that is represented by the governance debate. This can be considered from many sensitive and political angles. While this is not the place to consider such matters in depth it would be a disappointing omission if this context were not reflected upon briefly, especially bearing in mind the use of proxy votes referred to above in **4.1** and the dynamic steps taken by NGOs to influence the business environment.

In the US the Securities and Exchange Commission (SEC) has repeatedly broadened the permitted subject matter for shareholder resolutions as it has perceived subjects like social responsibility and corporate governance to be of growing concern to investors generally.

According to the Chairman of the Sustainable Development Commission and the Forum for the Future in the UK Jonathon Porritt, when delivering the Garner Lecture at an event held by the United Kingdom Environmental Law Association (UKELA) in 2002: 'the theme of governance is bound up with the issue of sustainable development'. He has described the concept of governance as:

> 'the way in which organisations manage or govern themselves, and the way in which they account for their performance to their respective 'stakeholders' – those constituencies or interests to whom they are accountable in one way or another'.

When considering drivers for due diligence and corporate governance it should be recalled that the private sector has become increasingly powerful from the point of view of ownership. This has been seen in the UK through:

- the outsourcing of many 'delivery' activities to arms-length agencies or quasi-autonomous non-governmental organisation (quangos);
- the systematic privatisation of what were once public assets; and

- the emergence of public/private partnerships as a core element in the 'modernisation' programmes.

Another broad driver may be described as the 'anti-globalisation movement'. Jonathan Porritt commented:

> 'It is in fact misleading to describe this oppositional force as anti-globalisation; what brings the disparate elements of this movement together is its opposition to global capitalism, to global institutions (such as the World Trade Organisation, the International Monetary Fund, G8 etc, which are perceived to be in thrall to the power brokers of global capitalism), and often to the United States whose principal mission as the world's sole remaining super power would appear to be absolute global hegemony for its own particularly unforgiving variation of capitalism.
>
> Wherever you may stand on these issues personally, the emergence of such a movement has already sharpened the debate about global governance systems and accountability...It seems hard to believe that the prevailing model of economic globalisation will survive the trauma of September 11 untouched. Globalisation in itself is not the culprit – most development economists have concluded that there is a positive relationship between international trade and economic growth, and that over time economic growth reduces rather than increases poverty. But as the most ardent promoters of the global economy, America and other OECD [Organisation for Economic Co-operation and Development] countries have systematically overlooked the economic, cultural and human costs of globalisation, 'cherry-picking' the commercial benefits of that economy without accepting enough responsibility for shared social obligations...to enjoy the fruits of globalisation in future, we must ensure that they are produced more sustainably and distributed more equitably... it seems equally obvious that efforts to establish more socially and environmentally responsible forms of development and organisation will be greatly facilitated by open, inclusive and accountable systems of governance at every level.'

For business, part of earning the licence to operate is to act transparently and accountably, not just in one's dealings with one's owners, but with all principal stakeholders. In theory, many companies are now seriously signed up to this and more are beginning to do so. Any prevailing practice of minimum disclosure, providing privileged access to information solely on a 'need to know' basis has to change.

Corporate reporting **4.19**

Many companies have actively lobbied the Department of Trade and Industry (DTI) over the whole question of mandatory environmental and social reporting. After a three year consultation period, the Company Law Review was presented to the Secretary of State at the DTI, and the recommendations as to changes in company law are mentioned below – in **4.40** to **4.42** It covers a huge range of issues, many of them touching on the issue of transparency and accountability. Companies will be required to produce a new Operating and Financial Review to improve flows of information with shareholders and other stakeholders. This requirement is scheduled to be introduced in January 2005. Nevertheless the debate continues over obliging companies (above a certain size or in certain sectors) to publish an Environmental and Social Report on an annual basis.

Further comment on reporting is made below in the section on socially responsible investing (**4.32** and **4.33**) while the discussion of governance and non-profits is developed further in **CHAPTER 17.**

Corporate social responsibility

Ethics and risk/reward issues 4.20

Corporate social responsibility (CSR) has been defined as 'essentially a concept whereby companies decide voluntarily to contribute to a better society and a cleaner environment' (see the European Commission Green Paper *Promoting a European Framework for Corporate Social Responsibility*. See also **APPENDIX 1.**) The Commission published the Green Paper in July 2001 to promote CSR. In addition, the Commission has voted in favour of social and environmental legislation reporting and indeed there are several initiatives debating and investigating the issue of CSR.

In the UK the government attempted to promote CSR by the Prime Minister's challenge to the top 350 FTSE companies to voluntarily report on social and environmental issues. Since business did not respond positively and since the EU has put pressure on the UK the government is considering new legislation on CSR and the concept is expected to feature in consultations for any proposed legislation. In addition an initiative regarding an international standard has taken place.

Business pressure 4.21

Locally, businesses do experience more and more pressure from stakeholders including public sector bodies and NGOs. All of the trends in the context of risk management support this and investors have begun to influence significantly the way in which businesses operate to the extent that the image and brand are increasingly vulnerable to external pressure (see **CHAPTER 7**). In addition, the requirements on listed companies to embed risk management in their corporate policy has meant that the Board must be aware of and evaluate the risks and their management for communication to stakeholders.

It has become clear that while some aspects of corporate responsibility are measurable, for example progress toward environmental management and enhancement, many others are not. Various strategies are referred to below to raise their profile with business decision makers. There are both corporate and government initiatives taking place. For example, adherence to the standards of the government backed Ethical Trading Initiative (ETI), which has 29 corporate members, is an extension of the idea of fair trade into other ethical procurement policies. These and other similar approaches are useful in their respective ways to help to promote high standards of business ethics within corporations and among their stakeholders.

Evidence of value 4.22

One of the most demanding tasks for CSR professionals is to prove that they add value to their organisation. Increasing sophistication is required to manage the intangible assets of a complex business in order to meet or exceed stakeholder expectations, sustain corporate reputation and enhance brand value. Wider groups of stakeholders scrutinise the way in which risk/reward decisions are made and communicated. Measuring the benefits of CSR is difficult, but not impossible. There are three components worth measuring:

- risk avoidance;
- reward enhancement; and
- volatility reduction.

Risk avoidance and reward enhancement may sometimes be measured, but CSR also reduces volatility and therefore adds measurable value. In a listed company, volatility reduction can be estimated and the value calculated.

The Enterprise Commissioner, Erkki Liikanen, delivered the following message to the European Multi-stakeholder Forum on CSR at the end of 2003:

> 'The evidence is getting more solid that it pays off for companies when they invest in corporate social responsibility.'

CSR in this context means companies behave responsibly towards their stakeholders in a strategic way and in their own long-term interest. The European Multi-stakeholder Forum on CSR, launched in October 2002, aims to improve knowledge and consensus among businesses, trade unions and civil society and to explore whether common European principles on CSR practices and tools would be appropriate. The discussions are also intended to increase trust and understanding and include reference to broader schemes such as the chemical sector's Responsible Care Programme and the UK's Ethical Trading Initiative. The potential benefits for companies are an important aspect of the Forum's work. For example:

- enterprises can gain competitive advantage when they invest in sound relations with their stakeholders and attend to their environmental performance;

- this can boost their reputation among employees, neighbours, customers and suppliers;
- sound relations with stakeholders can support an organisation's 'licence to operate' as societal actors and public authorities that have a positive perception of the organisation will be more in favour of its projects;
- it can spur innovations; and
- ultimately it can potentially improve bottom line performance.

Such assessments are in line with a growing number of scientific findings supporting the business case for CSR. The latter demonstrate vividly that being a good employer, neighbour and protector of the environment actually increase, rather than reduce, profits.

Since business continues to stress that the positive effects of CSR depend upon the ability of companies to tailor activities to their specific situations, the Forum has noted that the following factors can contribute to the success of CSR:

- firm commitment to the concept from the very top to the very bottom of a company and among all stakeholders;
- the integration of CSR into all day-to-day business;
- the nurturing of skills and the provision of educational tools; and
- the adherence to reliable transparency, including financial details.

While the matter of the necessity of a European CSR framework remains under discussion, the Institute of Business Ethics (IBE) has also considered various related questions. If it were indeed possible to produce evidence that ethics pays – that companies which undertake their business ethically perform better over the longer term – might it help convince more business leaders that it is a worthwhile thing to do – as well as the 'right' thing? According to the report 'Does Ethics Pay' published by the IBE in 2003, the management expert, Peter Drucker, is quoted as saying:

> 'What cannot be measured, cannot be managed.'

Measurement of the measurable is clearly worthwhile if management is to be effective. There are aspects of business which whilst difficult to measure accurately, do need to be managed. Whereas, as noted, some aspects of corporate responsibility can easily be measured, such as enhanced environmental management and performance (see **CHAPTER 16**), many others are not. Business ethics programmes are the latter. These may be considered as the soft areas, as was the case with environmental issues some years ago when they were considered to be separate from core business issues or policies. It is better to find indicators of 'tendency' or 'direction' rather than absolute numbers. In addition, some things can only be measured by their effects.

The IBE survey has found that corporate officers will question or even veto spending on programmes which cannot be shown to yield a benefit to the 'bottom line'. On the other hand, many chief executive officers and boards of directors are conscious of areas of business activity not susceptible to measurement, yet if not addressed, leave the organisation vulnerable. Their problem is that good policies aimed at lowering risk but which incur cost, are difficult to make convincing because they can only be justified by negative criteria such 'we did not have a disaster'. An example of this is corporate safety policy, the basics of which are required by law in many countries. A significant proportion of any pecuniary losses due to breaches of policy can be covered by insurance. However, can the same be said of security? In recent years, protection of premises from infiltration by unauthorised people has moved up the corporate agenda. Employees, customers, sub contractors, etc now expect some measure of security from the risk of random attack by intruders. Expenditure for minimising this risk has become a necessity and therefore has been built into budgets. Its is also true of integrity risk. There are enough examples of unethical, and sometimes illegal, behaviour by individuals or groups in corporations which have severely affected the reputation, if not the viability, of a business. Some of the examples that have been cited by the IBE and are well documented and include, for example, the collapse of Barings Bank as a result of losses made by trader Nick Leeson.

As is true of other areas of corporate governance, while it is not possible to guarantee the prevention of such unethical behaviour, it is possible to introduce preventative steps which will help to minimise (but not eliminate) behaviour which can lead to loss of corporate integrity. For example, by the implementation of an ethical policy. However, the IBE has found that:

> 'the need for such a policy is not yet widely accepted. It is still mainly the largest corporations that take business ethics seriously. One reason is of course, the cost of developing and embedding the policy. Different approaches have been tried with varying degrees of success'.

Ethical policy 4.23

There are a variety of approaches for businesses to incorporate and maintain ethical policy.

1. The IBE has found that one approach is for 'companies to make explicit their ethical commitment and incorporate it into their marketing or brand strategies. For instance, The Co-operative Bank claims that its 2001 profits rise was due, in part, to its policy of doing business only with those it considers ethical. The chairman estimated that between 15 and 18 per cent of the £96m pre-tax profit was attributable to Co-op's ethical values stance. The Bodyshop sets out to associate its products with environmental concerns, as do Ben & Jerry (now owned by Unilever).' Indeed as is seen in **CHAPTER 8** some businesses can also develop a managed risk culture.

2. As another approach, the IBE further found that some businesses use a 'strategy of associating the company and its products with concern for organisations (often in a developing country) which produce the raw materials or finished products. Members of the fair-trade movement, especially in primary products like coffee, emphasise ethical standards in purchasing policies eg purchasing from renewable sources, paying 'fair' wages and not employing child labour. In so doing, they seek to attract the sympathy market to their products ..Related to fair trade is the growing popularity of cause related marketing (CRM). This is the name given to a strategy that associates branded goods with a 'good cause' ..'

3. A third approach is to make it clear in public announcements by company CEOs and chairmen that they expect the highest ethical standard from their staff and that the company's long term sustainability depends on this. Examples of these that the IBE has referred to in its report are:

> 'Responsible behaviour makes good business sense ...At Rio Tinto we have found that maintaining the trust of local communities is essential for the long-run success of our operations. A sound reputation on ethical issues also helps us to recruit and retain high-calibre employees.'

Sir Robert Wilson, Chairman, Rio Tinto plc (minerals and metals)

> 'Each of us [in Sara Lee] – in every job, everywhere in the world – has a responsibility to do the right thing and to help our peers do the same. Our success and our future as a business depend on it. Companies that compromise their values and fail to do the right thing are sure to fail.'

C Steven McMillan, Chairman, President & Chief Executive Officer, Sara Lee Corp (food processing)

> 'Perhaps more than anything else we do, furthering our company's values and standards will have the greatest effect on the future success of our company.'

Ray Gilmartin, Chairman, President & Chief Executive Officer, Merck Corporation (pharmaceutical)

> 'Good leadership means doing the right thing when no one's watching. Values governing the boardroom should be no different from the values guiding the shop floor.'

Carly Fiorina, CEO Hewlett Packard (computer systems)

As has been indicated, these and other such approaches are valuable in the interests of high standards of business ethics and good corporate governance within corporations and among their stakeholders, as well as with the public.

Clearly, the way a company does business is critical to its own sustainability, wherever it operates. Studies by the International Chambers of Commerce

(ICC) have addressed this aspect in the context of environmental concerns[1]. A management that adopts a 'when in Rome, do as the Romans do' approach will soon realise that the inconsistencies that result are damaging to long term success. The US corporation Union Carbide's experience with the chemical plant disaster in Bhopal, India, illustrated that even if a business is not 100 per cent owned, basic standards are expected to be maintained wherever it operates. The Bhopal incident in 1984 has been described as one of the world's worst chemical disasters. It was caused by a gas leak at the Union Carbide plant in Bhopal, India. The incident killed over 3,000 people and adversely affected the health of many more thousands. It cost the American group over $US 270m in punitive expenses alone. In addition to monetary losses, all Indian assets were seized and the company was forced to leave India at a time when the market was most attractive to investors. The CEO at the time evidently regarded this event as his most important lesson in corporate management. It set the organisation on a new course of enhanced environmental awareness world-wide. (See also **CHAPTER 16**).

Since autumn of 2002, the New York Stock Exchange requires that having a code of ethics is to be a condition of listing. This should persuade those in the US and others listed on the Exchange who do not have codes to produce one. However, of course, a code is not sufficient in itself to ensure that a high standard of ethics is maintained. For instance Enron in the US had a 'standards of conduct' code. Such a policy, as with others, has to be embedded within the culture of the company to have any real effect in terms of ongoing due diligence, risk management and corporate governance.

One approach that has demonstrated that good CSR can assist good business performance is the SERM approach.

Accountability and SERM 4.24

The purpose of SERM's socio-environmental risk assessments is to make a judgement as to how companies are actively managing the socio-ethical factors associated with the industry sector and type of business in which they operate. The risk reduction factor derived from the ratings is one measure of the effectiveness of a company's management in identifying and lessening a set of direct and indirect social, ethical and environmental risks. SERM ratings indicate which companies are assessing and acting on their exposure to direct and indirect socio-environmental risk, that is putting into practice a commitment to ethical business. The ratings are calculated in such a way that they take account of industry sector differences and therefore provide a platform for cross-sectoral comparison of corporate performance. Information used to calculate these ratings, unlike other measures in this study, is not all in the public domain. The direct and indirect risks covered in the SERM ratings are:

- unrestrained corporate power;
- lack of corporate community involvement;

- adverse marketing practices;
- adverse business practices;
- engagement in bribery and corruption;
- poor human resources management;
- abuse of human rights;
- natural resources degradation; and
- negative impacts of technology.

One of the most crucial concerns for modern business is a good reputation. Without it success and profits are at risk and the confidence of stakeholders is lost (see **CHAPTER 7**). The evaluation of such risk is therefore a vital tool for business in a climate of media interest, rapid exchange of information and dynamic competition. A reputational risk valuation is used to:

- quantify the value that CSR adds and produce a direct benefit calculation by showing, for listed companies, the volatility reduction and, through option pricing, the shareholder value-added of CSR initiatives;
- direct CSR efforts to the areas of highest return and exit from CSR areas that are of little benefit;
- produce a cost/benefit analysis for future CSR activity; and
- develop a prioritised CSR plan that delivers shareholder value more rapidly.

Reputational risk valuation begins with an assessment of inherent risk – public profile, level of public awareness, longevity of public concern, focus of blame, stakeholder reaction. This is followed by an assessment of residual risk – entry or exit, operational controls and mitigations and managing stakeholder expectations. Both assessments are grounded in SERM's proprietary ratings, permitting statistically valid CSR/value benchmarking against peers or competitors.

Sustainable and responsible investment 4.25

Sustainable and responsible investment (SRI) is a generalised subject heading for different forms of investment strategy. SRI sometimes goes by the names of 'socially responsible investment' and 'ethical investment' (see also **4.32** and **4.33**).This goes some way to explaining the initial confusion over the different strategies. The shift in title also goes someway to explaining how and why it has developed.

A single question comes to mind that is essential to understand SRI. It is simply to ask who is investing? It is ludicrous to think that religions, charities, individuals and institutions have been grouped together under one nice comfortable banner, but yet they have. Here, therefore, begins the distinctions between some sophisticated investment strategies.

What is SRI? 4.26

The flippant but tempting answer to this is many things to many people. Research has revealed no fewer than five different definitions, all of which share one common theme. That theme is investment which in some form takes into account social, ethical and environmental issues. The problem with this definition is that it is too vague. For example, it makes it very hard to assess the size of the SRI market as under such a loose definition investors have been included who exercise their right to vote as shareholders, or write a letter to a company regarding a management practice. This constitutes engagement – central to some SRI strategy. Writing a letter to a company or exercising a right to vote is not enough to constitute what most professionals in SRI would consider sufficient to be an SRI strategy.

Therefore if we take this into account we can reform our definition by saying that it is action or information on social and or environmental and or ethical issues that informs the investment decision. The step change here is that it has in some way informed the investment decision differently than if the action or information had not taken place.

Development of SRI 4.27

In the last decade SRI has been growing at a tremendous rate. The tables below detail the rise of SRI in the UK and then the global rise in SRI assets. There has been a revolution of markets, revolution of technology and of demographics and values. These have all served to contribute to the growth of SRI.

Historically SRI developed from moral issues, hence its association with ethical investments. The Methodists saw alcohol as causing social problems in the nineteenth century and refrained from investing in companies that they saw as promoting social harm. In the 1970s opposition to the Vietnam War saw the creation of the PAX World Funds which did not invest in companies that were profiting from the conflict. The PAX World Funds represented the financial expression of new investment independence and values. When the fund started to outperform their peer group it was realised that maybe there was more to stock selection than pure financials.

Mainstream investment used to look at financial fundamentals to determine whether or not to invest in a company and by how much. However, there are strong suggestions that this was never solely the case. Investments would be made also on the belief that continued successful delivery of those financial fundamentals was possible, that the management was capable of delivering on predictions, that the business was sustainable and would still have a market for its product and finally for the belief that the business was capable of managing its potential risks. These were all measured and classed as financial risks. As business and economy has developed and grown ever more complicated in terms of regulation, operation and delivery these once clearly tangible business issues became broader and market behaviour more complex.

Growth in UK SRI Investment Universe £bn			
	1997	1999	2001
Church investors	12.5	14.0	13.0
SRI unit trusts	2.2	3.1	3.5
Charities	8.0	10.0	25.0
Pension funds	0.0	25.0	80.0
Insurance companies	0.0	0.0	103.0
Total	**22.7**	**52.1**	**224.5**

Global Revolution, SRI Assets 2001 £bn	
US	1603.2
UK	224.5
Canada	21.6
Europe	12.2
Japan	1.3
Australia	0.7

The market context **4.28**

Before detailing any further the specific aspects of SRI and of the types of investments held. it is important to understand the context of the market.

Institutional investors can be broken up into insurance and pension funds for the most part. The pension funds in the UK being required by law as of 3 July 2000 to report to what extent, if any, they take into account social, ethical and environmental (SEE) issues. Changes set out in the Pensions Bill in 2004, will require that pension fund trustees now need to be adequately trained in order to understand all investment decisions, including those relating to SEE issues.

It is important to note that the investments covered are representative of the majority of SRI, namely they are equities. However, SRI is much more than just equity investment. There is alternative investment in the way of venture capital and other private finance initiatives. There is property and, additionally, there are fixed income bonds.

Why and how is SRI used? 4.29

Distinctions between types and styles of investors need to be made to understand SRI better. These distinctions include differences between:

- passive and active investors;
- positive screening and negative screening;
- best in class; and
- engagement.

The complexity of the distinctions is furthered by the fact that none of these distinctions are mutually exclusive except for passive and active investors.

As a religious or moral investor, SRI is necessary to exercise a belief system. Typically this will take the form of screening against investment that is considered contradictory to the belief system, this is referred to as negative screening. Examples of this are green investors who choose not to invest in companies that contribute to environmental degradation, destruction of biodiversity or contributing to climate change. Religious investors tend to screen against investments in tobacco, alcohol, pornography and armaments. This represents possibly the most well known side of SRI and also the smallest financially.

Depending upon the size of investments involved, the funds will have a different risk structure. The larger the investment the more conservative the risk profile tends to be. Therefore, larger religious or moral investors tend to use a system of positive discrimination in stock selection, referred to as positive screening. For example, by only investing in companies that have International Organisation for Standardisation (ISO) environmental standards where appropriate, or companies that satisfy international standards.

There is also a system of selecting a best in class. In many ways this can be seen to a more risk adverse positive screen. A best in class strategy recognises the dangers of avoiding a sector entirely in terms of financial return and also the right of a business to exist in terms of satisfying legal requirements and fulfilling a market need. The financial year 2003 in the UK markets saw a strong out-performance of tobacco stocks. Nevertheless, this strategy believes that those companies that are better managed in terms of SEE issues will perform better and therefore selects quite simply a best in class in terms of SEE issues or CSR management.

As far as responsible investment goes, Hermes Pensions Management Limited (Hermes), the fund manager for the largest UK pension fund – the BT Pension

Scheme – is an excellent example. In 2003 it published ten principles committing itself to the idea of responsibility for its investments. As a large pension fund it is forced to invest across the entire stock universe of the FTSE 100, it is has a passive indexed fund. This means that they do not actively pick and chose stocks to invest in, instead they are passive and they pick a stock index (the FTSE 100) and invest their money across the whole index. The reason this approach is taken is because of the size of the pension fund and the risk allocation.

An active investor will change their investments according to their method of stock selection. Whereas a passive investor invests according to an index and may change how much of a particular stock is held will still hold the entire range of stocks across an index, this is usually adopted by funds that are of a very large size and need to adopt a low risk profile.

Passive funds are much lower in risk than active funds. As a pension fund their first duty is to provide their members with adequate remuneration. Hermes, however, believes that well managed companies will perform better, therefore companies that pay attention to managing their environmental and social risks will make more money. With this in mind from the entire stock universe within which it invests, Hermes will make financial assessments as to how much of that particular stock they will own, some stocks being overweight, as they believe they will perform well, and some underweight (as they believe they will perform poorly). Nevertheless, Hermes, as a universal investor, believes that it ought to influence the companies within which it holds stocks to 'internalise its externalities'. By this, Hermes means that companies ought to pay attention to potential risks and rather than wait and see, deal with them before hand and appropriately. Thus encouraging a company to internalise any potential external risk. Why? Because over time these 'externalities' will have had to be paid for somehow, they will cost, if not that company then the government, and then consumers, eventually the economy as a whole will be negatively impacted upon because of a rise in taxes to deal with these poorly managed risks. Hermes is very much on the periphery of the SRI universe but is nonetheless a useful example of the responsible attitude of a long term and conservative investor.

Additionally, large pension funds do not just have a fiduciary responsibility to its members but the re is an increasing examination on responsibility on a quality of life for its members. As such, the influence of large pension funds is increasingly being exerted to ensure that companies are run for the greater benefit of their stakeholders.

Engagement **4.30**

Increasingly, SRI funds have taken on an attitude of engaging with companies rather than disinvesting. Fund managers have been seeking added value through dialogue and improvement in companies that they found lacking. At the very least this will mean an active exercising of the voting rights afforded to the fund manager as a shareholder. Hermes engages with companies over issues of corporate governance and issues that are relevant to their two principles above. It is through engaging that Hermes hopes to add value, or minimise potential loss, to the company that it owns shares in.

ISIS Asset Management applies an engagement overlay to all their SRI funds. This overlay is intended to add value and to mitigate risks in essentially the same way that Hermes intends. The essential difference is that ISIS does not solely believe that financial return is the only aim of investment. ISIS, like other asset managers, believe that as an owner of a company, you have a responsibility as a steward, to ensure that the actions of a company are in keeping with at the least normative behaviour.

Insight Investment, the asset management arm of HBOS plc, has a two fold argument as to why SRI issues must be taken into account. Their argument is expressed in demanding high standards of corporate governance and corporate responsibility.

1. Financial benefits

   The first reason is that we believe that investment returns can be enhanced if the companies we invest in maintain high standards of corporate governance and corporate responsibility. Conversely, there are many examples where weakness in these areas has contributed to poor financial performance and even, occasionally, corporate collapse.

2. Responsibility

   Secondly, we believe that investors, especially in their role as shareholders, have moral responsibilities with regard to the companies in which they invest. It is widely accepted that individuals are responsible, not only for their own actions, but also, under certain circumstances, for the actions of those who act on their behalf. We believe this principle extends to the corporate context. As a general principle shareholders approve the composition of the board. The directors have a fiduciary duty to conduct the business on the shareholders' behalf and in their interests. Thus it is reasonable to accept that shareholders have a wider responsibility, under certain circumstances, for the actions of companies. While company law (see **4.4** above) strictly limits the legal liability of shareholders, it does not thereby limit their moral obligation.

This is not to say that shareholders' responsibility in this regard is open-ended. Effective control of companies, in law and in practice, lies with the directors. This is therefore where the main burden of responsibility for the company's actions should rest. However, given that shareholders approve board composition, the primary role of shareholders is to support, encourage and, where appropriate, challenge the directors in their efforts to achieve high standards of governance and corporate responsibility.

Some of the most effective engagements have come about through collective actions groups – in the US, for instance, a group called the Interfaith Center on Corporate Responsibility (ICCR). They are a membership organisation representing institutional funds of religious groups. The purpose of the group is to engage companies on SEE and management issues. Their most high profile and successful engagement was bringing a shareholder resolution against the energy company Talisman Oil which resulted in Talisman selling their stake in Sudan, which had been a concern for investors because of the human rights atrocities amongst other issues.

The concept of engagement itself is somewhat confusing. It may mean to simply write a letter or exercise voting rights. However, it is often taken to mean a prolonged campaign of communication including in all probability face-to-face meetings. More often than not once engagement has been started by a fund manager a company will usually agree to an examination of the issue highlighted and then, if it agrees, develop policy and implement change. Taking engagement as far as shareholder action is the strongest line any investor can take short of selling shares. It is essentially a constructive dialogue with a company in order to achieve what is seen as a mutually beneficial outcome.

Climate change **4.31**

Climate change (CC) stands at the gateway to the mainstream investment world and it stands on the shoulders of SRI. There have been individual companies and specific issues that have been picked up on in the past that have served to demonstrate that SRI has identified material risk. Never before has an environmental issue promised to be so all encompassing that it cannot be ignored by any sector. The threat posed by CC to our physical world can be seen with biblical numbers of avalanches in the Alps, glacial destruction, increased risks of flooding and of drought. CC needs to be addressed now.

Collective action from investors has been slow to organise. However, in the last few years both the Institutional Investor Group on Climate Change and the Carbon Disclosure Project have been established as positive forces in raising awareness amongst investors as well as companies. The more investors pressure companies, the greater the ability to force companies to give up carbon intensive practices and start to make a transition to a low carbon economy.

SRI has for the most part focused only on CC as another environmental issue. Tackling the root causes and mitigation of CC has not been properly explored, except possibly by the tiny market of alternative investment. Mitigation of CC and producing a low carbon economy is the best way to address CC. There needs to be a paradigm shift to a zero carbon economy.

Collective action looks to raise all members to at least a basic level of addressing CC. However, there are still many investors, mainstream investors, who do not address CC at all. The mainstream fundamentally needs to measure issues in terms of finance. The faster government produces allocations, that carbon trading establishes a more certain figure in terms of cost, the faster all investors will be in a position to take CC into their investment decision process.

Tempered against what is needed for the environment is what is needed for investors. Even in SRI there is the need to establish a sound financial return and for managers to satisfy their fiduciary responsibility. Therefore, CC is looked upon as being an environmental risk, but more often than not only insofar as it is a financial one too. Positive action is needed in addressing CC. Investment is needed for mitigation and adaptation. To achieve this there needs to be greater communication, greater understanding and the identification of those who don't make it a priority.

Socially responsible Investment

New reporting requirements **4.32**

It is indeed clear that sustainable or socially responsible investment (SRI) and social, ethical and environmental disclosure (SEED) are becoming increasingly important issues for business in the UK. In terms of reporting, therefore, it is appropriate to introduce and outline the main driving forces for these changes, and then give a more detailed breakdown of the most recent development – the White Paper: Modernising Company Law – and associated implications for business.

Trends **4.33**

Since the mid 1990s there have been a number of changes and trends in the arena of SRI, and the type and quantity of SEED made by companies, especially so in the last couple of years. SRI and SEED are receiving increasing levels of attention from business, government and society as a whole and with a move towards broader and more integrated disclosure due, in part, to the general erosion of trust in big business, not least following highly publicised and documented financial and accounting scandals such as Enron (see **4.23**) and Worldcom, and pressures from stakeholders to improve transparency. SRI, which can offer some protection for investors, is becoming increasingly valued and sought after, illustrated by the current 30 per cent annual growth trend in the UK SRI retail market. If this growth trend continues, it is likely to be worth at least £15bn in 2004 according to market sources.

In the year 2001–2002, 50 of the FTSE 250 companies reported for the first time compared with only 18 new reporters in the previous year, with 103 reporting in total and 95 of the reports including ethical and social information. However, it is worth noting that 91 companies produced nothing of substance, and concerns have been expressed about the scope and quality of disclosures made as well as the extent of variance in type, length, rigour and verification of reports.

Additionally, as is noted above, there is an ongoing debate centred on whether social, ethical and environmental disclosure should be a voluntary or mandatory requirement.

The Combined Code **4.34**

In the UK, the government has been putting increasing pressure on companies in terms of improving disclosure. Tony Blair challenged the top 350 companies to produce social and environmental reports by the end of 2001, (of which only a quarter did so) and in September 1999 the Combined Code, commonly known as the Turnbull Report, was issued, to which all UK listed companies must comply. This forced UK listed companies to take account of internal risk control, including reputational risk and the management of environmental, ethical and social responsibilities. Further comment on Turnbull is also made in **CHAPTER 10**.

The Pensions Review **4.35**

Following the introduction of the Turnbull Report, measures were introduced to encourage SRI in the form of the Pensions Review. The Pensions Review, effective as of July 2000, called for institutional investors to consider their position on SRI, and introduced a legal requirement to disclose a statement of investment principles to articulate their stance on the matter. This has led to a mixed response with a particularly insightful report published by Justpensions (*Do UK Pension Funds Invest Responsibly?*) indicating that poor practice is the norm and examples of best practice few and far between.

Corporate Responsibility Bill **4.36**

There have also been pressures from other sources in the UK and in June 2002 a coalition of NGOs and campaign groups joined up with Linda Perham, MP, to put forward a Corporate Responsibility Bill, drafted in response to the failure of the voluntary approach to SEED. This Bill required companies to publish reports on their social, environmental and economic impacts, and even though it failed, it succeeded in further raising the profile of CSR.

The Association of British Insurers **4.37**

The Association of British Insurers, also in the UK, has recently issued guidelines and called on institutional investors to give greater regard to SRI principles and for companies to report on important SEED issues that affect their business.

Pressures from Europe **4.38**

On a wider scale, in 2001, the European Commission published a Green Paper on 'Corporate Social Responsibility' and launched a multi-stakeholder forum for CSR in 2002. Some comment about the forum has been made above in **4.22**. This forum will consider and advise on standards for reporting and assurance and company codes of conduct, with recommendations likely to be made in 2004. Even though at this stage, it appears that neither the European Commission or the UK is prepared to make SEED a mandatory requirement, it is evidently viewed as an important issue for business.

A1000 standard and the Global Reporting Initiative 4.39

Verification of company disclosures and the quality of the verification are concerns that have been noted by commentators, and there are some prominent organisations within the SRI community, such as the Co-operative Bank, that back tougher verification as a means to improve the situation. A fundamental question is how and by whom standards of measurement might be developed and agreed upon.

In response to the problem of verification, the Institute for Social and Ethical Accountability has formulated the AI000 standard for social reporting and a new professional group of social auditors is emerging. Also the Global Reporting Initiative (GRI), set up in 1997 and established as a permanent institution in 2002, has developed a core set of metrics which are sector specific and applicable to all business, and has released a template for SEED. Even though the template is still underdeveloped and part of an ongoing process, it is a move in the right direction. However, the GRI has not been welcomed by all and there has been some resistance to the framework from some companies who see it as setting the standards unrealistically high.

Modernising company law 4.40

More recently, the UK Government published its White Paper: Modernising Company Law (Cm 5553) (see also **4.4**). The Paper implies greater environmental reporting on the part of large companies and contains numerous proposals for the simplification of modernisation of company law. Its aim is stated as the provision of 'a legal framework for all companies which reflects the needs for the modern economy..' The government is said to be committed to legislation following consultation on the White Paper as soon as parliamentary time permits. Along with the White Paper comes the publication of over 200 draft clauses. However, it must be stressed that this is not yet a draft Bill and the indications are that any eventual Bill will be longer still.

Drivers for reform 4.41

The introduction to the White Paper indicates a number of drivers behind the suggested reform. One of these is European Commission activity in pushing forward changes to rules on the annual reports of companies. In addition, the European Commission has recently put out a consultation on a proposal on continuing obligations of listed companies to make information available to the market. Whilst these proposals will largely cover financial reporting, the proposal makes clear the need for ongoing disclosures in other areas including corporate governance information. In the body of the report itself, the White Paper makes the following statement:

> 'The government believes that all the components for the corporate governance framework will continue to be important. In particular, it takes the view that whilst legislative and regulatory requirements have an essential role, there will also be a continuing need for a code of best practice and other guidance.'

Operating and financial review **4.42**

The report goes on to say that good company reporting is essential. As part of this process the White Paper promotes an operating and financial review (OFR). The proposal is to demand such a review from public companies where they meet two of the following three criteria:

- a turnover of more than £50m;
- balance sheet total more than £25m; and
- number of employees more than 500.

Private companies would also have to produce an OFR and in their case the relevant criteria are:

- a turnover of more than £500m;
- balance sheet total more than £250m; and
- number of employees more than 5,000.

The Accounting Standards Board first published a statement on OFR in 1993 and it is currently in the process of revising that statement. The government has stated that it wants more qualitative and forward looking reporting because companies are 'increasingly reliant on intangible assets'. Information should also cover future plans, opportunities, risks and strategies. In particular it proposes new requirements for environmental reporting which it describes as:

> 'A major contribution to both corporate and social responsibility and sustainable development initiatives.'

The government states that there is a business case for these for which the OFR provides the opportunity for company directors to respond.

The specific duties require directors to be able to identify relevant impacts or issues, processes must be in place which both identify new, and monitoring existing, issues. For example, environmental impact assessments for all new developments will identify potentially relevant environmental issues.

Whether issues are relevant to an understanding of the business depends on two variables.

1. Magnitude – single large environmental incidents are relevant.
2. Number – one small environmental incident eg effluent leak may not be relevant as an isolated incident – however the cumulative impacts of both environmental and financially (through penalties) may be.

Failure to appreciate the cumulative impact of many small incidents may lead directors to erroneously omit relevant information.

There is an exemption of medium sized companies from reporting non-financial indicators. The most pertinent other legislation is the EU regulatory framework which is set to dovetail with national legislation on OFR requirements. Companies fundamentally need to be acting now to improve management and reporting of non-financial issues. For instance, the review shall include information about:

- the employees of the company and its subsidiary undertakings;
- environmental matters; and
- social and community issues.

Page 45 of the White Paper states:

> 'It is important to note that a balanced and comprehensive analysis is not designed to require directors to cover all possible matters. They must not omit significant but uncomfortable information, nor should they include matters which are not necessary for an understanding of the company.'

In addition, the government has indicated that, in acting in the best interests of shareholders, the directors must recognise the company's need to foster relationships with its employees, customers and suppliers, its need to maintain its business reputation, and its need to consider the company's impact on the community and the environment.

The term 'significant' is paramount in the OFR. It is used as a dynamic concept and certain issues will become more or less significant to a company over time. The constant monitoring of a wide range of issues particular to a company and its wider industry sector will identify significant issues for inclusion and issues which are no longer appropriate for inclusion. Additionally a peer group review is an important element of this process as is a wider understanding of the changing political and economic environments in which companies operate.

Legal backing **4.43**

At the moment, as noted, any environmental reporting is voluntary. The Bill will leave it open for directors to decide what material information is relevant to their particular business and should be included in any report. However, at the same time, the White Paper states unequivocally that the Bill will make directors responsible for how these factors are covered in the OFR and goes on to state that:

> 'Ultimately the directors may need to defend the process behind their reporting before the courts.'

Preliminary draft clauses to implement the OFR are included in an appendix to the White Paper. In addition, because auditors will be required to report on the OFR, the major accounting bodies will need to determine how they will take into account the types of environmental and social factors that are now said to become part of the formal reporting process. Company directors will have to determine how important issues of corporate social responsibility and sustainability are in both the short term and long term futures of their business.

Conclusion 4.44

Currently, social, ethical and environmental reporting in the UK is not a mandatory requirement. However, times are changing. Over the last few years pressure has been building from a variety of sources to change the situation and improve disclosure, and is gaining momentum. The Modernising Company Law White Paper is perhaps the strongest signal yet that SEED is well on its way to become a mandatory requirement and no longer an option. This also has repercussions for small business: with this in mind a view from the perspective of small business is discussed in **4.45**.

Small business risks, due diligence and risk management 4.45

Environmental and related governance considerations will not go away for the owners of micro businesses (0–9 employees) however much they wish they would. It is said that the talk about jumping on the green bandwagon has died down; being green is no longer fashionable or glamorous, it is a fact of life with a direct impact on profitability.

It is important that the owners of smaller businesses consider the impact of environmental issues on their businesses but it is equally important that the regulators understand the problems experienced by businesses and adopt a pragmatic and realistic approach. Smaller businesses adopt a risk-based approach; they take priorities in order of apparent importance owners of small businesses have to be knowledgeable on all types of legislation. They have to monitor requirements that include:

- employment law;
- taxation;
- health and safety;

- consumer issues;
- product liability;
- IT; and
- the environment.

Small business does not have the resources in time or money to be, or to employ, an expert on all aspects of legislation.

Regulators and enforcers should try to adopt the 'carrot' rather than the 'stick' approach. Nevertheless it will become increasingly more important for small businesses to adopt a thoughtful approach to such issues and accept responsibility for their own environmental impacts. (For example, the *Landfill Tax Regulations 1996* (*SI 1996 No 1527*) relating to landfill tax, *the Special Waste Regulations 1996* (*SI 1996 No 972*) (as amended) relating to Special Waste and the *Producer Responsibility Obligations* (*Packaging Waste*) *Regulations 1997* (*SI 1997 No 648*) relating to packaging waste affect only a small percentage of small businesses but the implications of the *Environmental Protection* (*Duty of Care*)(*Amendment*)(*Wales*) *Regulations 2003* (*SI 2003 No 1720*) incumbent on every business, large and small, are still not understood.) Small businesses need to make an honest appraisal of where they stand in relation to environmental and related issues and develop a simple policy to address their potential problems over the short and long terms. Many regulations exist and environmental enforcement can result in fines – and a criminal record. Small businesses need to understand that the simple approach of good environmental housekeeping can save money.

It is now recognised at European level that, although the environmental impact of individual manufacturing small businesses is small, the total impact could be significant because of the vast number. Lord Strathclyde, when a Minister in the (then) Department of the Environment, said:

> 'Environmental problems are the consequence of millions of decisions taken by everyone in the course of their lives. The solutions to these problems lie in the same hands – individual action.'

That statement is still appropriate for us all in our business and domestic lives but it must always be balanced by the fact that many businesses are smaller than households.

Supply chain encouragement must replace supply chain pressure; there are examples of good practice and these must be extended. Banks have developed environmental criteria within their lending and credit policies and insurers are becoming even more vigilant about pollution problems.

Simple actions can make all the difference but the information must be the public domain. The 90:10 rule applies (the 80:20 has not yet been reached!); ten per cent discusses what they think 90 per cent should do, but the 90 per cent do not even know there is an issue.

Similar arguments can be applied to many aspects of corporate social responsibility and corporate governance. The trend for all business – whatever its size or location – must be to work toward best practice at all levels. Practical initiatives can change the performance of the workforce and the business as a whole, as well as improve public perception. Risk management strategies are not just the domain of big business. For instance, IT has proven to be one of the weakest areas for businesses of all sizes when implementing the Turnbull guidelines (see further **CHAPTER 9**). A practical priority is to invest in effective disaster recovery and business continuity plans. Early reports from the DTIs *Information Security Breaches Survey 2004* found two-thirds of businesses surveyed had to restore significant data from backup due to a computer failure or theft. For small business too, it is wise to ensure that a robust backup procedure has been implemented and that data can be restored efficiently in a worst-case scenario.

Finally, once the business has developed and implemented the risk management strategy it must be vocal about having such a policy in place. Investing in the development of a comprehensive risk management strategy improves the visibility of processes and demonstrates stability, which is particularly important for high growth SMEs and start-ups that are seeking grants and funding. One may even be able to negotiate more favourable insurance premiums for the business. In many respects all business should consider a risk management strategy as an extension to insurance arrangements, which will serve to boost employee, investor and stakeholder confidence in the business as a whole. Some of the requirements discussed in **CHAPTER 10** can also be applied to assist small firms to identify the main risks to the business and put plans in place to minimise them. In turn this will enhance the benefits of ongoing due diligence and corporate governance.

References

1. Spedding, Linda, *Environmental Management for Business,* Wiley & Sons, Chichester, West Sussex, 1996

2. Sparkes R. *Socially Responsible Investment: A Global Revolution.* Wiley & Sons, Chichester, West Sussex, 2002

Appendix

Corporate Sustainability and Responsibility Research

A standard has been drawn up by independent Corporate Sustainability and Responsibility Research (CSRR) groups, CSRR-QS 1.0, with the objective of promoting confidence in those bodies performing the research. The standard aims to improve quality management systems, to stimulate transparency, to facilitate assurance processes and to form a basis for further verification procedures.

This standard covers the functions of bodies whose work may include the collection of CSR data and subsequent SRI activities on the level of research, analysis, evaluation, rating, ranking, screenings, risks and opportunities assessments, all related products, processes, work procedures, services and subsequent reporting of results of these activities to clients and other stakeholders.

Although these CSRR activities may have numerous outputs and clients, the CSRR-QS 1.0 focuses mainly on the operational requirements of SRI related products and services. The initiation of the standard has been granted by the European Commission, Employment and Social Affairs DG as the outcome of the project 'Developing a Voluntary Quality for SRI Research'. It has been drawn up in the light of the European Commission's aim to build partnerships for the promotion of CSR, seen as a contribution to achieve the publicised strategic goal of becoming, by 2010, 'the most competitive and dynamic knowledge-based economy in the world, capable of sustainable economic growth with more and better jobs and greater social cohesion'.

Both the Green Paper *Promoting a European Framework for Corporate Social Responsibility* and the EC's Communication *Corporate Social Responsibility, a Business Contribution to Sustainable Development* call for 'more convergence and transparency of SRI rating methodologies', request that 'quality and objectivity should be ensured, not only on the basis of the information submitted by the management, but also by the stakeholders' and suggest that 'external audit and internal quality procedures should be used to assure accuracy in the research and assessment processes'.

Prior to the set-up of the standard a detailed survey has been executed in order to make an inventory of existing practices on the level of quality management, and related clients and stakeholder demands.

Chapter 5

Key Due Diligence and Corporate Governance Organisational Areas

Introduction 5.1

CHAPTER 4 has demonstrated the evolving nature of the definitions of due diligence and corporate governance, having regard to the business environment and the business drivers mentioned in that chapter. Evidently management should aim to have a clear policy that not only prioritises the positive aspects of running a successful business but also has considered the more traditional concerns that embrace such matters as:

- contracts;
- customer care;
- marketing;
- employment;
- health and safety; and
- other risks.

All of these can expose the business to potential conflict and dispute. While the better policy is to be proactive and protect the business as far as possible through careful drafting, with appropriate advisers inside and outside the business, there is no doubt that the risk of litigation is a day to day concern. In terms of ongoing operations, the threat of litigation looms large in the business lives of many and can devastate relationships with customers, financiers and suppliers. This has an impact upon the internal due diligence and corporate governance organisational issues of the business.

Since this is a potentially huge subject, for the purpose of this discussion certain topical examples have been selected to demonstrate its relevance in this handbook. The reader should bear in mind (the premise stated in **CHAPTER 1**) that the approach to due diligence and corporate governance being taken here is that both concepts are intended to enable the establishment and development of a sound healthy business in which sustainable decisions can be made. Accordingly, it should always be borne in mind that litigation can also in any event:

- prevent the success of a proposed deal or transaction;
- cause an unexpected blow to growth;
- affect morale;
- distract the human and financial recourses of a business, whether starting up, growing or established;
- impact upon reputation; and
- potentially cause insolvency and/or the end of the business.

Since the litigious environment of today's business world means that it is difficult to avoid confrontation, an understanding of the alternatives that are now available to mitigate the effects of such confrontation is very important. This can assist toward a 'litigation policy' so that management can aim to escape the disastrous impact on a business, small or large, through the distraction of resources. Many businesses try to avoid the reality of looming conflicts and potential legal battles. It is vital that a business ensures that someone in the business takes responsibility for this aspect of running a business from the outset and that they have a clear approach to litigation management. This is all the more true in this era of increased focus on corporate governance matters that requires transparency in all operations. Since many businesses in the UK do not in fact have in house legal expertise to assist with such a policy, a practical overview of the UK alternatives follows. It is noteworthy that the trend is toward dispute resolution and that in one major area of concern, employment law, there are new regulations, the Employment Act 2002 (Dispute Resolution) Regulations (due to come into force on 1 October 2004), giving rights to employers and employees, aiming to encourage both parties to resolve disputes internally through discussion, thereby reducing the need for further action and keeping the business moving. Both employer and employee must follow the minimum three stage process in the event of dismissal, disciplinary or grievance procedure before resorting to an Employment Tribunal.

Litigation 5.2

Prosecuting or defending civil litigation before the English courts is an expensive business. Frequently the legal costs overtake the amount originally at stake. Costs include lawyers' fees and expenses and in some cases the opposing party's costs. Even a party awarded its costs can expect to recover less than the full amount of those costs. There may also be costs that cannot be recovered if the other party has become insolvent in the meantime, which is why it is critical to monitor the solvency of the opposing party. In limited circumstances it may be possible to obtain an order requiring the prosecuting party to provide security for the defending party's costs at risk.

The Woolf Reforms **5.3**

In April 1999 sweeping reforms were introduced changing the way in which civil cases are prepared and conducted before the English courts. The Civil Procedure Rules (CPR), also known as the Woolf Reforms, provide a unified body of rules that apply to the County Courts, the High Court and the Court of Appeal. They are intended to ensure that the civil justice system is accessible, fair and efficient. The rules themselves are subject to an overriding objective which may be very briefly summarised as enabling the court to deal with cases justly, which includes so far as is practicable:

- ensuring that the parties are on a equal footing;
- saving expense;
- dealing with the case in ways that are proportionate to the amount of money involved, the importance of the case, the complexity of the issues and the financial position of each party;
- ensuring that the case is dealt with expeditiously and fairly; and
- allotting to it an appropriate share of the court's resources, while taking into account the need to allot resources to other cases.

The parties are required to help the court to further the overriding objective. The court may further the overriding objective by actively managing cases. Previously case management was almost entirely in the hands of the parties, or rather, in the hands of their lawyers. The courts could be asked to make orders giving directions to parties who were slow or uncooperative. Now the initiative lies substantially with the court to determine how the matter shall proceed and the pace of progress. There is now little opportunity for the parties to opt out and there is reduced opportunity for the lawyers to play games at their clients' expense. It is reasonable to expect that an average commercial dispute will take 18–24 months to come to trial from the date of issuing proceedings, often longer.

Without doubt, many of the reforms introduced by the CPR were overdue, are welcome and operate for the benefit of litigants, and in some cases for lawyers. Litigation can now progress to its conclusion quicker. It is now more difficult for the parties or their lawyers to abuse the system or delay. However, the cost of litigation remains high and the reforms seem to have discouraged a lot of litigation of smaller value.

How to reduce costs **5.4**

Costs can be reduced, or prevented from spiralling out of control, either by the client's effective management of the legal team or by finding alternatives to litigation. What is important is effective management. There is no cheap way of conducting litigation (straightforward debt collection excepted), but business people involved in disputes can take positive steps to ensure that their money is spent to best effect. The following are a few suggestions that should be helpful to any business, not necessarily in order of importance.

Using effective lawyers 5.5

At a commercial level at least, solicitors' firms tend to specialise. Some of the larger law firms (the majority of which, for the UK, are in the City of London) have more than 1,000 fee earners plus support staff. They often have a very wide range of expertise, all under one roof, but however much they may deny it publicly they do tend to be more expensive than smaller firms. Nevertheless, they have certain advantages. Sometimes the required expertise is not available elsewhere. Sometimes the matter requires a team, and may even demand working 24 hours a day seven days a week. Smaller firms can rarely provide this level of urgent service. If the matter does not require the expertise and service that only a major firm can provide, consider a less expensive alternative. Furthermore, many businesses find that it is not always necessary or expedient to use one firm exclusively and it can be useful to introduce a little competition (without overlooking the value and benefits of loyalty and trust).

Saving energy 5.6

There is a significant hidden cost of litigation which many lawyers fail to take into consideration. Complex litigation involves ongoing teamwork between the lawyers and their client. A substantial and ongoing input is often required from the client. This can take up a great deal of the client's time, a lot of the client's energy and there are cost factors involved. The client should consider whether that necessary input of time and energy might not be better spent pursuing the client's business. Critically set out a plan at the outset. The client should make clear what the litigation is expected to achieve (and at what cost) and make sure everyone has signed up to that. Ensure the lawyers explain their strategy for conducting the litigation, when they might be seeking settlement or mediation, what the costs plan is and how they will keep the client updated on progress.

Co-operation 5.7

If the matter is to be pursued, the client should ensure that his or her own input is well managed. This means making sure that staff are available or appointed to provide the lawyers with the answers to their questions and to provide all relevant documents. Documents should be presented in an understandable form and usually in chronological order. This reduces the many chances of mistakes and the time that the lawyers have to spend assimilating complex ideas, facts and evidence. One of the criticisms made of lawyers is that they have little idea of how they (and their work) are perceived by their clients. They think they have done a great job. The client thinks they have made an expensive mountain out of a molehill. Recognise these difficulties. Let the lawyers know what is wanted and expected.

Agreeing a budget **5.8**

Some of the financial uncertainty of litigation can be mitigated by agreeing a budget with the lawyers in advance and perhaps during the course of the case. Although each party to the dispute may say that its objective is to seek justice, what each really means is that it wants to win. It is hardly surprising that there is seldom continuous co-operation between the opposing lawyers. The pace, direction and cost of the litigation may be largely determined by steps taken by the opposition. It is therefore difficult for lawyers to accurately estimate in advance how the opposition will react and behave and as a result what a case will cost to fight and win or to fight and lose. The best the lawyer can usually do is either provide an estimate that takes into account all of the things that might go wrong (and in doing so risk frightening the client) or provide adjusted estimates for each stage of the litigation as the matter progresses, giving likely maximum and minimum figures. If the client wants an estimate that allows for no increase, he or she must expect to cover the costs risks. This means that he or she will have paid over the odds if the litigation progresses without difficulties.

It is a little different where lawyers tender for bulk work which might involve hundreds of disputes over a period of time. They can apply the swings and roundabouts principle. It is not in the lawyer's interest to take on work that turns out to be non remunerative and it is not in the client's interest to tie the lawyer down to a fee structure that tempts the lawyer not to spend as much time on the matter as it deserves. The lawyers should always be able to provide realistic estimates on an ongoing basis of the likely costs ahead.

Insurance **5.9**

Some domestic and motor policies provide additional insurance against legal fees including fees that may have to be incurred resolving disputes. This is now known as 'before the event' insurance. Business insurances are less likely to contain such benefits. Insurance is also available, provided by specialist insurers, to cover the expenses of a particular piece of litigation. The insurance can either cover the whole of the cost or the excess over an agreed sum. It is known as 'after the event' insurance. It provides an element of financial certainty but of course it can only be bought at a price, which can be, by way of premium, up to 40 per cent of the coverage required. The insurers do not expect to lose. The insurance can also cover the risk of having to pay the opponents' legal costs if the case is lost. Such cover is normal if the lawyer is working on a conditional fee basis, but not all cases can be dealt with in that way. Not all lawyers are prepared to work on that basis.

Legal expenses insurance is a useful way to cover the risk of a costs disaster, to contain litigation costs within a budget, or as a necessary adjunct to a conditional fee arrangement. One must be very careful about the terms of such policies. One key area of concern is the pyrrhic victory. This may be where a favourable judgment is given, but there are no funds to satisfy that judgment.

This is defined as a win in the terms of the policy and therefore any funds extended to cover own side costs become repayable regardless that the insured has not received any funds from the losing party. Secondly, the definition of win and lose needs to be carefully examined particularly where one is suing more than one defendant. One may lose against one defendant and have to pay their costs, but succeed against other defendants. Policies often define win as winning against any of the defendants, thereby preventing the insurer being liable to pay out against those who successfully defended the claim (which is more common than one might think). The insured nonetheless remains liable for the costs of the successfully defending party.

Litigation funders **5.10**

The *Access to Justice Act 1999*, coupled with a number of legal authorities, has opened the market to those wishing to provide funding for pieces of litigation, which they typically do either by taking an assignment of the action outright or in return for a share in the proceeds of the action. The old legal principles of champerty and maintenance make it clear, despite the developments welcoming the role of funders in providing liquidity to good claims which fail only for lack of funds, that such funders are not entitled to interfere in the conduct of the litigation. To do so would be to invalidate the funding arrangement. This is to say, therefore, that funders may not dictate the course of the litigation: that is the role of the client as advised by its lawyers.

Typically funders will be looking for a case to satisfy three key criteria.

1. That the case has a 70 per cent or better chance of success on the legal merits.
2. That the proposed defendant has the means to pay any judgment ordered against it.
3. That the costs of pursuing the matter are proportionate to both the size of the claim and the percentage which the funder will take by way of its return (typically 25–50 per cent of the net proceeds of an action, which needs to equate to a 3:1 return on costs spent by the funder).

Again, clients need to check the terms of such arrangements (eg when any funds might be repayable, what level of uplift the lawyers are entitled to under their Conditional Fee Agreement (CFA), that adequate adverse costs insurance is provided for, that adequate indemnities are provided for in any funding agreement), but the advantage of such arrangements is that there are no costs to be paid as these will only be taken out on recovery.

Overseas lawyers **5.11**

International businesses can find themselves with international (or cross border) legal problems. Contractual disputes may be subject to interpretation according to another country's law, and perhaps by that country's courts. Some of these problems can be avoided by taking legal advice before the contract is made, or by anticipating the problem before it arises and making sure that the applicable law and jurisdiction are agreed in advance.

When problems arise it may become necessary to involve foreign lawyers. Their costs may not be so easy to control. They may be less expensive than their UK counterparts but their basis of charging may be entirely different. If another language comes into the equation, the client should expect to pay a little more for having the luxury of a foreign lawyer reporting and receiving instructions in the client's own language. If text has to be translated professionally, the cost will be significant. Unless the client already has a good working relationship with an overseas lawyer in the country in question, he or she should consider instructing UK solicitors with an international practice to instruct overseas lawyers on the client's behalf. Some of the larger firms have overseas offices. Some large and small firms belong to one of the international associations of lawyers that provide access to trusted overseas colleagues. Instructing UK lawyers will mean incurring additional fees, but the UK lawyers will have some experience of the pitfalls to avoid. They should be able to ask all the right questions on the client's behalf and prevent nasty surprises when it comes to fees and expenses. They will endeavour to be cost effective. Sometimes part of their fee can even be recovered as recoverable costs in successful foreign proceedings.

Early settlement **5.12**

If the matter is capable of being settled before trial, then from the point of view of savings costs, the sooner the better. Costs tend to increase the closer the matter gets to trial. By then there will probably be increased reliance by the solicitors on the advice of counsel (barristers) and the need to instruct expert witnesses. These all add to the costs and to the costs at risk if the case is lost

Commercial lawyers in the UK generally consider it to be part of their job to resolve the matter as quickly and economically as possible and to avoid the heavy costs of a trial whenever possible. They will usually be skilled negotiators and be able to start settlement negotiations without the risk of their doing so being seen as a sign of weakness on the part of their client. The lawyers can usually broach the subject of settlement without it necessarily being implied that they do so on their clients' express instructions. Nevertheless, it does not hurt to remind the lawyers from time to time that settlement is preferable to trial if there is no point of principle or important point of law (that is, important to the client) to be determined.

Lawyers are often accused of dragging out litigation in order to increase their fees. No doubt this happens but most commercial lawyers take the view, quite apart from their professional obligations, that there is more money to be made by having happy clients who come back again and again than by milking a case for all it is worth.

Part 36 offers to settle 5.13

Part 36 refers to the particular section of the CPR governing the procedure described below in this paragraph. It would be unfair if a claimant were free to start legal proceedings, ignore a perfectly reasonable offer of settlement, press on regardless, be awarded less than was offered and but nevertheless recover all of his or her costs. To balance this, a defendant in litigation has always been free to make a money offer. If the right sum is offered and is backed by a payment of the offered sum into court, the claimant is put in a quandary. If he or she accepts the offer he or she is entitled to recover the costs he or she has incurred up to the date of the payment into court. If he or she ignores the payment into court he or she does so at his or her peril. If when the matter is later resolved at trial he or she recovers less than the sum offered (making due allowance for interest) he or she will recover the sum awarded and the costs he or she has incurred up to the date of the payment into court. However, he or she will have to pay his or her own costs and the defendant's costs (at the higher, or what is known as 'indemnity' level, by way of recognition that he or she has acted perhaps rashly) incurred after the date of payment into court.

Since the costs of the trial are likely to be substantial, he or she may well lose much more than he or she stands to gain. Incidentally, the fact of the Part 36 offer and of the payment into court is not revealed to the trial judge until after the matter has been decided but before the judge proceeds to deal with the matter of costs. The CPR takes this procedure one step forward. It now enables the claimant to make an offer by stating a sum that he or she would be prepared to accept in settlement. If the defendant fails to 'beat' the offer, he or she is not only penalised in the same way as to costs, but he or she will pay a punitive rate of interest on the sum awarded. The procedure is also available in respect of non-money claims.

It is of course the lawyer's job to advise the client of the availability of this procedure, when to use it and how much to offer, but it is no bad thing for the client to be aware of this from the beginning. An early Part 36 offer focuses the other party's mind on the risks being borne and often leads to an early settlement, although not necessarily on the terms of the offer.

Early settlement with or without the assistance of a Part 36 offer will probably do more than anything to bring about a saving in litigation costs.

Legal costs consultants 5.14

The rules relating to the calculation of lawyers' costs, especially the costs that can be recovered by the successful party are complex. Not surprisingly, it has created a specialised industry for independent law costs draftsmen. Lawyers sometimes prefer to use consultants rather than keep costs draftsmen on the staff payroll. Some of them also now offer their services to larger consumers of legal services. They can help the client to reduce fees. If they are engaged as a matter of routine, rather than as a result of a disagreement about the bill, they can prevent problems arising by keeping the solicitors in check.

Summary 5.15

Do not be too hard on the lawyers. A working relationship based on trust and respect for each other's role goes a long way to achieving the desired result. However, do:

- take steps to avoid disputes arising and treat litigation as a last resort;
- do not be fooled into believing that the Woolf Reforms have made litigation more financially viable;
- choose appropriate lawyers for the job;
- do not overlook the input that will be required from you as the client;
- be aware of the lawyers' problems and help to meet their requirements;
- agree a budget in advance;
- consider whether legal expenses insurance might be appropriate;
- if using foreign lawyers, consider using UK lawyers as an intermediary;
- be alive to settlement possibilities and opportunities;
- be aware of the possible advantages of a Part 36 offer; and
- consider the use of legal costs consultants.

Since litigation is almost always expensive, particularly so if you lose and in consequence you are ordered to pay your opponents' costs, alternatives to High Court or County Court litigation should also be considered.

Alternatives to litigation

Avoidance of disputes 5.16

It cannot be emphasised enough that in terms of due diligence and corporate governance, the avoidance of disputes is generally the best policy. This seems like obvious advice but avoidable commercial disputes occur all the time even when business on both sides is conducted in exemplary fashion. Disputes usually arise from misunderstandings. Typically the parties enter into a business venture, focusing on all the positive aspects of the venture but without much thought as to how matters will be resolved if things do not go according to plan. Carefully drafted agreements, contracts and business documents can legislate for the sort of misunderstandings or difficulties that may arise. It is usually much cheaper to pay a lawyer a fee to help draft something that will reduce the risk of problems arising, than to pay a lawyer later to sort out the mess.

Cost management by early settlement 5.17

It is surprising how often sophisticated business people who know the high cost of legal services automatically rush straight to their lawyers as soon as a dispute arises. When lawyers are seen to be involved, legal costs are incurred and the parties tend to distance themselves from each other and from reconciliation. It is worth considering for a moment what steps may be taken to bring about an early settlement. It is usually appropriate to seek legal advice on the merits of the prospective claim or defence at an early stage. It may not be necessary or appropriate to involve lawyers further in any initial attempts to achieve settlement. The client can be sure that if the first move is made by his or her lawyer, he or she will drive the other party into the arms of his or her lawyer too and the chances of an early settlement will probably be lost, at least for the time being.

Often in business, although the parties might not admit to it, there is an emotional obstacle to settlement. One party may feel that he or she has been bullied or cheated or there may simply be a clash of personalities. If this has happened and feelings are too strong to allow settlement, the matter should be taken out of the hands of the people involved, ie change the negotiating team.

An offer to settle a matter, or an offer to accept less than the full amount, may be perceived as a sign of weakness but not if the gesture is expressed properly. Care should be taken, however, to ensure that any settlement proposal is carefully expressed in such a way that it cannot later be used against you if negotiations fail. For example, up to a point, the negotiators can make it clear that their proposals are without prejudice. These words will generally ensure that the letters so marked or conversations expressed to be 'without prejudice', or 'off the record' will be excluded from anything placed before the court if the matter cannot be resolved. Negotiators can also hide behind the board of directors or their managers for formal authority to settle until such time as agreement in principle is reached. Lawyers can assist in negotiations and still be kept in the background. Bear in mind the value of what is sometimes called a 'commercial settlement' where the agreed debt or obligation is repaid by continuing or additional business between the parties.

Alternative Dispute Resolution 5.18

The term Alternative Dispute Resolution (ADR) is usually used to encompass two quite different means of resolving disputes – arbitration and mediation. Arbitration is an alternative to litigation before the courts. It is a less formal means of adjudication, and is often but not necessarily cheaper than court litigation. Mediation is a process that often takes place after litigation or arbitration has started, rather than as the first means of resolving the dispute. Mediation is directed to settlement, not adjudication.

Arbitration **5.19**

Arbitration as a means of resolving disputes has been employed in England for hundreds of years. Arbitration is generally understood as 'a reference to the decision of one or more persons, either with or without an umpire, of some matter or matters in difference between the parties'. Arbitration can be effectively used to resolve simple and very complex disputes. It is supported by a body of law, most recently the *Arbitration Act 1996* which sets out the following general principles (partly summarised).

1. The object of arbitration is to obtain the fair resolution of disputes by an impartial tribunal without unnecessary delay or expense.
2. The parties should be free to agree how their disputes are resolved, subject only to such safeguards as are necessary in the public interest.
3. The court should not intervene except as provided by (the Act).

Legislation such as this, together with the rules of natural justice provide an outline set of rules and principles which also generally limit the scope for interference or intervention by the courts. The courts remain in the background, to intervene only when permitted and when absolutely necessary. An arbitration award may be enforceable as such overseas in a number of countries. It may be enforced in the same way as a judgment of the court. It may be 'converted' to a court judgment if appropriate for the purposes of enforcement, for example if it is to be enforced overseas in a country where a foreign judgment but not a foreign arbitration award may be enforced. It is often easier to enforce an arbitration award overseas than it is to enforce a court judgment overseas.

Advantages/disadvantages of arbitration **5.20**

One person's advantage is often another person's disadvantage so rather than set out advantages and disadvantages it is more appropriate to set out some of the features of arbitration.

1. Freedom of choice

   Arbitration takes place only with the agreement of the parties. That agreement may be embodied in the contract giving rise to the dispute, or the parties faced with a dispute may subsequently agree to refer the matter to arbitration. In the absence of an arbitration agreement, the parties cannot be forced into arbitration. Conversely if there is agreement, the courts (at least in the UK) will set aside court proceedings in favour of arbitration. Many business people in the UK and overseas feel more comfortable having their disputes resolved by members of their own business fraternity whom they have chosen rather than by the judiciary. Not all courts overseas are competent to deal with complex business matters.

2. Cost

Arbitration is generally perceived as being cheaper than litigation but that is not necessarily the case. Arbitration procedure is usually less formal than court procedure. Even when arbitration takes place under the auspices of a professional arbitration body according to that body's arbitration rules, those rules will be very much simpler and more straightforward than the courts' *Civil Procedure Rules 1998* (*SI 1998/3132*) (*CPR*) although the arbitration rules themselves may be interpreted and applied in the same way as the CPR. In more straightforward matters, it is quite normal for the parties to conduct the arbitration reference without the aid of lawyers. Sometimes the parties will have help from lawyers, but manage without advocates. The saving in lawyers' costs may to some extent be outweighed, however, by the additional executive time devoted to the reference.

Heavy commercial arbitration can involve specialist solicitors, senior junior counsel and leading counsel briefed for the hearing together with an army of expert witnesses. The cost can be as high as the costs that would be incurred in court. Added to this the parties have to pay the daily fees of their arbitrators. It is quite common for the tribunal to consist of three arbitrators. In addition there may be the cost of hiring a room and facilities for the hearing. Although litigation in court involves the payment of fees to the court, the judge will sit for as long as the case takes to conclude, at no extra cost. No charge is made for the court room. At least one international arbitration body charges substantial administration fees in addition to the fees of the members of the tribunal. Some of the trade organisations that offer an arbitration procedure to their members (and others) make a modest charge for providing an administrative framework within which the arbitration reference can take place. Others make no charge, leaving it to the parties and to the tribunal to attend to all administrative matters, usually with some standardisation of fees. The cost of arbitration can be significantly reduced if the parties can agree to a single (sole) arbitrator. Of course when the parties are at loggerheads, they cannot always agree who should arbitrate for them. Sometimes, however, they can at least agree who shall appoint the arbitrator for them if they cannot agree between themselves and if there is no organising body with a set procedure for this event. The High Court has powers of appointment in these circumstances but even with the CPR reforms, the procedure can be fairly slow and costly when one is faced with an uncooperative opponent in a foreign country.

Some organisations provide a special procedure for small claims. The London Maritime Arbitrators Association (whose arbitrators routinely deal with complex disputes) have a small claims procedure for claims involving less than $US 50,000 (or equivalent). This procedure provides a straightforward low cost resolution service at a fixed cost. Under this procedure arbitration awards are made on documents alone, that is without an oral hearing taking place, and the matter is decided by a sole arbitrator.

3. Speed

The speed with which an arbitration reference may proceed is substantially in the hands of the parties. If the reference is to a sole arbitrator and the matter is to be determined on documents alone, it can proceed to completion in a matter of weeks; sometimes less with everyone's co-operation. If the tribunal consists of three heavily booked professional arbitrators, busy counsel and solicitors, heavily booked expert witnesses and witnesses of fact with similar problems, the hearing date may have to be fixed months, possibly a year or more in advance. It is probably true to say that with goodwill on both sides, arbitration is usually quicker than litigation before the UK courts and certainly much faster than litigation before some overseas courts.

4. Hearing

More often than not it is for the parties to agree whether an oral hearing should take place at which witnesses may or may not be called to give their evidence. Britain has a tradition of advocacy. The advocate's role in civil litigation is an important one. In many European countries, the advocate's role in civil matters is much reduced. The written submissions of the parties and their written evidence are largely supposed to speak for themselves. British arbitration tribunals are reluctant to deny a party his or her right to be heard, but it is probably true to say that the majority of arbitration cases are decided without advocates, on documents alone, that is on the basis of written submissions and written evidence. Significant cost savings can be achieved by dispensing with an oral hearing. The saving on advocates alone will be significant as will the saved cost of expensive expert witnesses standing by. There will also be a saving of executive time as well as the cost of room hire, catering, interpreters, messengers etc. Whether a formal oral hearing is really appropriate will depend on the nature of the case.

5. Representation

Some arbitration bodies exclude lawyers from any oral hearings, for example the arbitration rules of the Grain and Feed Trade Association (GAFTA). Lawyers may routinely be employed however in the preparation of the parties' cases, and the cost of lawyers may be recoverable by the successful party, even when lawyers are not permitted to take part in the hearing.

6. Challenge/appeal

As a general rule, UK arbitration awards are not subject to challenge by the courts except on the grounds of serious irregularity that has affected the outcome, or of the absence of the tribunal's jurisdiction to make the award. An award can be made the subject of an appeal to the courts but only on a question of law and then only (in the absence of agreement) with leave of the court. There is no automatic right of appeal. Some arbitration bodies provide an appeals procedure within the arbitration framework itself. This will not affect any eventual right to appeal by a court.

From the point of view of overall cost, restricted access to an appeals procedure can cut short what might otherwise be an expensive ongoing legal procedure. The parties can set the parameters in advance.

7. Confidentiality

Civil court proceedings are, with very limited exceptions, public. The decisions of the courts are reported not only in the press but in the numerous law reports that line the shelves of lawyers' offices. The decisions are, after all, the law itself. Arbitration proceedings are, with limited exception, private and confidential. If the matter becomes the subject of an appeal to the court, the facts and the parties are likely to become a matter of public record.

8. General suitability

One of the perceived advantages of arbitration is that the arbitrators need not be lawyers. They may be drawn from the trade or profession in which the parties have conducted their business and fallen into dispute. Arbitrators can be chosen who may have spent most of their working lives in the particular trade or business concerned. They are usually chosen by the parties themselves. They will be familiar at once with the technical aspects of the matter before them. Their expertise and experience may be unsurpassed, particularly for example in quality disputes involving agricultural produce or manufactured goods and in construction or shipping matters where the dispute is likely to be more one of fact (or rather facts) than law.

However, the substantial commercial knowledge of the judges of the Commercial Court of the Queen's Bench Division and of some of the county courts should not be overlooked. A judge of the Commercial Court may even be appointed to act as a sole arbitrator or as an umpire.

9. Getting to the truth

Where the choice whether to proceed before the courts or by arbitration remains open at the time the dispute arises, it may be appropriate to consider whether the dispute is likely to involve oral evidence from witnesses who may be tempted to be untruthful..

Arbitration is big business in the City of London. Disputes that have no connection with the UK are brought to London as a trusted neutral forum for dispute resolution. There is a good supply of competent professional and part-time arbitrators with surprising degrees of expertise. There are expert witnesses at hand with expertise in all manner of subjects. There is also a ready supply of lawyers and barristers routinely involved in arbitration work so the future continues to look bright for London arbitration. Strangers to arbitration should give the arbitration option serious thought.

Mediation 5.21

Mediation is a facilitative process by which the parties in dispute engage the assistance of an impartial third party, the mediator, who helps them to try to arrive at an agreed resolution of their dispute. The mediator has no authority to make any decisions that are binding on the parties, but uses certain procedures, techniques and skills to help the parties to negotiate an agreed resolution of their dispute without adjudication.

Thus, mediation is quite different to arbitration. Unlike arbitration, mediation does not involve making a finding of fact or law or the rendering of a final and binding award. The mediator has no authority to make a binding decision. At the moment, unlike arbitration agreements, an agreement to enter into a mediation process will not be enforced. As yet there is no substantial body of mediation law, but that may change in time. The rules of natural justice probably have little or no application to mediation.

How does mediation work? 5.22

Mediation takes place when the parties agree to try to resolve their differences by mediation or, perhaps more commonly, when during the course of litigation (or even arbitration) the parties are encouraged to try to reach a settlement by mediation. This applies equally to commercial disputes as, for example, family matters. The UK courts now encourage mediation by allowing adjournment of proceedings for that purpose. The court may require the lawyers for the parties to confirm that they have brought the possibility of mediation to the attention of their clients.

Mediation will not work if there is no will to settle – the skill of the mediator lies in helping both sides to reach agreement about how a dispute should be settled. Sometimes the parties will realise that at least some of the issues between them can be resolved, leaving fewer or shorter issues for the court to resolve afterwards.

There are no hard and fast rules. Different mediators have different ways of working. What often happens is that everyone involved meets in a room in the presence of the mediator. The mediator explains how the matter will proceed. It is up to the parties to decide whether they want their lawyers to accompany them. The parties then take it in turns to summarise what their case is about and to state what they are looking for. A time limit may be imposed. Then the parties go to separate rooms where they will be visited, probably several times, by the mediator who will discuss the case and try to determine where the common ground may lie or what are the underlying obstacles to settlement. The mediator will not disclose to the other party what has been discussed except to the extent that he or she is specifically authorised or requested to do so. The mediator will carry forward ideas, suggestions and, hopefully, offers. The parties are likely to be subjected to a deadline for completion of the process to make sure that their minds are concentrated on bringing the matter to a conclusion.

What happens if no agreement can be reached? 5.23

If no agreement can be reached it is the end of the matter. Nothing that has taken place during the mediation process will be referred to in the resumed proceedings or any subsequent proceedings brought in respect of the dispute. The process is confidential. The parties may even agree to a further mediation session at a later date.

What happens if agreement is reached? 5.24

Agreement may be reached as to the whole of the dispute, parts or aspects of it. The mediator has no powers to make any judgment or award but may assist the parties to draw up the terms of their settlement in such a way that it becomes legally binding upon both parties. If the mediation has taken place during ongoing court proceedings (or arbitration) the parties may ask the court or arbitration tribunal to make an order or award on the terms of the settlement so that it becomes enforceable as such. This may be important for enforcement overseas. If there are no underlying proceedings then the agreement will simply be of a contractual nature.

What happens if the settlement terms are repudiated or ignored? 5.25

If the settlement has become an order of the court or an arbitration award, it can be enforced in the same way as a judgment or award, possibly overseas. If it is purely contractual then it may be necessary to sue on the agreement. To some extent this means starting all over again but the issues this time should be clear and capable of being dealt with by a court quickly and cheaply.

What does it cost? 5.26

Arbitration can cost very little, especially when compared to the alternatives. If the mediation is conducted by a court (such as the Central London County Court) or a professional mediation body, modest fees are payable to cover the services of the mediator and the provision of facilities for the mediation to take place. The service offered by the Central London County Court is extremely reasonable. If the parties appoint their own mediator (anyone can be appointed but someone with experience is more likely to achieve a result) the fee is a matter for prior negotiation. Since the process is likely to last less than a day and will not generally involve the engagement of advocates or expert witnesses, the only other cost is the cost of the parties' lawyers (if they are to attend) and the price to be put on executive time for the parties themselves. The parties are usually required to agree beforehand that each will pay half of the costs of the mediation whatever the outcome.

Does it work for commercial disputes? 5.27

The answer is yes, provided the parties enter into mediation with the intention of trying to reach a settlement. The rate of take up is not particularly high – this may be the fault of lawyers unfamiliar with its possibilities and concerned that it might appear too soft or informal to their commercial clients. Indications are that when lawyers are present at the mediation, the success rate is lower. Perhaps this will change in time. Mediation is particularly appropriate where no point of principle is involved or where the parties may wish to do business together in the future.

Mediation should be kept in mind for all commercial disputes, if not at the beginning then during the course of litigation or arbitration. It costs little, arrangements can usually be made for mediation to take place quickly and there is really nothing to lose (other than its shared cost) if it does not work since neither party is prejudiced. Furthermore, a recent High Court decision (*Dunnett v Railtrack plc [2002] EWCA Civ. 303), [2002] 2 All ER 850*) made it clear that the courts will penalise on costs, those parties who, having been offered mediation, deny it and then go on to win. Where ordinarily they would get their costs, if they have denied the use of mediation they will not automatically get their costs back – this is a powerful weapon for getting the parties to mediate.

The introduction of the CPR has demonstrated that the cost of court litigation will remain high at least for the foreseeable future and that there is a real need for alternatives. Whilst there is a strong tradition of arbitration in Britain, other countries are well ahead in taking up the advantages of mediation and that is good news for the lawyers. Following the CPR reform – there was a boom in meditation according to experienced mediators. This was because there was a formal realisation that mediation was here to stay and due to it being made a formal part of the system. In comparison with court proceedings, the evident advantages of mediation are that it is:

- quicker;
- relatively cheaper; and
- can enable the resolution of disputes without damaging the commercial relations between the parties.

Some practitioners see mediation as a tool to be used in conjunction with – rather than as an alternative to – litigation. In any event, mediation does indeed seem to be on the increase along with the number of organisations providing mediation services.

Other ADR procedures 5.28

The ADR procedures that are discussed in the list below are also sometimes used. There are other procedures and, no doubt in time, more will be devised.

1. In some jurisdictions the court can refer the case or aspects of it to a referee (or expert) chosen by the parties to decide some or all of the issues. This is to be distinguished from the practice by which the court appoints an expert who reports his or her findings to the court. There is nothing to prevent the parties to a dispute agreeing to adopt this procedure.
2. In some industries (in particular the construction industry) the contract will provide for the appointment of a neutral adjudicator to make summary binding decisions without following litigation or arbitration procedures. This happens in the UK.
3. Sometimes the parties may themselves simply appoint an expert to consider the issues and make a binding decision without going through the motions of an enquiry followed by adjudication.
4. Statutory (or administrative) tribunals may be appointed (as in the UK) to establish fair rents, the price of freeholds, compensation awards, social security benefits etc. These tribunals have somewhat limited powers but they tend to be formal. A short comment on these is made in **5.30** as a result of the Leggatt Proposals.

   In the UK certain sectors such as legal and financial services may be investigated with public findings by an Ombudsman who may even be able to award compensation. In some jurisdictions a 'mini trial' may take place at which the lawyers for the parties present their cases to the parties and a neutral apointee who helps to clarify the issues and evaluates the merits. The neutral appointee may also play the part of mediator. Although the findings are not binding, settlement sometimes follows.

There are a number of variations of these alternatives but the least exotic and therefore perhaps most attractive at the moment seems to be mediation.

Summary 5.29

The following is a summary list of hints and tips to be considered.

1. Reduce the opportunities for disputes. Let the lawyers earn their fees by helping to avoid disputes arising.
2. If the dispute can be resolved amicably, the sooner the better if legal costs are to be reduced or avoided.
3. Consider the benefits of involving lawyers behind the scenes.
4. Change the negotiators if appropriate.

5. Frame settlement proposals carefully.
6. Consider including an arbitration clause in contracts and agreements.
7. Always be aware of the benefits of resolving the matter at any stage by mediation. Do not rely on the lawyers to suggest it.
8. Consider whether one or more of the other ADR procedures might be appropriate.

The role of tribunals 5.30

Another consideration for business is that they may be involved in tribunal proceedings.

There are over 85 tribunals in existence in England and Wales, generally under the supervision of the Council on Tribunals. These range from the Agricultural Land Tribunals to the Wireless Telegraphy Appeal Tribunal and deal with disputes between citizen and state and between private parties. Currently there is far from a unified system, for instance, in education and pensions there are several different bodies with different titles. Just as there was a massive reform of civil procedure in the courts it now appears that tribunals in England and Wales are likely to have a huge reorganisation. It has been reported that the government intends to implement proposals known as the Leggatt Proposals. The Leggatt Review (which led to the Proposals) identified about 70 tribunals in England and Wales, employing about 3,500 persons and hearing about one million cases each year. Whereas some deal with hundreds of thousands of cases per annum others were more or less defunct. The proposals would involve the Lord Chancellor's Department (LCD) bringing tribunals – currently also diversely scattered among various government departments – under one unified service. Such a reorganisation could take place in 2006–2007 provided ministers agree.

The Leggatt Proposals 5.31

The thrust of the Leggatt proposals is to provide tribunals with a more coherent structure. The report submits that tribunals will only acquire a collective standing that matches the court system once all of them, that is the administrative tribunals concerned with disputes between citizen and state and those concerning private parties, are brought under one administration. Within that structure there would be first-tier tribunals, for example education, health and immigration. These would have corresponding appellate tribunals. Some argue that employment tribunals have become so significant in terms of workload, however, that they should be the subject of a separate structure. They refer to the employment tribunal system task force recommendations. Similarly education tribunals are facing an expanding role and workload in the face of issues over disability discrimination. While the Leggatt proposals would remove the perception that tribunals are not independent from their sponsor government departments there is also concern that a unified tribunal system would be too cumbersome and may lead to a dilution in expertise.

Points to consider 5.32

While reform is clearly required to enable true independence for tribunals in England and Wales so that the *Human Rights Act 1998* is properly complied with there are several other issues that need resolution. In view of the fact that the Lord Chancellor's Department (LCD) is already preparing to assume responsibility for a unified courts system in 2004–2005 under the Courts Bill, to implement the Leggatt proposals over a similar period would require a vast increase in resources with the LCD more than doubling its current 12,000 staff to 25,000.

Due diligence and late payment issues for business 5.33

Recent legislation in England and Wales has assisted with the ongoing problem of late payment of invoices and accounts and, therefore, cashflow.

The *Late Payment of Commercial Debts (Interest) Act 1998* (*LPCD(I)A 1998*) came into force in August 2002. It makes provision for the payment of interest at a compensatory rate on the late payment of certain debts arising under commercial contracts for goods or services. Its main provisions (summarised) are that:

- interest becomes payable on the debt at the rate of eight per cent over the Bank of England's official dealing rate when payment becomes overdue;
- when no payment date has been agreed, interest starts to run after 30 days from the date of performance; and
- any term that purports to contract out of the statutory right to interest is null and void in the absence of a substantial contractual remedy for late payment.

The extent and nature of the late pay problem 5.34

It is well known that the provision of credit by suppliers to customers is an established feature of business transactions in most jurisdictions. Such a facility is essential for the efficient operation of the global economy. However, the provision of goods and services in advance of payment means, of course, that the supplier can be exposed and vulnerable to payment delays.

Debtors may be late meeting credit payments for several different reasons. These reasons were cited during the period of consultation as including:

- deliberate unjustified delay for financial advantage;
- disputes over the provision of goods and services;
- temporary cash flow difficulties;
- administrative errors by either debtor or creditor;
- misunderstanding or uncertainty over the agreed credit period;
- 'pay when paid' contract terms;
- breakdown in payment systems; and
- inability to pay due to insolvency.

Nevertheless, whereas it is a fact that small businesses have in the past regarded and continue to regard late payment as a serious problem, any quantification of the effect of the late payment problem is difficult. Some relevant surveys have, however, been carried out. By way of example, the Cork Gully Report of 1991 for the Confederation of British Industry (CBI) found that nearly 60 per cent of businesses regarded late payment as a significant problem. A more recent survey, which was undertaken in 1996 for Intrum Justitia (receivables management services company), found that a majority of UK companies surveyed said late payment caused a problem or serious problem for their cash flow, profit, growth and/or survival. Moreover the Forum for Private Business (FPB) has particularly lobbied the government on behalf of its 28,000 members – many of whom are within the definition of micro business as defined (0–9 employees) to deal with late payment issues as a matter of priority for small business. The FPB has been extremely proactive in this area.

It should be understood that actual payment performance against contractual credit periods would appear to support the concerns of small businesses. The 1996 survey undertaken on behalf of Intrum Justitia found that in the UK commercial debts were paid on average 18 days late (ie after the contractual payment date). Moreover, the existence of a statutory right to interest on late payments was supported by 80 per cent of the UK respondents in this survey.

As in many other parts of the world, small business plays a key role in the economy of the UK. The UK government has demonstrated its concerns that actions should not unnecessarily hinder the competitiveness of small business. There are countless examples that demonstrate the connection between late payment and business failure. It is therefore vital that larger businesses – as well as the smaller organisations – should keep abreast of the developments and observe their contractual obligations as part of their operational due diligence and corporate governance.

The late payment of commercial debt is recognised by most stakeholders in the business community as being a very serious problem, especially for small businesses (defined below and in **APPENDIX 1**) who are least able to carry the additional costs arising from payment delays. In particular, the recovery of debts can impose further expense in terms of diverted resources and actual costs, especially where recourse to the courts is required. Delays in payment are generally more harmful to a small business than to a large business. Small businesses are often highly geared, relying on short-term loans and overdrafts for working capital. Cash flow problems caused by late payments can, therefore, have a significant negative impact on the ability to trade of many companies and businesses, especially the smaller players. Small business, particularly in supply chain situations, can be very exposed to the problems of insolvency (some of which are mentioned briefly in **5.35** below). It has been acknowledged that any late payment, like any other breach of contract, should attract appropriate sanctions.

Some international examples and comparisons: applicable law 5.35

The problem of late payment is not, of course, limited to the UK. Most European Union (EU) Member States report payment delays. In general, legislation in Member States of the EU tends to apply to all commercial debts. A notable exception is Ireland where a right to claim interest against public sector bodies exists, though the right may be extended to allow claims against the private sector in the future. The following list illustrates other comparisons to be made.

1. The right to interest on overdue payments already exists in Spain. Legislation there was enacted in 1985 and the average overdue payment period is only six days. By comparison, Greece has no statutory right to claim interest and the overdue payment period there is 19 days.
2. In Sweden, legislation was introduced in 1976 to provide a creditor with a statutory right to interest. Since then the payment culture in Sweden has been cited as providing a leading example for Europe. Whereas in the UK the average overdue payment is 18 days late, in Sweden it is only seven days late. Evidently the legislation is used by all businesses who sell on credit as there is an appreciation among creditors that to tolerate non-agreed credit without any form of economic sanction amounts to being an unofficial bank that does not charge interest, causing, inter alia, unfair competition.

What is of obvious practical importance to business is not only the existence of a statutory right to claim but also the existence of the necessary legal systems to enable a creditor to enforce the right efficiently and effectively. As indicated above, one key reason for businesses not exercising their right to interest is the cost involved in pursuing the interest. It is evident that the more expensive and time consuming the legal process, the less likely that interest will be claimed and the less effective the legislation will be in improving the payment culture. By way of example, in Sweden, interest can be claimed automatically and, where not paid, can be pursued through the courts even if the principal debt is paid. A summary court procedure for undisputed claims exists. If disputed, the claim is referred to the court for litigation. Notice of this is sent to the debtor, seeking payment or objection/defence within eight days, in the absence of which a summons will be issued. The debtor must pay interest and compensation for the cost of pursuing the claim where a claim for interest is upheld.

It is noteworthy that the US has also introduced legislation to address the problem of late payment by Federal Government bodies. The *Prompt Payment Act of 1988* provides small businesses with a right to interest on overdue payments by Federal Authorities.

The matter of applicable law is therefore also vital. The UK legislation, the *Late Payment of Commercial Debts (Interest) Act 1998* (*LPCD(I)A 1998*) applies to any commercial contract, including imports and exports, written under the UK law unless foreign law expressly applies. Where the choice of law is a foreign law the Act applies if, but for that choice of law, the applicable law would have been UK law in general.

Implementation of the Late Payment of Commercial Debts (Interest) Act 1998 5.36

The implementation of this recent legislation has given rise to various practical issues for small business. Evidently as a result of the representations made to the government during the consultation period, it is of the view that many small businesses are currently addressing issues of credit management and should be given sufficient time to develop and implement their systems. This is why the government proposed the following timetable.

Claimant	**Timing**
Small businesses against large businesses and the public sector	From enactment of legislation
Small businesses against all businesses and the public sector	Two years after enactment
All businesses against all businesses and the public sector	Four years after enactment

Evidently the government believes that many small businesses will take time to adjust to the legislation. Accordingly, the interim stages should enable their credit management systems to be in place and checked before they face claims on possibly their largest bills. In this connection the Better Payment Practice Group was formed in 1997 as a partnership between the public and private sectors. Its publicised aim is to improve the payment culture in the UK business community and reduce the incidence of late payment, using the *LPCD(I)A 1998* as a catalyst for change and offering practical assistance to the business community in improving credit management. This is most useful bearing in mind the deadlines noted above.

Some outstanding issues 5.37

Meanwhile, certain areas of concern relating to the application of the interest rate defined by the *LPCD(I)A 1998* (Statutory Rate of Interest (SRI)) to various common circumstances often found in insolvency were raised by the small business policy representative body, the FPB, and outlined below.

1. A main concern relates to insolvency caused by cash-flow problems, caused by large debtors with partially disputed bills, or deliberate delay. This is a particular problem in construction and some other areas, and sometimes on insurance claims as there is no case law at present. It is unclear whether insurance claims are, or should be, covered by SRI.

2. Is the amount of a finally agreed invoice subject to eight per cent above base from the date the original invoice was due? At present, interest applies to tax assessments by companies or individuals, backdated to the original assessment – does SRI apply also in similar manner?

3. In an insolvency situation, it should not be necessary to commit to that process in full without an initial assessment of the viability of the sales ledger. This could well be done in smaller companies for a fixed £500 fee and pro rata for larger companies, which would look at the percentage of bills regularly paid in full, and also (relevant to industries which have a practice of dispute and adjustment) what the average percentage settlement is. Should that assessment be that most of the sales ledger will be realised, an assessment of the impact of the likely return under the *LPCD(I)A 1998* should be made so that it is sufficient to cover bank overdraft interest in the interim. If that is the case, the rights to SRI could be assigned to the bank (whether on part or the whole of the debtors), who would pursue it, and have increased security on their claim. This would add weight to have the outstanding amounts settled promptly, yet leave the company trading. Acting in this manner would avoid the normal costs of an insolvency practitioner to realise assets that reduce the return to unsecured creditors.

Therefore while the steps taken to date toward better payment practice and awareness of other approaches referred to above are helpful to business, it is clear that it is only through further debate and experience that such practical issues can be resolved.

Business interruption and recovery 5.38

As comments made in other chapters demonstrate, such as **CHAPTER 9** on IT, business interruption is one of the greatest risks that faces an organisation. This is even more true in the fast moving world that we inhabit: it can involve litigation, lead to insolvency and cause real business crisis. In **CHAPTER 9** this area is discussed from the e-commerce perspective since the repercussions are clearly dramatic in that context. Companies should invest in effective disaster recovery and business continuity plans as a priority. This is so regardless of the size of the business or its location, as with other risks. Unfortunately the Institute of Chartered Accountants in England and Wales (ICAEW) in their recent policy papers 'Entrepreneurship: the Key to Growing the SME Sector' found that small firms place too little emphasis on risk management. Although many companies focus on financial and insurable risks the approach that risk should be managed throughout an organisation remains relatively novel. Moreover, despite the fact that IT has been proven to be one of the weakest areas for businesses of all sizes, when implementing the Turnbull Guidelines (see **CHAPTER 10**) the DTI's 'Information Security Breaches Survey 2004' found that two thirds of businesses surveyed had to restore significant data from backup due to a computer failure of theft This is despite the increased reliance on business intelligence (see **CHAPTER 10**) and the growing sophistication of internal and external threats. See also **5.44**.

Global events **5.39**

The ramifications of the terrorist attacks on 11 September, 2001 to the World Trade Center in New York have been discussed extensively at many levels – political, commercial and personal – in many forums and no doubt they will continue to be debated for some time yet to come. What has been demonstrated in such discussions in no uncertain terms is the practical reality of the 'shrinking world' in relation to business activity. Global events really impact on national activity and the consequences for global business really impact on both large and small business. The broader implications and the increasing involvement of international law are common threads for business and initial observation should be noted here. When considering the events of 11 September, one executive summary has noted that the incident 'hit when the world economy was already weakened. As a result, global growth will be even weaker than formerly expected and the anticipated recovery postponed'.

Moreover since mid–2000 global growth had been on a downward trend due to:

- a collapse in investment spending;
- a tight monetary policy; and
- high energy prices.

Even before the terrorist attacks, the economic outlook for the US had deteriorated and the American consumer was beginning to 'retrench'. The business disruptions due to the attacks probably expedited the process however and the repercussions are widespread due to the high level of interdependency in the global economy. It is now understood that the downward trends at a high level can be disastrous and can disrupt micro business, bringing about an increase in insolvencies, pending upward trends.

Small business **5.40**

As this handbook aims to demonstrate that impacts can be felt regardless of size in the light of today's business relationships, the concerns of small business must be addressed. Having regard to the global events referred to in **5.39** above, some insight is given into the recent developments regarding insolvency in the context of corporate governance. Their impact is, of course, an important issue for small business at times when the global economy is highly volatile and affects all business and the supply chain directly. It should also be noted that the contribution of a healthy small business sector is dramatic in its positive impact on the economy in view of the size of its membership. Wide scale insolvency and bankruptcy can therefore have damaging lasting effect throughout the business community as a whole.

Insolvency

Meaning of insolvency 5.41

There has been extensive debate regarding the appropriateness of the UK's regulatory framework relating to insolvency and business failure. It is not intended to go into the present laws of insolvency (which is a vast subject) but rather to consider the meaning of insolvency and its application by some reference to other relevant countries that are driving change. What, then, is usually meant by insolvency? Generally:

- the inability to pay debts as they mature;
- under the American *Bankruptcy Act of 1898* the insufficiency of assets at a fair valuation to pay debts; or
- under various other laws the insufficiency of assets at a fair saleable valuation to pay debts.[1]

The second meaning is sometimes referred to as the balance sheet insolvency test and is the predominant meaning in civil law jurisdictions. Non-lawyers are accustomed to using the term 'insolvent' as an adjective, such as an insolvent debtor whereas lawyers sometimes use the term attributively as a noun, that is 'an insolvent'.

Also relevant is the concept of bankruptcy, which generally refers to:

- the fact of being financially unable to carry out one's business and meet one's engagements, especially to pay one's debts;
- the fact of having declared bankruptcy or having been adjudicated bankrupt under a bankruptcy statute; or
- the field of law dealing with those who are unable or unwilling to pay their debts.

In this respect the relevance of the American approach has been widely debated in the UK. In the US the phrase 'Bankruptcy Act' refers to the law of 1898, which governed bankruptcy cases filed before 1 October 1979. The phrase 'Bankruptcy Code' refers to the *Bankruptcy Reform Act of 1978* (frequently amended since then), which governs all cases filed since 1 October 1979. What is well known in American law – and increasingly understood here

– is what is called 'Chapter 11'. In American legal usage Chapter 11 has become synonymous with corporate reorganisation for the purpose of handling debts in a structured way, under the protection of a federal bankruptcy court. The phrase is often used attributively. By way of example it has been noted that:

> "The purpose of a Chapter 11 filing is to give a chief executive an opportunity to reorganise a financially troubled business by putting its creditors on hold. When the money problems have been straightened out and the company restored to health it emerges from the protection of the bankruptcy courts and picks up where it left off"
>
> (*John Taylor, 'Bankruptcy Was a Disappointment', N.Y.Times 10 December 1989, p 11*)

Although to 'go Chapter 11' does not appear to have the stigma often attached to insolvency and bankruptcy evident here, certainly the consequences for business – as well as personal consequences – may in practice be devastating. This has been witnessed in many very recent cases, such as Kmart and Global Crossing and including the controversial Enron case study that has been so publicised and analysed. This led to calls to make US directors more accountable and to require a new team of managers under Chapter 11 rather than leaving the existing management in place. In some ways the US is moving closer to UK thinking, to make it more difficult for bankrupts to be forgiven their debts and to be rehabilitated without sanction. It is recognised that insolvency may be a result of mismanagement, misfeasance or fraud.

Nevertheless is must be appreciated that the stigma that attaches here, together with the ongoing practical repercussions of business interruption and failure – insolvency or liquidation – often means that a valuable contribution to the business economy is damaged in circumstances that are often beyond the control of the small business at risk and are disproportionate in effect. In some ways the UK is moving closer to the US by attempting to make it easier for companies to be rescued and make it less of a crime for an individual to go bankrupt. This is where there is no blame involved. It can be appropriate in circumstances in which small business often finds itself, that is with cash flow problems in circumstances beyond its control through the late or non-payment of debtors, which is the focus of the discussion below (see **5.42**).

Key areas of concern 5.42

At an EU seminar on business failure cash flow was reputed to be the cause of 30 per cent of insolvencies. The European representative body for small business, UEAPME, noted that similar concerns were being debated across Europe. In the UK the representative body the FPB has similarly noted key areas of concern. Discussed below are the key issues under debate in Europe, the recent insolvency proposals of the government and some comparative analysis of proposals from other European countries as they relate to small business. Some points in response to the Department of Trade and Industry (DTI) review of company rescue and business reconstruction mechanisms are also made. What have emerged as key areas of concern in the insolvency debate are:

- the role of the banks;
- the appointment of an insolvency practitioner;
- the priority of debts;
- the costs of insolvency; and
- the problem of cash-flow and causation of insolvency.

While the debate continues, certain recommendations set out below may assist in reaching a satisfactory conclusion for stakeholders when any insolvency occurs, particularly as regards small business.

Methods applied in Europe: a brief synopsis 5.43

It is helpful to consider the highlights of other approaches in place and proposed in Europe in a short overview. Four jurisdictions have been selected and discussed alphabetically below.

1. Austria
   - aim: restructure and continuation;
   - all creditors ranked equal, no advantage to preference or secured;
   - experienced lawyer/business person appointed, procedure supervised by court under the provisions of the most recent legislation and is unpublicised;
   - creditors who do not attend creditors' meeting can take no further part;
   - costs to society are taken into account;
   - employees who are made redundant are paid by National Insurance fund to which all contribute.

2. Belgium
   - legal settlement (and proposed simplified legal settlement);
   - court supervised procedure, which is not published in official journal;
   - all creditors ranked equal (no preference), individual creditors prevented from acting;
   - creditors vote on insolvency plan;
   - court can impose moratorium, and pay suppliers first to stop domino insolvencies;

- costs to be lessened and flat rate imposed. Priorities – identify companies paying utilities, rates, taxes etc late;
- have informal meeting with owner to prepare recovery;
- trained judiciary with great business experience;
- if plan does not work, then bankruptcy – dealt with as swiftly as possible.

3. Germany
   - since 1999 duty of insolvency practitioners to continue the company, if possible;
   - distinction between insolvency and liquidation removed;
   - recovery plan voted on by creditors, who are divided into four groups – real estate, mobile /fixed assets, employees and all other creditors. All groups have to be in favour by 50.01 per cent. If one is not, the plan can be overruled by the court;
   - there are no preferential creditors, other than creditors owed less than €1000 (who are paid in full), and employees.
   - the insolvency practitioner can demand restoration of supplies, change any contract or lease including re-entry by the company into premises closed by a landlord, but is personally liable.
4. Netherlands
   - proposed a moratorium of up to four months to include preferential creditors;
   - reduction in quorum to simple majority of creditors for more than 50 per cent of receivables, and compensation to third parties with a lien, with energy suppliers required to continue supply;
   - breach of conditions would result in removal of debtor or management;
   - there would also be a centralised register of insolvencies;
   - all changes were aimed at making recovery successful;
   - current proposals are for a low-cost single gateway for insolvent companies (bank, accountant etc) backed by court application if there is survival chance. Tax should have preferential status removed. Employees should be able to be released (now only possible in bankruptcy), and then re-employed as business works out.

Business continuity and operational risk management 5.44

Bearing in mind the above sections, it is also appropriate to consider business continuity and operational risk management at this stage. Since business continuity can be affected for an array of reasons, as has been noted (in **5.38**), management should be prepared as part of its internal due diligence and corporate governance. One preliminary issue is whether a business does in fact consider operational risk management and business continuity as being entwined. The basis for this discussion is the experience of many organisations where operational risk management and continuity planning are considered to be two entirely separate disciplines. So often, according to risk practitioners, the two departments never really work together to any significant extent. Yet according to many risk advisers they are one and the same. Continuity planning is simply one of the opportunities – and an increasingly important one – available to the modern risk manager.

Part of the problem, as ever, is understanding the terms. Ask a dozen people for a definition of risk and there will surely be at least 15 answers. We are talking here about operational risk; nasty surprises that come along and divert the organisation from its strengths and objectives. (See also **CHAPTER 8** regarding risk management culture.)

The science or art (depending on your point of view) of risk management has an unfortunate foundation in people who called themselves risk managers, but were in reality buyers of insurance programmes. Other 'risk managers', including company secretaries, treasurers, auditors, lawyers, facilities managers, continuity managers, health and safety managers, security managers and business directors, all developed there own risk approaches quite independently.

A business director could cheerfully accept an exposure of, say £10m or even £100m, and see doing so as a business profit opportunity – comfortably fitting within the assets and cash flows that are managed on a daily basis. The 'insurance buyer' in the same company however may be spending millions buying insurance for PCs and photocopiers. The lawyer may be cheerfully transferring risks to other contracting organisations without it being part of the job description to consider the residual risks on their own organisation if that party failed to meet those contracted responsibilities.

Continuity management, on the other hand, comes into the twenty-first century from a twentieth century foundation of IT and facilities managers owning the responsibility for, and developing continuity facilities for their services. Fortunately, the science of continuity management have moved on somewhat. Increasingly, the business impact analysis is a vital tool, and there is more ownership and exercising by business managers. It is a history, however, that is still not easy to forget considering the background and the reporting lines of some continuity managers.

The evolution of risk management in practice 5.45

Looking at risk management first, the ground has shifted noticeably under the old 'insurance buyers' and in different ways at the same time. Firstly, their employer's organisation is almost certainly undergoing such major change that the old organisation of just a few years ago, and the new, are barely recognisable as one. Following mergers, it is likely to be much, much, bigger, and much more international. Computerisation and communications have created different marketing, service delivery and cost saving opportunities. These developments have reduced the need for locations and people dramatically.

The focus on creating value at each individual stage of the supply chain has created new critical dependencies in third party organisations that are less easy to supervise in detail. These dependencies have frighteningly shorter and shorter periods where delay can be tolerated before destructive damage occurs. Entirely new risks – e-commerce, internationalism, media, and others – have evolved; as have customers' expectations been raised towards a seamless 24 hour/7days service. (see **CHAPTER 9**).

E-commerce – where basic entrepreneurial instincts are fuelled by ever more powerful computers – along with telecommunication and data mining tools, is one huge area where the rewards of the first pioneer are totally disproportionate to the rest. In that atmosphere of headlong sprint and laying bets so large that they will create, or kill, careers, the risk and continuity managers asking for time and resources to plan effectively can be ignored.

Sometimes it is an 'old' risk that, because of these changes, now has a new potential for total, organisation-wide and simultaneous destruction across individual business units – miles or even countries apart. What good is it being able to produce good products if the world has lost confidence in the product's name and won't buy them?

Conversely, these larger corporations have opportunities to absorb much more risk within the strength of their balance sheets and cash flows. They are large enough also to have some flexibility to keep them in their marketplaces whilst problems are being resolved without stakeholders feeling an unacceptable impact.

New challenges **5.46**

The challenges that are currently facing this new generation of risk managers therefore include:

- making best use of the new strengths within the organisation;
- a consistency when approaching risk evaluation, risk tolerance and risk management – leading to seamless risk decision making;
- both a bird's eye view and a detailed view across the organisation at the same time;
- managing second-hand, destructive risks or timescales in suppliers;
- risks that are beyond the ability of the insurance industry to support;

- communication on matters of risk and thus managing diverse expectations; and
- keeping up with change.

The concept of 'killer risks' (having such an effect that it literally kills the company) is increasingly emerging. The brand and stakeholder confidence concerns (see **CHAPTER 7**), are such killer issues; as is insolvency (see **5.41**), and business and financial control. Dependencies on central group-wide facilities such as computerisation and communications to deliver the products on time to an acceptable standard; also the intellectual assets within the organisation (see **CHAPTER 9**), are just a couple more amongst others. These are all not unfamiliar to the continuity manager.

As is mentioned in **CHAPTER 1** and **CHAPTER 4** in this handbook and discussed further in **CHAPTER 10**, the changing regulatory needs are also demanding a more 'holistic' approach to risk management. Stakeholders have no interest in internal organisational boundaries. They concern themselves only with the potential for unacceptable impact on the shareholding, or on any other relationship they may have with the organisation itself. These regulatory needs, Turnbull included, are driving organisations to consider enterprise-wide operational risks more formally. These organisations however are more comfortable with the clearer cut aspects of financial risk and indeed they have experiences, sophisticated strategies and controls, developed over many years. The evaluation and cost/benefit analysis of non-financial, operational risk, decision making, however, is not as simple to quantify. Some are clearly struggling with the commercial decision making that is being demanded (see also **CHAPTER 4** and **CHAPTER 7**).

Its worth considering at this point where the insurance programme fits into this new style corporation and its 'killer' risks. Insurance is indeed extremely useful eventually but where is that value in those crucial, threatening minutes and hours after a disaster where survival of the organisation and its dependencies is the only challenge? The organisation and its most crucial dependencies need to survive first to fully gain, later, the value of its insurance programme.

Stakeholders **5.47**

The risk manager will consider that the organisation is no more than the brand value, its intellectual and free value assets together with the combined influence and support of its stakeholders. These stakeholders, with their quite different interests, are at the centre of the risk manager's thoughts. This is true of the continuity manager too. They include:

- employees;
- suppliers;
- customers and distributors;
- regulators;

- the media;
- private and quoted shareholders;
- bankers;
- the public – via their impression of the brands;
- the environment, and others.

Risk evaluation **5.48**

When evaluating risk, the risk manager measures them against the agreed risk tolerance levels of the organisation itself. If the risk, and/or the potential impact, is not acceptable to the organisation, then the risk manager sets out to bring these aspects within that agreed tolerance level. It is rare that the risk manager can remove all risks or impacts altogether. In addressing unacceptable risks, the risk manager can consider the commercial realism of a range of options and the relative value of each option. They are:

- reduce the likelihood or the potential frequency;
- ensure that the impact is reduced to an acceptable level – whether that be in human, operational or financial terms;
- transfer the impact to another organisation eg a counterparty, an insurer, a captive insurer, the financial market or another; and
- prepare for the incident by way of continuity planning of business critical issues.

The suggestion here, therefore, is that contingency planning is just one of the options available to the risk manager. If credible, tested plans can be in place so that the organisation can get through an incident without serious damage, then surely that is one of the options alongside risk expenditure and resources. This is especially so where risk management constrains the organisation from doing what it is best at doing and when dealing with low frequency, high impact exposures. Resultant expenditure incurred after the disaster can often be an insured expense – a benefit close to finance directors' hearts. What is important here is to highlight credible, tested plans. This is all familiar to the continuity manager, who sets out to identify risks and evaluate them within the context of the impact on safety, and the urgencies, survival needs and responsibilities of the organisation. However, one should also include here, not just business continuity plans, wherever contingency planning is needed, for instance:

- kidnap;
- extortion;
- bomb threat;
- suspicion of major fraud;
- succession planning;

- media criticism;
- product recall; and
- others.

These have common denominators, of course, but the needs of each must be met.

It may not be cost effective, or just unachievable, to remove risk altogether by risk management. Continuity planning may be the only answer left when all that is realistically preventative can be done. All of these measures though, ones that can include business decision making, security, health and safety, resilience in production lines, etc. are, with continuity planning, best effective when all are part of a relatively seamless process of risk and impact understanding and management.

Even the challenges to risk managers and to continuity planners, are similar. Firstly, how to get the attention of the board to the point that the right level of priority is given? How to gain resources for risk and continuity management in competition with projects that are about what is happening today not what may or may not happen sometime in the future? How indeed to get the directors to pay more than superficial attention and concede fully, that not only this thing may happen, but it may happen within the ever-shorter tenancies of that particular top job.[1]

Risk management and continuity management are both commercial business issues which incorporate the special challenges of acceptability and urgency. Each discipline is evolving within itself. There is real value in them working much more closely together and each providing valuable support for the other.

Crisis management – a view from the US 5.49

Today's business environment requires that crisis management is prioritised, especially since the terrorist attacks on 11 September 2001. Events that span from the threat of Y2K to terrorist attacks mean thatserious threats to the business exist in many ways. It is useful to consider the approach in the US in terms of important business risks such as the environment and health and safety from the perspective of the external adviser.

Managing an environmental, health and safety crisis 5.50

The world collectively breathed a great sigh of relief on 1 January 2000, when the much anticipated and greatly feared Y2K bug failed to bite, and the long predicted worldwide crisis never happened. In the weeks leading up to the new millennium, even those of us who had convinced ourselves that everything was under control and that no crisis would actually occur, had quietly put aside some extra cash and had stocked up on grocery supplies, just in case. In hindsight, many questions are being asked about why we survived the stroke of midnight unscathed. Was it all just a hoax? Was it misinformation? Or perhaps, was it due to the unprecedented planning and testing of computers and computer-related equipment that had taken place in preparation for the new millennium?

Those who believe that the Y2K bug was exterminated by the enormous planning efforts of corporate America think there is a lesson to be learned from this experience; namely, there is no substitution for good planning. That concept is certainly true when it comes to planning for and managing a crisis in the environmental, health and safety (EHS) arena. Unfortunately, despite an organisation's best efforts, it is almost inevitable that somewhere, sometime, a very serious EHS accident will occur. Experience has shown that companies caught without a comprehensive EHS crisis management plan, suffer both severe public relations troubles and significant legal liability (see also **CHAPTER 16**).

Heightened expectations **5.51**

Due to the rapid pace of technological advances, a timely response to an EHS crisis is more critical than ever. Because of the widespread availability of electronic mail, mobile/cell phones and other forms of immediate communication, there is a growing belief that there is no excuse for delayed response. In the US both the public and regulatory agencies, such as the Environmental Protection Agency (EPA) and the Occupational Safety and Health Administration (OSHA), have raised their expectations considerably. Regulatory agencies have also stepped up enforcement activity in the wake of such accidents. Moreover, computer technology has assured that, not only will a company's EHS crisis make the news, it will also be posted on numerous websites, complete with all the details and pictures.

These reasons make it an absolute necessity for companies to have in place, well thought out, usable and responsive EHS crisis management plans. Because of the myriad complex issues involved in crisis management, it is extremely important that a multidisciplinary team develop the plan. Too many companies make the mistake of having a plan developed solely by lawyers, a public relations firm, or engineers. Such plans are inevitably missing critical elements. Once a plan is complete, it is also important to make certain that all of the players involved in EHS crisis management, from top to bottom, are familiar with it.

How crisis leads to liability **5.52**

In order to properly prepare for an EHS crisis, it is important to understand why and how companies can be subject to both public relations nightmares and legal liability in the event that such an unfortunate accident occurs. First and foremost, a company that is unprepared to deal with an EHS crisis, eg a chemical spill or industrial accident, will very likely not be in a position to contain the spill or minimise the consequences of the accident. Obviously, to the extent the problem gets worse, the more expensive the cleanup bill becomes and the higher the agency penalties and claims in third party law suits will be.

In addition, there are myriad requirements under environmental, health and safety laws to immediately report a spill, release or an accident. For example, both the *Comprehensive Environmental Response, Compensation and Liability Act of 1980* (*CERCLA*) and the *Emergency Planning and Community Right-to-Know Act of 1986* (*EPCRA*) or the *Superfund Amendments and Reauthorisation Act of 1986* (*SARA Title III*), contain provisions for immediate reporting of chemical releases, which a company knew, or should have known, are in excess of reportable quantities set out in the regulations. EPA has taken the position that reporting such a chemical release more than 15 minutes after the release occurs is a per se violation of law subject to the imposition of penalties. In fact, the reporting requirements of these statutes require that the caller provide details such as the type and quantity of the released chemical and its health effects, as well as exposure and precautionary information. In order for a company to comply with that requirement, it must have a sophisticated record keeping and reporting protocol in place, before a spill occurs. A company that is unprepared to appropriately respond to such a release will undoubtedly incur significant penalties from an agency.

Moreover, history has shown that EHS crises inevitably lead to additional, and sometimes far-reaching, government regulation. Consider the track record. In the 1970's, flaming rivers led to the enactment of the *Clean Water Act of 1977*. Hazardous waste dumped in Love Canal in New York led to the enactment of *CERCLA* in 1980. More recently, major air releases in Bhopal, India and Institute West Virginia, led to the enactment of *SARA Title III*, as well as a number of provisions in the *Clean Air Act of 1990* requiring companies to provide significant volumes of information to the public, as well as increased record-keeping requirements. Finally, the Exxon Valdez oil spill in Alaska, one of the most publicised oil tragedies in history, directly led to the *Oil Pollution Act of 1990*, which required additional planning and which contained many new requirements for the prevention of releases of oil to waters.

The public relations trap **5.53**

Often public relations issues feed the legal liability problems after an incident such as a spill or an industrial accident. Naturally, an injured party, or someone who believes they are injured, is more likely to bring a lawsuit when they perceive that the situation was not handled properly. A good public relations strategy is key to altering that perception. Simply responding to an EHS crisis is not enough. One must manage the crisis.

A company that is not prepared to deal with the public and that does not have senior management responding in a way that assures the public the company has things under control, is more likely to become the subject of intense scrutiny. That scrutiny can come in the form of government enforcement, including criminal investigations and prosecution, as well as third-party suits such as toxic tort and citizen suits; the latter becoming increasingly relevant to the UK as a result of the impact of European environmental legislation (see **CHAPTER 16**).

A crisis and the crisis management plan 5.54

Some companies have invested considerable sums developing EHS crisis management plans with the assistance of able public relations firms and lawyers. However, it is common for an organisation to merely take the plan and put it on the shelf, being satisfied that it is there when needed. This is a mistake. A crisis is no time to test a plan for the first or even second time. Moreover, during a crisis, there is simply no time to consult a plan, particularly a voluminous one, with which you are not already thoroughly familiar. Crisis management requires almost instantaneous reaction by companies, both plant personnel on the scene and senior management, wherever on the globe they happen to be.

Everyone involved in managing an EHS crisis, from the second shift process operator, to the chief executive officer (CEO), must be familiar with both the contents of the plan and, more importantly, their role in the crisis management process. The kind of organisation required to effectively manage an EHS crisis is something that requires careful thought, planning, testing, practice, and updating.

Updating plans 5.55

For many reasons, it is very important to make sure that the EHS crisis management plan, once developed, undergoes frequent periodic review and updating. There is the growing public expectation that companies will not miss a step during a crisis. Therefore, periodic review and updating of plans will assure that they are both correct and familiar to those who are tasked with the responsibility to execute them. Outdated data, such as contact or health, affects information, can cause serious delays and either over or under reporting of information to agencies and the public. Even sophisticated organisations with well-prepared crisis management plans end up with serious liability due to outdated data. By way of example, if the material safety data sheet (a form required to be maintained by both OSHA and EPA) for a new process chemical does not make its way into the plan this can mean an incomplete report to EPA during a process release and a significant penalty. If a release drifts to a nearby neighbourhood, the outdated information could bring about far more devastating consequences.

In addition, in these days of frequent corporate mergers and takeovers, EHS managers are often faced with responsibility for entirely new facilities and divisions. Because of personnel changes and other reasons, EHS accidents are more likely to occur during such transition periods because EHS might be overlooked for a short period. Ironically, most EHS managers are simply too overwhelmed by the integration of day to day EHS functions to focus on integrating crisis management plans. That can unfortunately lead to serious problems in the wake of EHS accidents.

Finally, there are many new federal, state and local laws, regulations and policies that can come into play in any EHS crisis. While most companies track new EHS requirements and incorporate them into operations, many forget to make sure those new requirements are integrated into updated crisis management plans and systems.

The lawyer's role in EHS crisis management 5.56

As stated in **5.54**, an effective crisis management plan is the work product of a variety of professionals bringing their particular expertise to bear on the process. The experienced environmental/OSHA lawyer, particularly one who knows your business well, should be an important member of the development team. In most cases, use of specialised outside environmental/OSHA counsel for this task makes sense regardless of the in-house legal structure of the organisation.

Even though many companies have specialised in-house EHS counsel, many are already overwhelmed with day to day regulatory issues, briefing management on significant issues, overseeing litigation, and reviewing EHS issues in transactions. Despite their best intentions, consistent participation of these individuals in EHS crisis management planning is simply not feasible. Moreover, many in-house lawyers spend significant time on the road. Therefore, even assuming they are immediately reachable during a crisis, they may be unavailable to effectively participate in the necessary efforts.

Due to the many legal issues and requirements involved in a crisis, the expertise of a lawyer is necessary in both planning and during a crisis. For example, an EHS lawyer can help assess the extent of a crisis and the applicable legal reporting requirements (including reporting to insurance companies). In addition, the lawyer can assist in (although not participate directly in) the dissemination of information during a crisis, and help prepare statements by the company to the public and regulatory agencies. Also, the lawyer can assist in the event inquiries lead to criminal investigations. A lawyer can also assist in, and potentially protect through privilege, internal investigations of root causes and can begin to develop a record for use later. Finally, to the extent the lawyer was involved in the planning and response, he or she can be invaluable in responding to the host of government inquiries in the days following the event.

Ongoing considerations 5.57

EHS crisis management is extremely important, particularly in these days of immediate and prolonged media coverage. Involvement of a experienced team of professionals, including outside environmental/OSHA lawyers, in both planning and during an actual crisis, can have payoffs far beyond the costs involved. History has shown that few managers look back on a crisis and wish they had prepared less.

References

1. James A. Machachlan, *Handbook of the Law of Bankruptcy.*

Appendix

Glossary

Key definitions

For the purpose of this discussion, certain key definitions may be set out – some of which are explained further below. It may be noted that the government always intended to retain the ability to amend definitions by means of secondary legislation.

Late payment

This is the receipt of payments beyond the contractually agreed credit period or the default credit period as defined in the legislation.

Commercial debt

In principle, this means any debt entered into by business, eg a registered company, body corporate, partnership, trust, sole-trader or public organisation for the provision of goods and/or services, where both creditor and debtor are party in the course of their business.

Small business

This has the meaning as defined in the *Companies Act 1985*. This provides that a small business meets two of the following criteria:

- a turnover of no more than £2.8m;
- a balance sheet total of no more than £1.4m;
- no more than 50 employees.

Unincorporated and new businesses, until their first set of accounts are due to be lodged, will only be required to satisfy the employee criterion of that definition under the *Late Payment of Commercial Debts (Interest) Act 1998*. For the purpose of this discussion, a large business means any business that does not fall within the definition of a 'small business' as defined above. A brief overview of the proposal is as follows:

Territorial extent and applicable law

The legislation applies across the UK and to any contract written under the law of England and Wales, Scotland or Northern Ireland.

The size of the debt

Small businesses should be able to claim interest on any commercial debt, irrespective of monetary size.

Credit periods

Where no credit period is defined in a contract, or no contract exists, the credit period should be thirty days from date of the invoice for payment, or the delivery of the good or the performance of the service, whichever is the later.

Interest rate

The objective was to recompense creditors for the cost of payment delay rather than to impose a penal rate of statutory interest. A rate of interest of base rate plus eight per cent, this being the average bank lending margins to small businesses, was proposed. The interest should be open to separate pursuit and assignment.

Chapter 6

Commercial Due Diligence Case Studies

Introduction 6.1

In a traditional due diligence exercise commercial due diligence forms the practical aspects of the transaction, some of which have already been considered in **CHAPTERS 1** and **2**. The target or candidate company is assessed in the context of:

- its competitors;
- the market;
- brand value;
- product image;
- the integration of the workforce and their interaction at different levels; and
- business goals and objectives.

As a matter of ongoing due diligence and prudent commercial practice a business must, of course, constantly monitor its competitors and the market in general. Media headlines are often filled not only with stories of failure in this respect but also likely acquisition moves, as well as sensitive sector comment. This handbook as a whole deals with commercial due diligence and corporate governance, having regard also to the management of risk – this Chapter focuses on the latter by way of practical case studies in selected sectors. Many of the issues are, of course, relevant for many business sectors. Nevertheless, it must be emphasised that this Chapter does not intend to address how a business should deal with the competition. Instead it approaches the topic by considering sectors that are both highly competitive and risk driven.

It has been noted in earlier chapters, specifically in **CHAPTER 5** that in today's fast moving business world a business can be set up, grow, be acquired or fail in a short period of time. A vivid illustration is the information technology (IT) sector, which was the subject of so many 'boom and bust' stories and examples. Some sector specific comments regarding e-commerce are made in **CHAPTER 9**. Moreover, in **CHAPTER 7** consideration is given to the reputation risks that can affect business as a result of decisions taken, having regard to the sector in which they operate. For the purpose of this chapter, the two sector case studies that have been selected demonstrate a more modern area of extensive business interest and a very old traditional industry in the context of commercial due diligence. They show that whatever the sector, this is clearly the age of risk management, internal due diligence and corporate governance. With this in mind an attempt has been made to consider relevant macro and micro issues in this chapter in accordance with the theme of the handbook.

The choice of two sectors, the biotechnology sector and the shipping industry, is also topical since both often attract the attention of the media, non governmental organisations (NGOs) and the public in general. In order to compete successfully, businesses operating in both sectors need to demonstrate a sensible approach to legal risk management. While the first case study highlights, in particular, the importance of legal risk management, the second emphasises the need for the implementation of good governance to enable this sector to be more competitive.

Case study 1 – the biotechnology industry

Introduction 6.2

Managing legal risk exposures is an area that all businesses face throughout their life cycle. Given the unique nature of the biotechnology business with its very long product development cycle, high financial investments and multi-faceted risk environment, legal risk exposures can be particularly accentuated. The need to evolve a structured and proactive legal risk management framework is an imperative that stakeholders in biotechnology ventures cannot ignore.

The process that a biotechnology company goes through from initial venture funding to product release and facility expansion is long drawn and fraught with risks. This case study seeks to provide an understanding of the legal risk management process and the tools available for identifying, managing, controlling and ideally eradicating, if not, mitigating such risks for biotechnology ventures at different stages of development from the start-up to the market roll-out phase.

The biotechnology industry has seen rapid technological advances over the last decade, often led by small start-up companies that focus on innovative, early-stage technologies. While the returns in investing in these companies can be lucrative, the risks are equally high. When investors are confronted with a decision as to whether they should fund an early growth company, a very thorough risk assessment on the investment proposition offered by early-stage technology companies is critical in the investor's decision-making process.

What is biotechnology? 6.3

Biotechnology can be defined as any form of technology that makes use of the natural biological processes or products of living things to modify and advance human health and the human environment. Although biotechnology has dominated the corporate agenda in recent years, the concept of biotechnology is not new at all. Traditional forms of biotechnology include brewing, bread making and cheese making. These processes use natural fermentation processes of micro organisms to either preserve food produced through agriculture or create new tastes and textures. Advanced or modern day biotechnology, for example in the area of genetic engineering, attempts to achieve the same goals of producing new or better products but in a more technologically advanced and efficient manner. Today these same processes of micro organisms are used for a wider range of applications, to produce valuable products such as antibiotics or enzymes for medicine or industry.

Areas of biotechnology 6.4

While, as noted in **6.3**, biotechnology embraces processes that are very old, from the point of view of due diligence and corporate governance it is the more modern applications that attract controversy and more risk. Modern or advanced biotechnology today encompasses broad areas of the life sciences and includes both products and processes. Examples of relevant products and processes are:

- products that include genes (including modified genes, expression vectors and probes); proteins (including modified proteins, monoclonal antibodies and receptors) and other chemical compounds; and
- processes or methods that include a new method of using a known compound and methods of medical treatment.

Biotechnology today, therefore, covers a wide range of industry sectors and typically involves the following areas of technology:

- medical biotechnology which uses micro organisms (such as bacteria or fungi) to make antibiotics or vaccines;
- industrial biotechnology which uses micro organisms to make enzymes (eg to add to biological washing powders), or to produce beer, cheese or bread;
- environmental biotechnology which uses micro organisms or plants to clean up land or water that is polluted with sewage or industrial waste; and
- agricultural biotechnology which aims to produce better crops, 'natural' fertilizers, or feed additives.

The life cycle of biotechnology companies 6.5

Comment has already been made in **CHAPTERS 4** and **5** regarding the extensive issues involved in establishing and maintaining a healthy business. For example, some consideration has been given to the choice of business vehicle, disputes and their avoidance, as well as other business issues bearing in mind the demands of good corporate governance. This consideration sought to address the concerns of a business throughout its life cycle, from idea stage, start up, operations, expansion and so on. Biotechnology companies typically go through the following stages in its business life cycle:

- conception of idea;
- research and development (R&D);
- sourcing of funds (including corporate finance);
- incorporation of business entity or formation of alliance (eg a corporation or partnership or joint venture);
- conduct of clinical trials;
- application for approvals, licenses and permits;
- early product roll-out in the market;
- market review;
- expansion of facilities and scale up of marketing; and
- ploughing back of commercial returns into further R&D to start a new or overlapping cycle.

This can be a long drawn out process: the whole growth cycle typically takes between six to twelve years. From the perspective of mergers and acquisition, evidently, as with certain other sectors, biotechnology companies may be acquired at different points in their business life cycle and the risk exposures vary according to:

- the stage of development that the biotechnology company has reached, for example a start-up drug discovery company as opposed to a drug company that has generated revenue from its R&D activities;
- the type of industry segment the biotechnology venture is in and its risk profile, thus a drug company that has to carry out clinical trials in a highly regulated and strict licensing regime would have greater risk exposure in terms of compliance than one where the regulatory regime is more relaxed; and
- the complexity of the transactions that the biotechnology company is involved in.

Biotechnology business **6.6**

Compared with other technology-based transactions, biotechnology ventures may be considered as unique in several ways.

1. At the heart or core of biotechnology ventures are inventions which define what the whole business is. Biotechnology business is about innovation and the quality of ideas and inventive flair will determine whether there is a potentially lucrative business to start with. Therefore the business is very much dependent on the quality of researchers and research managers. When such inventions eventually have legal protection, typically in the form of patents and sometimes as a trade secret, due diligence in the area of the adequacy of legal protection becomes central to the whole risk management exercise. Accordingly, in managing legal risks in biotechnology ventures, due diligence relating to patents and other forms of intellectual property tend to take centre stage.

2. Biotechnology ventures are by their very nature very capital-intensive and have an extremely lengthy turnaround time from R&D to the market place. For instance, it could typically take some twelve years and over $US 30m on average to bring one medicine onto the market. Given this turnaround time, the risks are far greater compared to those found in other technology ventures. Moreover, the risks are highest when the biotechnology ventures, which are often target companies, are at their R&D or early growth stage. Such a long investment cycle also means that there are distinct challenges in evaluating existing and pending patent claims which may be made at any particular point in time in the life of the biotechnology companies.

3. The biotechnology products and services which are typically the subject of acquisitions more often than not deal with matters that affect the medical conditions of humans, such as pharmaceuticals and modified genes, or things that humans consume, such as genetically-modified (GM) food. This means that the health and environmental risk factors are accentuated. These are particularly sensitive issues that are of continuing interest to the media and the general public, as well as government agencies and traditional stakeholders. It is vital that the business monitors such risks which can impact on vital business decisions, including the choice of location, very carefully. The GM debate is one that has aroused particular concern in many parts of the world.

4. Biotechnology related data analysis relies heavily on the use of IT and this raises new risk factors (as can be seen from the discussion in **CHAPTER 9**). Take for example the area of bioinformatics – the marriage between biotechnology and IT. The intensive use of electronic tools and processes, such as the application of software to process the huge amount of data in genomic analysis, creates new risk exposures. As such data are far more fragile and are more vulnerable to attacks from hackers and technical piracy, acquirers must take greater effort to protect such digital assets. The loss of such digital assets in turn may mean

greater potential liability or losses compared to physical assets such as land, equipment and machinery. In this respect the physical assets are relatively 'safer' compared to digital assets that are created out of bioinformatic transactions.

The key legal and business issues 6.7

As is the case in other sensitive sectors, the risk exposures that biotechnology ventures face are very much dependent on what stage of development they are in. These stages include different types of activities and may be summarised as follows:

Research phase 6.8

This includes:

- conception of idea;
- planning of research;
- conduct of research;
- research breakthrough;
- development phase; and
- clinical trials.

Corporate set-up phase 6.9

This includes:

- compliance programme;
- business and deal structuring;
- financing and funding of the business;
- document structuring;
- contract management; and
- human resource activities (including the employment of researchers and managers).

Operational phase 6.10

This includes:

- intellectual property (IP) audit;
- laboratory management practices;

- intellectual asset protection;
- IP licensing strategies; and
- IT security (particularly in relation to bioinformatics).

Marketing and product roll-out phase **6.11**

This includes:

- product safety and liability;
- environmental compliance;
- health and safety requirements;
- technology transfer; and
- dispute prevention.

Ongoing work **6.12**

This includes:

- audit functions;
- asset audit;
- corporate governance practices; and
- use of IT and its attendant legal risks.

Business and legal risk factors **6.13**

As is the case with all major commercial transactions, there are a range of legal and business risk issues that should be taken into account. In biotechnology ventures, the major transactional risk factors may be summarised and include:

- financial risks (for example where there is a huge investment in a research programme involving genetically modified crops where the market response is uncertain);
- compliance risks (for example some jurisdictions have laws that ban human stem cell research);
- asset creation, protection and exploitation risks (for example the failure to register an invention as a patent or failure to commercially exploit patents or to maintain a trade secret);
- legal liability risks including contract enforceability (for example there may be non-performance of ongoing contractual obligations for clinical trials of new drugs);

- operational risks (for example the inability to obtain live samples of micro-organism such as viruses in order to carry out R&D);
- business model risks (for example a biotechnology venture may be dependent on critical data from other parties thereby increasing their risk exposures);
- ethical and reputational risks (for example the failure to abide by public sentiments on moral issues of patenting living organisms);
- alliance formation risks (for example the teaming arrangements between private companies and university research institutions);
- third party risks (such as an overdependence on third parties for critical technologies);
- privacy risks (such as the protection of personal data of individuals involved in research projects);
- IT security risks (particularly in bioinformatics-related activities);
- product liability risks (for example when a genetically modified crop poses potential or actual health risks); and
- health and occupational risks (including employees and staff).

In managing a biotechnology venture, the above risks must be recognised, addressed and managed. As is the case in other businesses the board of directors and senior management of biotechnology ventures are usually responsible for:

- developing the organisation's business strategy; and
- establishing an effective management oversight over risks.

Having regard to the demands of good corporate governance also, the board and senior management can therefore be expected to take an explicit, informed and documented strategic decision about the organisation's overall risk management framework.

Legal risk management 6.14

In **CHAPTER 1,** legal due diligence was considered as part of the traditional due diligence process. In the course of that discussion risk management was mentioned (this is also examined in **CHAPTER 8** in the context of reputational issues and in **CHAPTER 10** when considering corporate governance issues). Moreover, the connection between due diligence, risk management and corporate governance is a general theme of the handbook.

Legal risk management is a process designed to help organisations identify, quantify, and control their legal risk exposures with the ultimate aim of protecting the organisation and creating greater value to the shareholders and other stakeholders. Legal risk management can be a complex undertaking. The failure to adequately address legal risks can subject businesses and organisations to virtually unlimited financial liability. In other words, risk management can be simply seen as a process to help an organisation protect its bottom line. The risk management process is significantly more difficult in the biotechnology sector than other industries due to the huge investment amount and the long turn-around time from ideas to the market place.

As has been stated earlier in this chapter that this is the age of risk management, therefore all organisations should adopt and uphold high standards of risk management and for those businesses operating in sensitive sectors such as biotechnology this is a priority. In developing this legal risk management framework, primarily biotechnology ventures are expected to:

- establish a sound and robust risk management process;
- strengthen their intellectual asset management programme; and
- establish their risk management programme as part of the organisation's good corporate governance practices.

Depending on the quality of corporate management, the risk profile of the biotechnology company would vary from company to company and requires a tailored risk mitigation approach. There is therefore no 'one size fits all' approach to risk management in biotechnology transactions. Moreover, if the intention is to dispose of the business, a careful grooming exercise should be carried out. This may also be the case in the event of corporate financing. As a general principle, biotechnology companies differ in their stage of development, size and their risk management approach. It is important that businesses operating in the biotechnology sector should have in place risk management processes which are appropriate for their:

- individual risk profile;
- operational structure; and
- corporate governance practices.

In biotechnology transactions, the people or 'brains' factor behind the technological products and processes tend to be more important than technology assets. Indeed in many of the biotechnology companies, the company would not have any valuable assets if it were not for the people behind the inventions, unlike companies that are rich in physical assets such as land, minerals or equipment. Therefore the board of directors and senior management in biotechnology companies and research institutions should establish rigorous and effective risk management systems and standards in their business operations to safeguard their human as well as technological assets. As part of this process, they need to continually monitor the adequacy and effectiveness of their risk management functions as well as implement compliance and audit procedures to ensure that their business and legal interests are protected.

In view of the varied business and legal risk exposures that the board of directors and senior management of biotechnology companies face, it is also important to establish a holistic and integrated risk management framework at the corporate level. The absence of such a framework suggests that the overall risk management approach is weaker and the stakeholders of the biotechnology venture should address this aspect of the corporate framework carefully. This should be considered in the context of the overall approach to due diligence and corporate governance.

The legal risk management process 6.15

A legal risk management framework at the corporate level that contains the key priority elements for the business is a necessary tool that every organisation should evolve, have and use. The absence of such a framework may weaken the organisation's capacity to better manage its risk liabilities and to transfer its risks to other parties, wherever possible. The overall intention of the risk management process is to minimise the financial impact of losses on the company's bottom line and the strategies and techniques must all be geared towards this goal.

The process of legal risk management may be summarised here and involves:

- identification of legal risk events (for example dangers of an unprotected biotechnology invention);
- analysing and weighing the gravity of legal risks and analysing their consequences (for example non-compliance with the laws banning test on human stem cells is far more serious than a weak indemnity provision in a licensing agreement);
- taking proactive, preventative, responsive or reactive steps – depending on the scenario– to prevent the problems from either taking place or worsening (such as entering into legal agreements or requiring parties to comply with legal obligations in ongoing projects); and
- reviewing the efficacy of the legal measures introduced to counteract against a problem or a potential problem in the future.

Governance aspects 6.16

Biotechnology ventures should play a proactive role in educating the researchers, business executives and partners on the benefits and risks of biotechnology services and products through a structured and enterprise-wide risk management framework. Once an organisation has effectively managed its legal risks exposures and the attendant financial risk liabilities through such a framework, it would then have a better frame of mind to pursue its corporate strategy and other regular business activities in a manner that would enhance its growth.

Case study 2 – the shipping industry 6.17

Another example of a sensitive sector in today's business world is the shipping industry. The shipping sector can be exposed to usual commercial risks, as well as sensitive ethical issues and concerns of governance and sustainability. Indeed the shipping industry can affect the performance of many other business sectors and, in turn, this affects the economy and society as a whole. Bearing in mind the age of this industry it is most important that it monitors the trends of good corporate governance and fulfils modern day responsibilities. Moreover, since this industry has to deal with both macro and micro issues shipping can provide a useful case study that addresses many aspects of risk management and due diligence and complements the earlier one. In order to operate a successful business in this area a company should not only carefully take account of the commercial and contractual issues but also be aware of the broader concerns to which it is exposed. This is intended to be a practical case study that:

- considers some of the main day to day business risks;
- offers some business hints and tips; and
- proposes some governance strategy for the sector.

Loss, damage and expense to cargo – avoiding and reducing the risks 6.18

When goods are moved by road, rail, inland waterway, sea or air in the course of trade, they are inevitably exposed to a wide range of risks of loss or damage. Below the usual and less usual risks are summarised.

Usual risks 6.19

Goods in transit are a target for opportunists and thieves. Fire can break out almost anywhere. More rarely, road and rail accidents occur. Goods carried across water are subject to the predictable risks of loss or damage during the course of loading and of discharging, by sinking, stranding, by the ingress of water, by contamination by other goods, sweating and damage through shifting, etc. Perishable goods suffer if delayed. Some cargoes, especially bulk cargoes, suffer a measure of inevitable loss due, for example, to natural shrinkage or to the difficulties of accurately measuring or assessing their true weight on loading and discharge. Cargoes carried by air tend to suffer less damage but as non-perishable air cargo is often of high value, air cargoes are a magnet for thieves. International carriage can involve partial carriage by sea, road and rail with transhipment and consolidation for the purposes of making up full loads. Each operation increases the risk of goods going astray or becoming damaged. The above are examples of just some of the many risks to which goods in transit are exposed.

Less obvious risks 6.20

There are some less obvious risks that fall upon the owners (whether shippers or receivers) of goods carried by sea. The following are examples.

General average 6.21

General average may arise when goods are saved from loss or damage by sacrifices made by the ship or by other cargo interests for the common good. For example, the sacrifice made by a ship owner who incurs emergency towage and repair costs when the ship's engines fail at sea or by the owners of other cargo jettisoned to make continuation of a voyage safer in bad weather. The owners of the ship and the owners of the cargoes on board and even freight at risk (freight that will only be earned if the cargo is delivered) will be required to contribute rateably to the cost of the sacrifice. Contributions are determined by a general average adjustment made by an average adjuster, much later. Cargo interests will be required to deposit security for their contribution before being allowed to take delivery of their cargo.

Marine salvage 6.22

When a ship in difficulties or distress receives assistance from another vessel or from a professional salvor, the salvor is usually entitled to a reward for his services. The salvor's reward will be based substantially upon the value of property salved. Professional salvors are awarded substantial sums as an encouragement to them to keep their expensive salvage vessels, equipment and crews on constant standby in busy waters. Cargo interests will be required to bear their rateable proportion of the salvor's reward. In practice the salvor (especially the professional salvor) will usually require the master of the vessel requiring salvage services to enter into a 'Lloyd's Open Form' written salvage contract. The salvor will have a supply of forms on board the salvage vessel. The 'Lloyd's Open Form' contract terms enable the salvor to take over control of the ship for the purpose of the salvage operation without discussion being necessary about terms and the price of the job. The price is determined much later in London, by specialist salvage arbitrators. The salvage operation can therefore usually take place without the risky delays that are likely to result from negotiation. Cargo interests will be required to provide immediate security for the eventual salvage arbitration award as a pre-condition of being permitted to collect their cargo or continue the voyage.

Unexpected taxes and charges 6.23

Taxes or other charges, direct or indirect, may be levied by acquisitive or desperate governments, without warning, at ports, airports and entry points. An import licence may be required. The importation of certain goods or commodities may suddenly be made unlawful with the result that goods may have to be surrendered or forwarded to another country for sale, or returned. The associated costs may well exceed the profit expected from the original sale.

Cash crisis 6.24

Sometimes the sea carrier or contractor simply runs out of money or claims to have done so and is unable (or unwilling) to pay the crew and pay for fuel and provisions to continue the voyage. If there is cargo on board, trans-shipment to another vessel, if indeed possible, is usually risky and always expensive. If the voyage is to continue, the cargo interests may have to pay the voyage costs all over again or risk losing their goods. Carriage, by any means, may be delayed. Markets may be lost. Perishable cargo may suffer. Similar problems might arise in relation to other means of transport.

Liens **6.25**

Where either a shipowner or charterer who has sub-let their chartered vessel have not been paid, the unpaid shipowner or charterer may be able to exercise a lien on the cargo to cover the unpaid freight or hire. A cargo owner may promptly pay freight or hire due, only to find that payment has not been made up the line to the head-owner or head-charterer who then purports to exercise a lien for unpaid freight or hire on the innocent cargo owner's cargo. That the lien may not be effective according to English law is no guarantee that it will not succeed overseas.

Delay – demurrage **6.26**

Demurrage is compensation in the form of liquidated damages payable to the shipowner when the charterer exceeds the time permitted by the charter contract for the loading and discharging of the cargo. The seller of the cargo (who may also be the charterer) may make the buyer similarly liable for delay in discharging his cargo at destination. Similar provisions may apply to other means of transport.

Dishonesty/fraud **6.27**

Documents, especially bills of lading or other documents of title to goods on board ship or conveyance, may be forged, tampered with or may simply misstate the nature or condition of the goods or the date of loading. Such events are likely to affect the sellers' right to payment and are likely be to the sellers' ultimate loss.

Absence of justice **6.28**

Justice, or the means to resolve problems quickly and effectively may not be available in the country where the problem has arisen. Even if justice is available, it may be slow or uneconomic. Local courts may be persuaded to seize jurisdiction to resolve disputes despite a contract stipulating that the courts or a tribunal of another country shall have exclusive jurisdiction.

Avoiding and reducing the risks **6.29**

As is the case when operating any business, some of these risks (and there are others) can be avoided altogether or reduced in a number of ways. Some suggestions follow by way of a practical guide that offers some hints and tips since it is not possible to go into extensive detail.

Avoid the risks altogether **6.30**

Certain methods of risk avoidance may be summarised as follows.

1. If goods intended for ultimate delivery to a place overseas are sold on ex-works (EXW), ex-factory terms or free carrier (FCA), free alongside ship (FAS) or free on board (FOB), the buyer will carry most if not all of the risks associated with carriage.
2. Will the seller necessarily make any more profit or will the seller necessarily gain an elusive sale by selling goods on delivered terms or similar? If not, the seller could end his responsibilities either at the factory gate or at delivery on or alongside the buyer's nominated ship or other means of transport.
3. Does the transaction justify the risks, and will the profit justify the cost of insuring against risks that cannot be passed on directly to the buyers? If the seller sells on cost and freight (CFR) terms or cost, insurance and freight (CIF) terms, risk in the goods will usually pass to the buyer on shipment. If the sale terms are CIF, the seller has to take out only the minimum insurance cover.
4. If the seller's skill lies in producing or procuring goods of a certain description, it may not be appropriate for the seller to get involved in the wholly different business of organising and carrying the risks of international carriage of those goods. Chartering ships is not for the faint-hearted and will require dedicated management. The profit potential can be substantial, but so can the risks.
5. Much of the burden of organising international carriage (but few of the risks) may be taken from the seller's shoulders by competent freight forwarders who can, for example, arrange a 'door to door' service.

Insurance **6.31**

All of the risks associated with carriage of goods can be insured, including provision for periods when, in the course of transit, the goods are likely to be stored. Of course the broader the insurance cover, the more it costs. The cost of insurance can vary according to the state of the insurance market at any one time. The cost of truly comprehensive cover can cut deep into the envisaged profit margin, particularly in relation to low profit bulk commodities.

Shippers of bulk cargoes often insure subject to Institute Cargo Clauses Form C or free of particular average (FPA). Such insurance is much cheaper because cover is limited to catastrophes. The cargo owner's liability to contribute in salvage or in general average is covered, but partial loss or damage is not covered. The insurers will take care of any general average or salvage security that has to be deposited.

Form B insurance provides extended catastrophe cover, but restricts the perils insured against. It costs more than Form C cover, but less than Form A, which is 'All Risks' cover.

If the seller requires the buyer to insure the goods (therefore saving the seller the possibly heavy cost of doing so) it is unlikely that the insurance will insure for the benefit of the seller. Until both property and risk in the goods pass to the seller (an event usually determined by the terms of the contract of sale) the seller is at risk and must consider whether to incur the expense of insuring against that risk. A good insurance broker will be essential.

Sale terms **6.32**

In terms of the usual carriage-of-goods risks, what cannot be adequately covered by cost effective insurance can usually be covered by sale terms that place the risk elsewhere. If there is a straightforward contract sale, the sale terms will determine when both property and the risk in the goods pass to the buyer. However, even well used standard form contracts can be vague about such matters. Property and risk pass when the sale contract so provides, which is when the parties intend that property and risk shall pass. The reality is that the intentions are often not clearly expressed.

The sale terms will identify with whom the risks lie. As is the case with all business dealings, careful drafting of contracts and the use of well tried and tested models are advisable. The party who takes the lead in drafting the sale terms is usually best placed to draft to advantage. If a lot of money is involved, or if there are likely to be multiple contracts on the same or similar terms, it may be appropriate to seek assistance from specialist solicitors.

Payment provisions **6.33**

These provisions are, of course, perhaps the most important part of all the agreement. The payment terms should be made very clear in the contract terms. If payment is to be by documentary letter of credit, care should be taken to ensure that so far as is possible the documents that trigger payment are in order before shipment is made. Careful thought should be given to what documents will be needed. Is there likely to be any problem about their coming into existence? Are they actually relevant to proper execution of the contract? The letter of credit should be confirmed by the sellers' own bank (preferably at the buyer's expense) if there is any doubt about the paying bank.

It the payment terms are not tied up carefully, the buyer may seize the opportunity to avoid payment. Recovering payment in a foreign country may be fraught with problems. It is likely to be expensive.

Documents 6.34

Vigilance should be exercised to ensure that the documents required to trigger payment under the letter of credit (or other means of payment) will not only be available but that they are genuine. The buyer should always have in mind that the documents that will trigger his or her payment should, so far as possible, evidence due shipment of goods corresponding with the contract terms. The seller should make sure that the required documents will be readily available. The necessary documentation to allow import and export of certain goods should be obtained.

Superintendence 6.35

It may be necessary or advisable for bulk cargoes (especially) to be subject to specialised superintendence by recognised superintendence companies. They will oversee the loading process (perhaps also the discharging process) and issue certificates certifying the quantities shipped and discharged. This will help to avoid fraud and disputes. Banks or even insurers may insist on it. The loading superintendence certificate would then be included amongst the documents required to obtain payment under the letter of credit.

Early action 6.36

If something appears to be wrong or shipment or delivery seems to be delayed, early action should be taken. Action will usually be supported by the cargo insurers, who will arrange for expert surveyors to investigate, attend and report. Sometimes it may be appropriate to instruct lawyers to become involved at an early stage, but first on the scene is usually the cargo surveyor. If goods are at risk, any delay usually compounds the risk.

Supervision 6.37

Sometimes the cost of engaging a surveyor (as opposed to a cargo superintendent) to supervise loading and discharging operations will be justified, even though the cost forms no part of the price of the goods exported or imported. Apart from reducing risks of damage by bad stowage and bad handling, the surveyor can sometimes reduce time spent loading and discharging, reducing the potential demurrage bill.

Summary 6.38

The following is a list of hints and tips for risk avoidance:

- identify the most likely risks;
- where appropriate, ensure that the sale terms pass responsibility for carriage risks to the other contracting party;

- insure, recognising the limitations of less expensive cover. There is little or nothing that the shipper or receiver can do to avoid a liability in general average or salvage, but marine insurance will take care of the financial and administrative consequences;
- the small risk of unexpected taxes or import restrictions in the country of destination is avoided if property in the goods has passed to the buyer;
- the occasional risk of falling into the hands of impecunious shipowners or charterers is difficult to avoid as the risk is often unpredictable. The risk may be reduced by avoiding sub-charterers altogether;
- be especially careful to ensure that the payment provisions are effective. The bank should be able to give some assistance;
- exercise vigilance to avoid fraud. Dishonesty appears in so many different forms that the best approach is probably to trust no-one;
- consider whether superintendence and the issue of superintendence certificates is appropriate;
- appoint cargo surveyors to arrange supervision of loading and/or discharging if the cost can be justified. Keep in mind that demurrage can be substantial, and is usually difficult to recover from the buyer/receiver; and
- take prompt action if something appears to be wrong.

Macro issues and the shipping industry 6.39

Many of the business risks associated with the shipping industry have been summarised above in **6.19** to **6.28**. This discussion has largely considered the risks form a micro perspective. Yet, as was mentioned at the beginning of this case study at **6.17**, there are also macro issues that the shipping industry should address. In this case study, therefore, the discussion proceeds to consider governance concerns.

Ship demolition – environmental consequences 6.40

Fortunes are made and lost in shipping. Freight rates are pulled in different directions by differing national economies – they seldom remain constant. When freight rates are high, ships command high values. When values are high, some shipowners may even feel confident enough to order the construction of new ships. As new ships are constructed, old ships are sold for demolition.

Governance trends and ship breaking 6.41

In recent years there has been a steady reduction in the volume of surplus over age tonnage. There has been increasing pressure on and from the maritime nations to make ships safer to navigate and environmentally safer. In the long term this makes for more profitable and stable freight rates, and rids the seas of ships that are either unseaworthy or in danger of becoming unseaworthy.

When ships reach the end of their economic or safe life, they can be disposed of in much the same way as an old car. They can be delivered to the breaker's yard or they can be dumped and abandoned.

Some parts of the ship can be salvaged for use on other ships, but the greater function of the breaker's yard will be to reduce the ship to her basic metal components, the most important being steel. The value of a ship for demolition is determined by the current market price of her steel. Anything else on board that can be salvaged is a bonus for the breaker.

As an alternative to breaking, a ship may be abandoned at sea, sunk, run aground or stranded in shallow waters. She may then become a danger to navigation, a source of pollution (she will have significant fuel and lubricant residues on board) and perhaps an eyesore. Fortunately the sheer size of ships means that even as scrap ships can be worth millions of dollars, so the majority are scrapped for profit rather than abandoned.

The greater the number of ships scrapped, the cleaner and safer the seas are likely to become. The construction of safer and cleaner ships is likely to reduce the number of environmental disasters to which our seas and coastlines are at risk. Ship breaking, however, has an ugly side that is likely to remain for years to come. Ship breaking is a dirty and labour intensive business. It cannot be neatly confined to a small industrial area. No significant ship breaking takes place in Europe, including Eastern Europe. Ship breaking cannot be undertaken profitably outside third world countries where labour is cheap and where health and safety and environmental controls are non existent or primitive, unenforced and unenforceable.

Not surprisingly, ship breaking has become an almost exclusive third world industry, with India and China leading the world as the major ship breaking nations. The operation is often conducted in the most appalling health and safety, as well as environmental, conditions.

A ship to be broken is likely to be more than 20 years old, perhaps built in a country not too sensitive at the time to the dangers of certain types of asbestos. The asbestos will be stripped by hand. The ship will be stripped with bare hands (and feet) sledge hammers and gas torches. Heavy metals and poisons will be released, contaminating the workers, the beach on which the operation takes place and the sea that washes it. Asbestos fibres will also be airborne. Partially empty fuel tanks may be explosive. Metals may be reclaimed by the process of burning and raking the ashes. Further air pollution will take place.

Governance – macro issues 6.42

Whereas awareness is growing regarding the environmental and health risks associated with the shipping industry, largely through the pressures from the media and non governmental organisations (NGOs) such as Greenpeace, the solution to these problems must be considered by the industry itself as a matter of good governance. The traditional approaches did not consider these issues and the responsibility for solving such problems has been ignored. It is a matter of importance not only for the shipping industry but also other sectors since the discussion also illustrates well the importance of earning the licence to operate in today's shrinking business world.

How can the dangers to health and the environment be avoided? Without the intervention of vast amounts of capital to subsidise the ship breaking industry and corporate governance drivers this is a global problem that must be resolved. Who will provide the capital or subsidy? If scrapping is not made economically attractive to the ship owner, there will be a disincentive to dispose of older ships. If a ship owner cannot be guaranteed a return after taking into account the running costs of getting the ship to a scrapping destination (which may be many thousands of miles away) the ship owner may be tempted to dump or scuttle the ship. Even ships forming part of a fleet tend to be owned by individual and often extremely anonymous companies based offshore. The companies will have no assets, from the moment they dispose of the ships. Individual persons can escape personal liability for the consequences of dumping or scuttling. Higher profile ship owners, anxious not to tarnish their reputations, can avoid any association with the messy business of scrapping by selling the ship to a third party. The third party will then in turn sell to the breaker leaving no connection between breaking and the original owner.

Since ship owning is a truly international business the imposition of, for example, a tonnage tax to provide the subsidy necessary to clean up the business of ship breaking will simply drive owners to register their ships in a different jurisdiction where no tonnage tax (or any other tax) is imposed. Ship owners have long since abandoned countries where taxes are levied.

If the imposition of safety controls on the demolition process had the result that ship breaking in India, for example, became more expensive (thus reducing the scrap value of the ship to be broken), the industry would shift elsewhere. Many of the workers in India are migrants. They or other migrants would migrate to the new scrapping base. Even if subsidies became available to entice the business to Europe where health and safety and environmental controls are more likely to be enforceable, few European countries would want the business in their backyards. It is likely therefore to remain a third world industry.

The provision of capital resources to implement effective health and safety procedures and environmental protection may not guarantee that those procedures would be adopted or that their adoption would be enforceable, but they would certainly improve the present state of affairs.

The solution to the pollution and health and safety risks must lie in a combination of heavy subsidy, implementation of appropriate controls and enforcement of rules and regulations giving effect to those controls. The countries in which demolition takes place cannot afford to provide their own subsidies. Those involved in the business of owning and operating ships cannot (and will not) be forced to pay. Ship builders will not agree to build deconstruction costs into the sale price, nor build for safer deconstruction. If they are forced to do so, they will lose business to other countries eager to subsidise the industry, without the burdens of deconstruction. The necessary subsidies, grants or donations will have to come from governments rather than directly from the shipping industries, or from businesses willing to accept part of the responsibility for cleaner seas, safer shipping and improved working conditions in poorer countries.

Ship breaking – environmental, health and safety implications in the third world **6.43**

The imperatives of corporate governance are drivers to consider the following concerns:

- the shipping industry is sometimes in crisis largely because the freight rates do not justify the cost, the risks and an deal level of investment in health and safety;
- it is in the interest of ship owners worldwide to encourage a programme of scrapping (breaking) older vessels. This would lead to better freight rates. There would be fewer ships chasing available cargoes. It would also stimulate the construction of new and better ships. Moreover, the recycling of materials by scrapping is generally agreed to be environmentally desirable;
- a ship's value as scrap will fluctuate according to the market value of steel. Its value can be millions of dollars;
- scrapping usually takes place in third world countries where this is an ample supply of cheap labour and minimum or absence of rules and regulations to obstruct the process and to diminish the price attainable by the ship owner;
- the bulk of ship breaking takes place in India and China often in appalling conditions for the labourers and with inadequate regard for the environment, bearing in mind the presence of the following risks:
  - oil residue – sometimes explosive, always environmentally demanding;
  - asbestos – especially around pipe work stripped manually without any or adequate protection for the labourers or environment;
  - dangerous metals and chemicals that are used in the construction of ship's equipment; and
  - pollution of beach/land, sea and rivers.

Bearing in mind the above a cost-benefit risk analysis is summarised below.

Practical problems **6.44**

These include:

- the failure of preferred scrapping destinations to provide legislation or pressure to protect:

  (*a*) the workforce; and

  (*b*) the environment;

- the necessity for ship owners to get the best price if a vessel to be scrapped;
- there is no economical alternative to scrapping in unregulated third world countries;
- the lack of social conscience in a desperate industry. The majority of the world's merchant ships are owned by one-ship companies domiciled in flag of convenience countries. This is not likely to change in the foreseeable future;
- ship owning and ship management are usually separated;
- ships tend to be registered in countries that do not charge taxes;
- offshore ship owning companies are difficult to track, trace and condemn. Legislation serves no useful effect. Penalising the ship owner will not work;
- whilst the compartments of tank ships can be rendered 'gas free' before delivery to the breakers, the industry does not consider it viable economically or otherwise for ships to be 'cleaned' prior to scrapping. Apart from cost, vessels usually use their own motive power to reach scrapping destination. Stripped of their offending parts, they would be reduced to a hazard to shipping and to the crew; and
- most ships are owned by one-ship companies, even those forming part of a major fleet. When a ship is sold the proceeds of sale generally disappear instantly and there are no other corporate assets to attack.

Positive action **6.45**

Positive action could incorporate:

- the education of breakers, improving the awareness regarding these major practical concerns;
- the co-operation of major fleets with breakers should be developed as a matter of good governance.
- an obligation on trading vessels over, say, 20 years to carry out a detailed inventory of hazardous materials used in their construction, identifying

presence of hazardous materials. (This could be enforced by International Maritime Organisation (IMO) regulations currently in force, relating to safety generally);

- the consideration of best practice from other industries can be helpful, for example the motor car industry, in the Netherlands for instance. There should be some co-operation with manufacturers to design for hazard free deconstruction and disposal; and
- as with any other industry, financial incentives should be considered to effect change. The carrot is more likely to succeed than the stick, again as with most sectors in business.

Conclusion 6.46

This form of industry response to collective liability can have relevance to other sectors. In the context of this case study it is important to note the risk of ships being dumped, thus polluting the sea, rather than broken and the commercial impact on the well known companies in terms of reputation.

In view of the increased interest and awareness, relevant members of involved sectors should be proactive in considering the state of the industry that takes account and bringing it up to date, in line with the drivers of corporate governance and due diligence.

Chapter 7
Reputation Risk Management

Why does reputation matter? 7.1

According to *The Aon European Risk Management and Insurance Survey 2002–2003*[15], loss of reputation is seen as the second biggest threat to business (after business interruption). These findings were based on the views of risk managers, insurance managers and financial directors of over 100 of Europe's largest companies. Aon's research also shows that the top 2000 private and public sector organisations regard reputation as their biggest risk. The results of a similar survey carried out in Australia revealed a very similar picture, with loss of reputation, business interruption and brand protection topping the list. It was also pointed out that, the key causes of concern for brand management were ethics, corporate governance, compliance and product quality.

'It takes 20 years to build a reputation and five minutes to ruin it.' Fortune Magazine reported Warren Buffett's view in 1991, arguably the world's most successful investor. It now takes more than just good public relations and a clever advertising campaign to successfully secure financial success – ongoing reputational due diligence and corporate governance exercises are needed. If a company commits to an idea, it does so in public and will be therefore subject to public scrutiny and examination.

Consultation with stakeholders is the best way to ascertain stakeholder perceptions and expectations about building credibility.

A good corporate reputation can influence:

- investors' willingness to hold its shares;
- consumers' willingness to buy from it;
- suppliers' willingness to become its partner;
- competitors' determination to enter its market;
- media coverage and pressure group activity;
- regulators' attitude towards it;
- its cost of capital;

- potential recruits' eagerness to join;
- the motivation of existing employees;

Corporate reputation is now very much defined in stakeholder terms. One of the definitions of corporate reputation sees it as the 'aggregate perceptions of multiple stakeholders about a company's performance.'[7] A good reputation is therefore achieved when stakeholders' expectations and experiences of the company are aligned. Stakeholder expectations represent the expectations of all conceivable parties interested or in some way involved in the workings and development of a company. (The stakeholders that really matter to the private sector are invariably customers, employees and investors. Others may include regulators, strategic partners, suppliers and the local community.) It is considered good business sense to perform market surveys on relevant consumers. The rationale behind this is that these consumers will have an effect on the corporate health of a company. If consumers like a product or brand, they will purchase it. This will result in profits. This is well understood because it can be seen to have an obvious effect on the economic bottom line. However, if consumers are examined more closely, they can be seen to be swayed by fashion, values and other outside influences. The risks a company faces is having a product boycotted, found unfashionable, of poor quality or not purchased because of bad press. To minimise these risks it is sensible to address stakeholders to ensure that there is minimal exposure to liability. Furthermore, it will also give the business the opportunity to be able to recognise market trends faster, change faster and predict the social effect on the economic aspect of their business.

The following are the most significant areas where a risk to reputation may arise:

- delivering customer promise;
- communications and crisis management;
- corporate governance and leadership;
- regulatory compliance;
- workplace talent and culture;
- financial performance and long-term investment value;
- corporate social responsibility.

Businesses cannot afford to ignore risks related to their reputation and brand. However, apart from business interruption, the top risks identified by the Aon survey are uninsurable. Therefore, mitigation of such business risks is largely an issue for corporate governance. The Cadbury Report[18] defined corporate governance as 'the system by which companies are directed and controlled' (paragraph 2.5) (see also **CHAPTER 10**.) The report's findings marked an important advance in the process of establishing corporate governance.

A broader definition of corporate governance is that it ensures that the Board of Directors develops, implements and explains policies that will result in an increased shareholder value and address their concerns. It will also reduce the costs of capital and diminish business, financial and operational risks.

However, the relationship between risk and materiality is not necessarily a straightforward one. Western case law seems to have defined materiality to mean those things that would be material to shareholder interests. For example, the current US regulatory regime (see **CHAPTER 11** and the discussion of the *Sarbanes-Oxley Act of 2002* legislation) requires directors of listed companies to develop a system of detailed business controls that enables them to report accurately on any financial or business risks that are material to shareholder interests. In other words anything that, if the information leaked out, could lead to a material decline in the share price or value of its bonds. This is remarkably close to what investors expect the purpose of good corporate governance to be. This implies a very much wider interpretation of risks than those usually reported on by companies in their annual accounts. This can be illustrated with respect to the well-known experience of Nike and the labour practices of its suppliers in Pakistan and Vietnam. Those local companies were found to be using unregulated child labour in the manufacture of Nike sports goods. Nike, of course, was aware of this. From a financial accounts perspective, these were not material risks or issues, other than it kept manufacturing costs low. However, once the news came out Nike's shares dived as a result of the ensuing consumer backlash. That decline in the share price was definitely 'material to shareholders' interests'. Being socially responsible and engaging in stakeholder dialogue with investors and consumer groups would have helped avoid this problem.

Some companies have been publishing Operating and Financial Reviews (OFR), providing more qualitative and forward-looking information on a wider range of issues than have traditionally been covered by company reporting, for some time. In fact the Accounting Standards Board first issued a statement of best practice for the OFR in 1993. Many listed companies do so as a matter of best practice or to enhance their reputation and profile in particular areas, such as the environmental impact of their business. A survey by Deloitte & Touche[22] reported that in 2003 over 60 per cent of listed companies prepared an OFR or adopted the broad approach set out in the Accounting Standards Board statement, and a further 30 per cent include some recognition of the OFR in their reporting. However, the general upward trend in the quantity of reports has not been matched by the provision of improved quality of information. The UK government's Company Law Review found that the content and rigour of reporting varies widely, and that a significant proportion of large companies fall well short of meeting the Accounting Standards Board's recommended practice, especially outside of the FTSE 100. Only few reports go further than explaining their 'policies' and contain any quantitative data about performance, mention progress on previous targets or strengthen future objectives. Moreover, only very few reports are audited by third parties. Even if this does happen reports are often mainly concerned with the methodology and not the verification of information or the even the investigation for why certain information has not been included.

However, in January 2005, reporting on 'intangible' issues is scheduled to become compulsory for listed companies in the UK by means of the Operating and Financial Review. This expansion of company law will make it much easier for shareholders to obtain information on how companies deal with reputational risk issues.

> 'We expect companies to create wealth while respecting the environment and exercising responsibility towards the society and the local communities in which they operate. The reputation and performance of companies which fail to do these things will suffer... For this reason, ... increased, high-quality shareholder engagement is vital to creating the modern economy that we all want. But, if shareholders are to hold the directors of their company to account for its performance, they need full and accurate information. This will allow them to act if they see a risk that a company may go in the wrong direction. And it will be a discipline on management, making them analyse and report accurately on performance and prospects. This consultation document contains the Government's proposals for a new statutory Operating and Financial Review (OFR) designed to give shareholders the information they need. ...The OFR will give shareholders information in an accessible form allowing them to make a full assessment of their company. It builds on best practice followed by a number of our larger companies, and follows the recommendations of the independent Company Law Review. It does not constrain companies' freedom of action. It simply demands that directors explain their stewardship to their shareholders.'
>
> (*Department for Trade and Industry, 2004, p. 5–6*[6])

Value of brands and their importance to reputation 7.2

When the FTSE 100 index was first calculated on 3 January 1984, the value of the 100 largest companies (by market capitalisation) listed on the London Stock Exchange was £100,145m. Exactly twenty years later this figure stood at £1,110,838m. According to research carried out by a major branding consultancy – Interbrand, over 70 percent of this value is accounted for by goodwill, or in other words reputation and brand. In 1994 goodwill, according to the same organisation, accounted for 44 per cent of total value. The above facts suggest that the eleven-fold growth in the value of FTSE 100 companies is largely attributable to their reputation or brand.

Widescale use of brands started in the late 19th and early 20th centuries. Improvements in manufacturing and communications allowed mass marketing of consumer goods. Consumer brands like Coca-Cola, Quaker Oats, Shredded Wheat, Kodak, American Express and Heinz date from this period. Trademark legislation was introduced to allow the owners of these brands to protect them in law. Since then, brand protection was little more than a legal issue.

Trademark law prohibits a company from using the same or similar names, logos, or marks if they are already in use by another company. However, a brand is much more than a name or a logo.

> 'Today's brand strategists say brands evoke distinct associations; they ascribe human personality traits to brands; they speak of long-term relationships with customers, rather than transactional exchanges. Brands carry emotional attachments; brands are your neighbours, your colleagues, and your friends. In short, brands have the potential to provide customers with a variety of pleasant, or unpleasant, experiences.'
>
> (*Schmitt: 2001, p. 236*[12])

Considering this view of the brand and the value of brands to their owners, there is a need for more than just legal protection and protection from competition. Companies need to manage brands in an integrated manner across the entire company. Brand is reputation, which has to be managed in the eyes of all of the company's stakeholders and the public at large. Via the Internet and other new media, consumers, concerned citizens and non-governmental organisations can mobilise powerful campaigns against corporate behaviour of which they disapprove. Anything a company does or says can add to, or destroy brand value and reputation.

> 'Brand owners are accountable for both the quality and the performance of their branded products and services and for their ethical practices. Given the direct link between brand value and both sales and share price, the potential costs of behaving unethically far outweigh any benefits, and outweigh the monitoring costs associated with an ethical business... The more honest companies are in admitting the gap they have to bridge in terms of ethical behaviour, the more credible they will seem.'
>
> (*Brands and Branding, 2003*[2])

To many companies, brands are their most important assets despite being intangible. According to research carried out by Interbrand in 2002, the contribution of such brands, such as Coca-Cola and McDonald's, to shareholder value are 51 per cent and 71 per cent respectively. In fact, it is possible to argue that the majority of business value is derived from intangibles like brand and reputation. This is certainly supported by the attention that they have been receiving from management in the recent decades. Other research shows that companies with strong brands generally outperform the market in respect of several indices.

However, despite all of the above, a number of large brand acquisitions in the 1980s have revealed the inability of most accounting standards to deal with goodwill in an economically sensible way. The purchase of Rowntree by Nestle for £2.5 bn, which was two and a half times the pre-bid price and eight times the tangible net asset valuation, sparked off the debate about the inadequacy of the accounting standards for such transactions.

In the UK, France, Australia and New Zealand companies can put acquired brands on their balance sheets as identifiable intangible assets, however they are not encouraged to do so. Nevertheless, the recognition of acquired brands by a number of companies led to a similar recognition of internally generated brands as valuable financial assets within companies. In 1988, Rank Hovis McDougal, a leading UK food conglomerate, defended itself from a hostile takeover by Goodman Fielder Wattie, by demonstrating the value of their brand portfolio and later accounting for it on the balance sheet. This was the first brand valuation in the UK and it established that it was possible to value brands not only when they had been acquired, but also when the company itself had created them. Since then there has been a number of accounting standards changes on the topic of treatment of goodwill on the balance sheet, including the introduction of FRS 10 and 11 by the UK Accounting Standards Board in 1999, and the IAS 38 by the International Standards Board.

Since the late 1980s, a number of brand valuation models have emerged. According to Jan Lindemann, the Managing Director of Global Brand Valuation at Interbrand the two main categories of these models are:

- research-based brand equity evaluations;
- purely financially driven approaches.

Research-based brand equity evaluations are usually insufficient for assessing the economic value of brands unless they are integrated into a financial model. This is due to their heavy reliance on consumer research and relative performance of brands. Financially driven approaches, can be further subdivided into four categories: see the table below.

Brand value assessment approaches	
Cost-based approaches	These define the value of a brand as the aggregation of all historic costs incurred or replacement costs required in bringing the brand to its current state: that is, the sum of the development costs, marketing costs, advertising and other communication costs, and so on.
Comparables	Another approach is to arrive at a value for a brand on the basis of something comparable. But comparability is difficult in the case of brands as by definition they should be differentiated and thus not comparable.

Premium price	In the premium price method, the value is calculated as the net present value of future price premiums that a branded product would command over an unbranded or generic equivalent.
Economic use	Approaches that are driven exclusively by brand equity measures or financial measures lack either the financial or the marketing component to provide a complete and robust assessment of the economic value of brands. The economic use approach, which was developed in 1988, combines brand equity and financial measures, and has become the most widely recognized and accepted methodology for brand valuation.

Interbrand's own approach to brand valuation, is one of economic use. According to *Business Week* (4 August 2003) it values brands the same way analysts value all other assets – on the basis of how much they are likely to earn in the future and then discounting these projected earnings to a present value based on how risky they are.

Reputation and corporate culture 7.3

Reputation management is a natural extension of brand management and some rules apply to both. Aligning an organisation with its brand is crucial for managing a brand in the eyes of consumers; and aligning an organisation and its reputation is paramount for managing the reputation in the eyes of stakeholders.

Aligning the organisation, its operations and culture around its brand values brings the 'promise' that the brand makes to life. A corporate brand stands for the relationship that an organisation has with its employees, as much as it represents the relationship that it has with its customers through its product and service offering. For a brand to 'come to life' with customers, the organisation must be internally aligned to deliver the brand promise through the organisation's culture, reward systems, key success activities and structure. In other words, employees must 'live' the brand values in their day-to-day interactions. Also management must demonstrate their commitment to these values through their behaviour as well as corporate communications, demonstrating sincerity – not just rhetoric[8].

A good reputation helps to sell goods and services, recruit new employees and reflects positively on a company's share prices. It is needed for creating shareholder value. Delivering on a 'brand promise' is crucial for reputation management. Companies that commit themselves to this promise generally demonstrate the following characteristics:

- effective use of internal communications to raise employee morale and commitment through shared beliefs and vision;
- managers and staff are given a deeper understanding of the brand promise and the behaviours and values the promise demands – they are trained to adapt their behavior;
- all employees understand how their own work processes and responsibilities contribute to delivering the brand promise to customers;
- such company policies as recruitment, training, and rewards are changed so that the organisation is also behaving in line with its brand promise. When employees understand and accept that the values are genuine, they align their attitudes and behaviour to the brand values. The result is greater satisfaction for both customers and employees, leading to employee and customer preference and loyalty.

The above approach is also consistent with the responsibility of organisations towards their employees and fits into the principles of corporate social responsibility as so many reputational risk issues do. At the same time, neglecting the human interface between the company and its stakeholders (especially customers) can destroy all the other attempts at reputation management. Implementing an integrated approach to reputation management across the organisation is thus critical to success.

The high proportion of mergers and acquisitions] failures can be largely attributed to the inability of managements to align corporate cultures. This fact supports the importance of having an integrated approach to reputation management for the creation and sustainability of shareholder value. Reputation management is necessary for protecting the long-term value of the brand and should not be confused with either just creating a marketing image of an organization's product or just a public relations exercise.

Examples of major losses of reputation 7.4

Since corporate reputation is increasingly defined in terms of the relationships that a company has with its stakeholders, transparency and accountability are vital to reputation risk management because it is not enough to *be* 'good', it is also important to be *perceived* as being 'good'. Through becoming transparent, companies engage their stakeholders in dialogue and thus become accountable. By engaging in dialogue, companies can address the concerns of their stakeholders thus mitigating the risks that arise out of legal action, bad press and other actions taken by stakeholders that may have a negative effect on the company.

Reputation risk management differs from traditional risk management in an important respect: reputation is largely about perception. Many management teams have been criticised for the way they handled a crisis – not because their strategy was ill conceived or clumsily implemented, but because they failed to tell the outside world what the strategy was. The Exxon Corporation received relatively little comment about its recovery plans after its tanker *Exxon Valdez* ran aground in Prince William Sound, Alaska, but was criticised for its communication with the local community.[3]

A highly relevant example of why companies should address the opinions of stakeholders is the well-known case in 1995 when Shell focused on the legislative aspects of disposing of its Bent Spar oil storage rig. Having assured itself that the procedure would comply with all relevant legislation, Shell decided that the best environmental option was to wreck the rig in the North Sea. Environmental activists protested vigorously but Shell paid no attention, it was content that its case was legally sound and therefore risk free. However, Greenpeace released a torrent of negative publicity against the company, based on what later proved to be an incorrect analysis of the environmental impact of sinking the Brent Spar. Nevertheless, Greenpeace's attack served to bring the proposal to the attention of the public so successfully that Shell had to back down in the face of public outrage demonstrated by a boycott of its products in Germany.[13] Addressing the opinions of the relevant non-governmental organisations (NGOs) in the decision making process of the disposal of the rig would have helped to prevent the damage done by the lack of transparency and subsequent boycott.

In contrast, British Midland Airways demonstrated a good example of effective crisis management after the 1989 Kegworth air disaster in which the airline's Boeing 737 crashed near East Midlands Airport. The chief executive of the airline handled media interviews in a convincing and honest manner that confronted the scale of the disaster and addressed the questions that the various stakeholders had. Not only did the airline avoid a decrease in sales, but it actually saw a short-term increase in sales after the disaster.

After 25 years of NGO campaigning, Nestle, is still the target of boycotts over its policies in promoting infant formula over breast milk. In the 1970s, when the company was accused of selling infant formula in developing countries at prices that could not be afforded and where clean water was unavailable, the company ignored allegations of irresponsible behaviour and contravening the World Health Organisation's (WHO) code. Nestle suffered reputational and commercial damage by refusing to debate the issues in public. By the time the company was prepared to negotiate with campaigners and other stakeholders, no amount of attempts to align with the WHO code made a difference.

The Cost of Crisis			
Corporation	**Event**	**Cost ($m)**	**Year**
Pan Am	Lockerbie	652	1988

Union Carbide	Bhopal	527	1984
Exxon	Valdez oil spill	16000	1989
Perrier	Product recall	263	1990
Occidental	Piper Alpha disaster	1400	1988
P&O	Zeebrugge	70	1987
Barings	Financial collapse	1200	1995
Ford/Firestone	Product recall	5000	2000
Coca-Cola	Product recall	103	1999
TotalFinaElf	Oil spill	100	1999
Monsanto	GM crops	2000–3000	1999
Source: Larkin, J *Strategic Reputation Risk Management*, Palgrave Macmillan, 2003.			

A study by Knight & Pretty[9] (pp 275–279) reveals further implications that catastrophes have on the share price of companies. In their research they focused on 15 major corporate disasters, similar to the ones in the above table, and traced their effects on stock returns and trading volumes. The selection of catastrophes was based on five criteria, which also make this research relevant to reputation risk management:

- man-made as opposed to natural disasters;
- involved publicly quoted companies;
- received headline coverage in the world news;
- occurred since 1980;
- the organisation had to be affected on a symbolic level as well as on a physical level.

In all of the cases studied, the catastrophe was found to have had a significant negative impact on the share price – on average amounting to eight per cent of stock value. However, after just over 50 trading days some stocks experienced an apparent full recovery with the exception of the non-recoverers whose initial drop in share price amounted to eleven per cent and up to 15 per cent one year after the catastrophe.

What also became apparent from the research was that for the first six months after the catastrophe, the stock market tended to judge the companies on the impact of the financial loss on its average market capitalisation and the number of deaths arising as an immediate consequence of the catastrophe. However, after six months the market paid more attention to whether the company was perceived to be responsible for the disaster. Investors' perceptions of managerial responsibility were the most important ones.

The negative impact of the catastrophes on the companies' cash flows is usually reduced by the extent of insurance recoveries, and the stock market usually forms a collective opinion regarding the cash flow impact when the exact figures are not known in the immediate aftermath. However, a lot more depends on managements abilities to deal with the consequences.

> 'Effective management of the consequences of catastrophes would appear to be a more significant factor than whether catastrophe insurance hedges the economic impact of the catastrophe. ...Evidence suggests that managers' immediate, honest and efficient disclosure of relevant information to all parties helps secure share price recovery. ...The lessons for managers are clear. They must do all that is reasonably possible to prevent catastrophes from happening.'
>
> (*Knight & Pretty, 2001, pp 278–279*[9])

Integrated reputation risk management 7.5

The above examples demonstrate the amount of damage that can be done to an organisation once its reputation is damaged. Reputation damage can harm organisations of all sizes and activities, however it is the most misunderstood and ill managed of all enterprise risk management activities. Once an organisation has done something that reveals or even suggests a possibility that its products or services are unsafe or unreliable, that the management was incompetent or corrupt, no amount of immaculate crisis response and highly paid public relations consultants can prevent the damage. Therefore, reputation risk management can only be effective if it operates in an integrated manner – not as a specialist function to be activated in an emergency but as a major influence on the organisation's actions, behaviour and standards. The key to this is to understand your reputation, matching its management to its needs.

In many cases, the potentially catastrophic consequences of not managing the crisis properly becomes apparent only when a reputation incident has already severely damaged the credibility of an organisation or one of its brands, or its standing in the eyes of its stakeholders. Many organisations make the mistake of assuming that all that is needed is crisis planning. However, a reputation crisis exposes to public and media scrutiny not only the organisation's competence at crisis handling, but also the values, standards and shortcomings that existed beforehand. Reputation risk management should have two main objectives: to prevent the causes that could damage reputation; and to minimise the impact if, despite the best efforts, a reputation crisis still occurs.

In her book on strategic reputation risk management[10], Judy Larkin argues that there are six steps to successful reputation risk management.

1. Establish early warning and monitoring systems – the reputation risk 'radar'.
2. Identify and prioritise risks (and opportunities).
3. Analyse the gaps; identify response options.
4. Develop strategies and action plans.
5. Implement strategies and action plans.
6. Keep the 'radar' tuned.

Early warning and monitoring systems 7.6

Establishing early warning and monitoring systems or surveillance systems designed to scan commercial, political/regulatory, social, economic, technological and other trends that may have an impact on business strategies is the first step. This may involve stakeholder profiling in relation to a specific risk issue, examination of look-alike and legacy issues, quantitative and qualitative stakeholder opinion polling and assessment of websites, chat rooms and scientific, public health, consumer and other relevant databases. The need for a finely tuned 'radar' that can scan for threats and track individuals, groups and organisations that may have an interest is critical.

Prioritisation of risks 7.7

Identification and prioritisation of risks provide the basis for developing and validating risk issue management strategies. This may involve facilitated scenario planning, auditing and benchmarking as well as gathering qualitative and quantitative data that can assist planning. The aim is to identify all risk issues that may have an impact on the business. It is essential to factor in the reputational risk dimension into integrated risk management and internal audit procedures (eg include in operational 'risk register').

Gap analysis and response options 7.8

Analysing the gaps and developing response options involves analysis of any gaps between current performance and stakeholder expectations. This provides a basis for determining anticipatory or response options that can contribute to closing the gap. Questions to ask are:

- Is there a gap between performance and expectation?
- If so, what is it and why is there a gap?
- Is the risk evaluation effective?
- Does the company really deliver the standards and values it claims?
- Which people determine how the company behaves?

- Which stakeholders can influence the company's reputation and performance on these issues?

Asking these questions helps a company realise the differences between how it sees itself relative to the perceptions of its key stakeholders.

Strategy development and action plans 7.9

Developing strategies and action plans involves selecting the most appropriate response options, deciding on the company's position, assessing resources, identifying stakeholders to be targeted and developing an action plan. The action plan would need to consider steps to be taken, responsibilities, timelines and measurement criteria.

A standard, bullet point template which describes the risk issue alongside a risk assessment, objectives, potential scenarios, strategic approach, key messages and summary operational and communication actions can help to filter out all but the most important information for implementation purposes.

Implementation 7.10

Implementation is about putting the approved strategy into action and communicating the response effectively to relevant stakeholders. Testing position statements and engagement techniques may be a necessary first step. Building a support base and using trusted third parties for stakeholder research and communication are also important considerations. The implementation phase will also require the preparation of supporting materials, which may include position papers, question and answers, press statements and websites. Throughout this process, the reputation risk 'radar' must be tuned to track media, Internet and trend and event data influencing the issue. Internal communication is also essential.

Keeping the 'radar' tuned 7.11

Keeping the 'radar' tuned involves evaluation and ongoing vigilance. Answering the following questions may be useful:

- has the issue become more serious in terms of its reputational impact and/or likelihood?
- is there support among key stakeholder groups?
- to what extent has the issue faded from the radar?
- can the company fulfill its objectives more effectively?
- what input can be provided to future strategies?
- what learning can be disseminated and built upon as part of this process?

Examples of reputational risk management 7.12

In-house teams for multinational companies have begun to try and establish systems to satisfy these points. At the Institutional Investor Group on Climate Change meeting in London November 2003 Lord Browne, CEO of BP announced that for an initial £20m investment they had a return of £650m. Other major companies have not been so adept at measuring these values and have turned to specialist rating agencies. The SERM Rating Agency, based in London, specifically rates companies on social, ethical and environmental risk and reputational risk. They do this solely from the point of view of the external stakeholders' perceptions and express the risks as a percentage of market capitalisations. Their ability to quantify and measure these risks is becoming ever more popular amongst the FTSE 350 companies who are increasingly being forced to examine them not just from a reputational risk perspective but also from a legislative one.

Figure 7.1: Suggested Risk Management Process

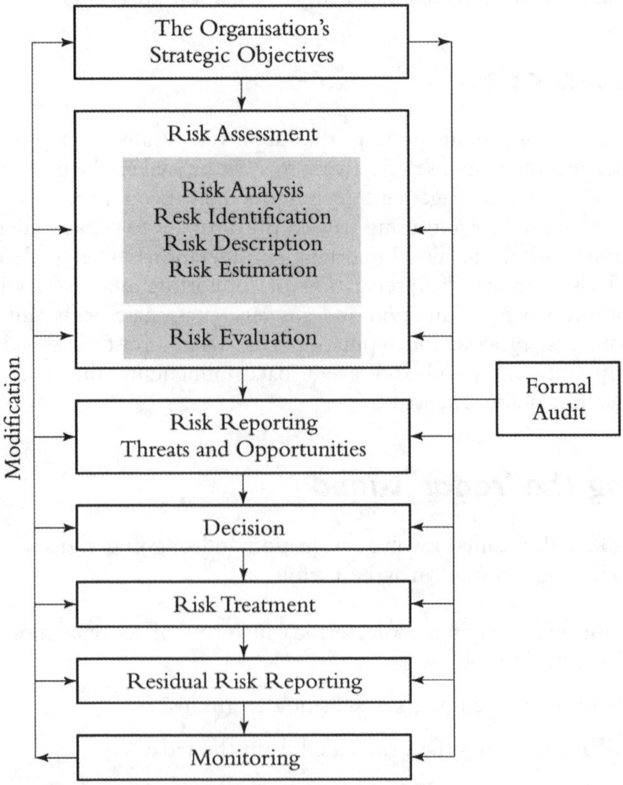

The Organisation's Strategic Objectives

Risk Assessment

Risk Analysis
Resk Identification
Risk Description
Risk Estimation

Risk Evaluation

Risk Reporting
Threats and Opportunities

Decision

Risk Treatment

Residual Risk Reporting

Monitoring

Modification

Formal Audit

Reproduced with the kind permission of the Institute of Risk Management, the Association of Insurance and Risk Managers and the Association of Local Authority Risk Managers.

The above diagram, taken from the UK *Risk Management Standard*[1], represents the suggested risk management process and broadly corresponds to the six steps to successful reputation risk management proposes by J. Larkin. Indicating that reputation risk can be managed just as any other risk facing organisations today.

Reputation risk management, the Turnbull Report and the new Combined Code 7.13

Following a series of relatively recent high profile corporate disasters such as Enron, Worldcom, Maxwell, BCCI, etc. and controversy over directors' pay, the general trend towards risk awareness as an integral part of corporate governance has been supported by such guidelines as the *Cadbury Report* (1991)[18], the *Greenbury Report* (1995), the *Hampel Review* (1998), the *FSA Listing Rules,* the *Turnbull Guidelines* (1999), the *Higgs Report* (2003)[19] and the *Smith Report* (2002)[20] that were all incorporated into the new *Combined Code on Corporate Governance* in July 2003. All of these recognise the importance of risk management in safeguarding and enhancing the investments of shareholders as an integral part of corporate governance. The relevant sections of the *Cadbury Report*[18] states that:

> '4.23 The basic procedural requirements are that the board should meet regularly, with due notice of the issues to be discussed supported by the necessary paperwork, and should record its conclusions. .. Boards should have a formal schedule of matters specifically reserved to them for their collective decision, to ensure that the direction and control of the company remains firmly in their hands and as a safeguard against misjudgments and possible illegal practices. A schedule of these matters should be given to directors on appointment and should be kept up to date.
>
> 4.24 ...such a schedule would at least include: risk management policies
>
> Boards should lay down rules to determine materiality for any transaction ...
>
> 4.31 Directors are responsible under *s 221* of the *Companies Act 1985* for maintaining adequate accounting records. To meet these responsibilities directors need in practice to maintain a system of internal control over the financial management of the company, including procedures designed to minimise the risk of fraud. There is, therefore, already an implicit requirement on directors to ensure that a proper system of internal control is in place.'

The Hampel Review (1998) states that the board should maintain a sound system of internal control to safeguard shareholders' investments and the company's assets not only by means of financial controls, but also risk management. The above principles have also been confirmed by the *FSA Listing Rules*, which require listed companies to supply the following information in their annual reports and accounts:

> '12.43A(a) a narrative statement of how it has applied the principles set out in Section 1 of the Combined Code, providing explanation which enables its shareholders to evaluate how the principles have been applied.'

Company directors increasingly have a corporate responsibility to identify risk and to ensure that adequate risk mitigation measures are in place. The *Turnbull Report* requires company directors of Stock Exchange listed companies to demonstrate a sound system of internal control to safeguard shareholders' investments and company assets. The Report recommends a regular programme to assess and evaluate all risks to the business – both internal and external. Historically internal audit procedures have concentrated almost exclusively upon confirmation of financial veracity, a position that is then verified by external auditors. The *Turnbull Report*, however, defines business risk in a much wider sense and requires that a company should consider and monitor its overall exposure to risk. Reputation is now so important that the *Turnbull Report* on corporate governance advises companies to treat it in the same way as all other assets.

The Turnbull Guidelines (1999) were produced to provide guidance to companies on what is required to comply with the *Combined Code* (1998):

> '17. In determining its policies with regard to internal control, and thereby assessing what constitutes a sound system of internal control in the particular circumstances of the company, the board's deliberations should include consideration of the following factors:
>
> - the nature and extent of the risks facing the company;
> - the extent and categories of risk which it regards as acceptable for the company to bear;
> - the likelihood of the risks concerned materialising;
> - the company's ability to reduce the incidence and impact on the business of risks that do materialise; and
> - the costs of operating particular controls relative to the benefit thereby obtained in managing the related risks.'

The above document also states that the board's statement on internal control should disclose that there is an on-going process for identifying, evaluating, and managing significant risks. The Appendix of the *Turnbull Guidelines* provides a list of possible questions that should be asked in order to assess the effectiveness of the company's risk control process.

Questions related to risk assessment include:

- does the company have clear objectives and have they been communicated so as to provide effective direction to employees on risk assessment and control issues?
- are the significant internal and external operational, financial, compliance and other risks identified and assessed on an ongoing basis? (Significant risks may, for example, include those related to market, credit, liquidity, technological, legal, health, safety and environmental, reputation, and business probity issues.)

Questions related to control environment and control activities include:

- does the board have clear strategies for dealing with the significant risks that have been identified? Is there a policy on how to manage these risks?
- does the company's culture, code of conduct, human resource policies and performance reward systems support the business objectives and risk management and internal control system?
- are authority, responsibility and accountability defined clearly such that decisions are made and actions taken by the appropriate people? Are the decisions and actions of different parts of the company appropriately co-ordinated?
- does the company communicate to its employees what is expected of them and the scope of their freedom to act? This may apply to areas such as customer relations; service levels for both internal and outsourced activities; health, safety and environmental protection; security of tangible and intangible assets; business continuity issues; expenditure matters; accounting; and financial and other reporting.

Questions related to information and communication include:

- do management and the board receive timely, relevant and reliable reports on progress against business objectives and the related risks that provide them with the information, from inside and outside the company, needed for decision-making and management review purposes? This could include performance reports and indicators of change, together with qualitative information such as on customer satisfaction, employee attitudes, etc.

Questions related to monitoring include:

- are there ongoing processes embedded within the company's overall business operations, and addressed by senior management, which monitor the effective application of the policies, processes and activities related to internal control and risk management?
- do these processes monitor the company's ability to re-evaluate risks and adjust controls effectively in response to changes in its objectives, its business, and its external environment?

- are there specific arrangements for management monitoring and reporting to the board on risk and control matters of particular importance? These could include, for example, actual or suspected fraud and other illegal or irregular acts, or matters that could adversely affect the company's reputation or financial position?

The above questions do not purely relate to reputation risk management but to risk management generally as an integral part of corporate governance. However, the recognition of reputational risk issues mentioned earlier is clearly apparent.

Some argue that following the recommendations of the *Turnbull Report* regarding boardroom responsibility for risk management and accountability for intangible assets such as reputation; and the Higgs Review of corporate governance recommending a more active and independent role for non-executive directors (NEDs), NEDs should be appointed as reputation guardians for the corporation – in much the same way as they now sit on audit, nomination and remuneration committees. This recommendation is given added weight by the proposed implementation of the Operating and Financial Review (OFR) in 2004, which, seeks to embed reputational enhancement and protection in corporate reporting. NEDs will need a powerful strategic management tool (stakeholder audits) to ensure that they can perform this task properly. They will also need the support of an experienced communications function, which will be critical in conducting, interpreting and advising on appropriate courses of action arising from the stakeholder audit.[11]

The OFR is designed to give shareholders the information that they need to make informed decisions about the companies that they invest in. The OFR is a narrative report by listed companies setting out the principal drivers of a company's performance. It should cover the issues traditionally seen as key to a company's performance – an account of its business, objectives and strategy, a review of developments over the past year, and a description of the main risks. However, it will also cover issues relating to the environment, employees, customers or social and community issues where that information is important for an assessment of the company. This will make reporting on these 'intangible' issues directly impacting corporate reputation potentially compulsory in January 2005.

References

1. The Institute of Risk Management. *The Risk Management Standard*. 2002.

2. The Economist, et al.*Brands and Branding*, Economist Books, 2003

3. Brotzen, D., 'Wise Words and Firm resolve when Times Get Tough' in Pickford, J. (ed), *Financial Times Mastering Risk Volume 1: Concepts*, FT Prentice Hall. 2001.

4. Business Week.'The 100 Top Brands', league table, *Business Week*, 4 August 2003.

5. Clifton, R., 'Editorial: Brands and Our Times', in *Brand Management*, Vol.9, No.3, pp.157–161, January 2002, Henry Stewart Publications. 2002.

6. Department for Trade and Industry, *Company Law: Draft Regulations on the Operating and Financial Review and Directors' Report: A Consultative Document*, May 2004.

7. Fombrun, C., Gardberg, N., & Sever, J., 'The Reputation Quotient: A multi-stakeholder measure of corporate reputation' *Journal of Brand Management*, Volume 7, p.241–255. 2000.

8. Interbrand Insights, *Aligning Your Organization and Your Brand for Performance*, No.1, March 2001, Interbrand.

9. Knight, R., Pretty, D., 'Day of Judgement: Catastrophe and the Share Price', in Pickford, J. (ed), *Financial Times Mastering Risk Volume 1: Concepts,* FT Prentice Hall. 2001.

10. Larkin, J., *Strategic Reputation Risk Management*, Palgrave Macmillan, 2003.

11. Murray, K., 'Reputation – Managing the single greatest risk facing business today', in *Journal of Communication Management*, November 2003, Vol.8, No.2, p.142–149, 2003.

12. Schmitt, B., 'Branding Puts a High Value on Reputation Management' in Pickford, J. (ed), *Financial Times Mastering Risk Volume 1: Concepts*, FT Prentice Hall. 2001.

13. Stone, H, Washington-Smith, J, *Profit and the Environment: Common Sense or Contradiction?*, John Wiley & Sons Ltd, 2002.

14. Survey of Risk, *The Value of Strategic Risk Management: Learning From Best Practice*, March, 2004.

15. *The Aon European Risk Management & Insurance Survey 2002–2003.* Extracts available on: http://www.aon.com/about/publications/issues/2003_uk_biennial_survey.jsp (06 June 2004).

16. *The Analyst: A Perspective on Business Risk*, Number One, October 2003, Aon.

17. The Financial Reporting Council, *The Combined Code on Corporate Governance*, July 2003.

18. Committee on the Financial Aspects of Corporate Governance. *Report with Code of Best Practice,* [Cadbury Report], London: Gee Publishing. 1992.

19. MORI Social Research Unit. *Review of the Role and Effectiveness of Non-Executive Directors*. (The Higgs Report). 2003.

20. Financial Reporting Council. *Audit Committees – Combined Code Guidance.* (The Smith Report). 2002.

21. Institute of Chartered Accountants. *Internal Control: Guidance for Directors on the Combined Code.* (The Turnbull Report.) 1999.

22. Deloitte & Touche. *From Carrots to Sticks: A Survey of Narrative Reporting in Annual Reports*, p.1, 2003.

Chapter 8
Cultural Due Diligence

Introduction 8.1

There are many lengthy publications, manuals and books that address the vital matter of corporate culture. In this chapter selected aspects will be mentioned and prioritised in accordance with the theme and discussion of due diligence, risk management and corporate governance as explained in **CHAPTER 1**.

Having regard to the trends in favour of outsourcing it is important that businesses should make it clear whether or not this is part of their corporate culture. If they do not do so misunderstanding within business relationships can arise. Indeed the approach of an organisation to, for example, diversity, gender, employment practice, the environment, health and safety and other key issues can all form part of the corporate culture. It should also be noted that much of what has been covered in most of the earlier chapters could be described in terms of ongoing cultural due diligence.

It has already been clearly shown that the success or failure of a business depends very much on the quality of the decisions it makes. Decisions are made by everyone within the organisation at all levels and range from those involving strategic direction, major investments or acquisitions to those evidently simple tactical judgments made on the shop floor of the business. It has already been seen that a brand is more than a product – it is a promise. No matter how globally aware consumers become they will always want to carry out business with brands they know and trust – brands that fulfil an actual or implicit promise made by the company. The key to the promise is the most powerful word in branding today: trust.

Building and nurturing brand trust have never been more important than in the current climate of suspicion of corporate behaviour. For example, in 2004 in the retailing sector Gap Inc was developing strategies to fend off any sweatshop image in its use of factories in places such as China, Taiwan, Saipan, Cambodia and sub-Saharan Africa by spending resources on training and helping factories develop their own compliance programmes. Gap is working with non-governmental organisations (NGOs) and local authorities on labour policies. Meanwhile in the banking sector Merril Lynch has been accused of having a culture of sexual discrimination. On the other hand the oil giant BP has announced its commitment to diversity that has its roots in a larger cultural shift. The CEO has stated that he intends to create a positive culture of 'inclusiveness'. In addition, competing marketing messages, brand pollution and consumers' propensity to shop around for the best price mean that familiarity and brand awareness are ever more precious. They are critical to building and maintaining sustainable success. The approach to all of this reflects the culture of the business.

A managed risk culture 8.2

All decisions present different levels of risk: sometimes the risk will be obvious and the decision maker will consider the risk formally or intuitively. At other times the risk will not be so evident and therefore will not be taken into account or prioritised. Clearly in the context of cultural due diligence one challenge is how to get risk management integrated within the decision making process so that risk is considered as a normal part of business and its internal due diligence. In this discussion of cultural due diligence it is intended to focus on the importance of risk management being part of that culture to enable successful business alliances. Creating a culture with the objective of 'no surprises' does not mean that the company becomes risk averse. Rather, it is free to take risky decisions with comprehensive and intelligent knowledge of what the risks are.

According to many risk managers a managed risk culture can be defined as creating an environment that:

- enables people to take more effective decisions;
- allows risks to be fully understood so that calculated risks can be taken; and
- encourages employees to consider the consequences of decisions and actions that they take.

It is true to say that such a risk management culture can be positive for the generally perceived culture of the business, as well as its brand/s and its reputation.

As has been discussed in the discussion of reputation in **CHAPTER 7**, a risk management tool is of particular use to the director of corporate communications because it provides a comprehensive and scientific basis on which to argue for increased focus on reputation management strategies. Given the constituent parts of the overall reputation story – brand, vision, values, media, public affairs and public policy, compliance, governance, regulation, corporate responsibility – a framework for improved co-ordination of the reputation management strategy is important. The tool is also valuable to the risk director since it provides hard data for inclusion in the company risk register, where it is possible that no entries have been made before, either in terms of categorisation or value. It will also become vital to the company secretary for operating and financial review (OFR)-compliance reasons referred to in **CHAPTER 10**. A value for intangible assets, and related values for the component parts, is of tremendous importance in today's business climate. The brand and the culture of a business are intangible assets that are of utmost value, yet rarely taken into account. Taken together, the increased visibility which results from the valuing of intangible assets, and the improved understanding of what is necessary to control, mitigate and eliminate the risks associated with a decline in asset values, should lead to strengthened risk management, reputation management and financial modelling procedures. The innovative approach that takes account of intangible assets beyond goodwill alone also presents the opportunity to refresh and redefine key parts of business process such as:

- improved techniques for risk and reputation management;
- the mitigation of threats/disaster recovery planning and crisis management;
- the supply value chain and other partnerships;
- alignment and incentives for all betterment initiatives including good corporate governance.

More powerful applications of the tool relate to issues such as:

- the implications for strategy;
- articulating and communicating the corporate character and 'personality' of the company;
- stakeholder engagement and management by the alignment of internal behaviours and external expectations; and
- change management programmes (see **8.3**).

Whereas **CHAPTER 7** has discussed brand and reputational risk issues in more detail here the connection between brand and corporate culture should be noted. In this respect, broadly speaking a brand is regarded as a particular product or a characteristic that identifies a particular producer. To be a successful brand one of the most important issues facing the Board – and the marketing director specifically – is to establish a customer-driven corporate culture. As has been noted above companies can decide upon positive strategies as regards their culture. This is particularly important when they are acting globally and deal with external concerns such as location, which involves language, regional and other cultural differences. Even the regulatory culture has to be considered, and flexible internal cultural policies implemented accordingly, along with some demonstration of, for example, a sensitive corporate social responsibility strategy (CSR) strategy along the lines proposed below. This is also clear from the study of outsourcing. Cultures can motivate and stimulate companies and increase productivity and profit. A positive culture – especially one that espouses CSR and many of the features described in the context of enlightened corporate governance – can be invaluable.

Creating a sustainable marketing culture has been described as one of the hardest tasks facing business. It requires:

- proper training and education;
- sharing good practice;
- awareness and commitment on all levels; and
- excellent corporate communications.

Marketers of a business also have to think and act as 'brand champions'. Brand strategies include consideration of the following questions:

- has a clear philosophy on the corporate brand evolved?
- has there been a thorough investigation of the brand portfolio?
- has it been established which brand needs boosting – perhaps by buying brands – which brand requires development and which should be pruned or sold?
- has the management of the brands been incorporated into the organisational structure?
- is there a logical system for brand naming?
- has the company tried to establish a financial value for its brands?; and
- have the local nuances in new markets been thoroughly investigated to try to decide which brands can travel?

In terms of corporate culture certain questions can form an initial checklist that can impact on several departments.

1. Is there a defined strategy for changing the corporate culture?
2. Has a corporate identity programme been carried out in place of noticeable corporate cultural change?
3. Is there real communication in the organisation or do employees tell superiors what they want to hear?
4. What is the first impression given by reception when calls are answered?
5. What is the role of the marketing director?
6. What is the role of marketing beyond spending on advertising?
7. Has restructuring been considered to integrate marketing with sales?
8. Do the various departments know what the others do?
9. Is there a philosophy of moving staff around to perform different functions?

Change in culture must be focused on:

- employees;

- leadership style; and
- organisational processes and functions.

Change management **8.3**

This is another vital aspect of corporate culture that requires ongoing monitoring in the context of ongoing due diligence and good corporate governance. As has been emphasised in earlier chapters change has become the only real constant in everyday business life. All kinds of changes take place: social, technological, political, economic and cultural. These changes develop at an accelerating pace. Every day we witness different products, new markets and different customers. Some of the resulting significant challenges have been referred to as the five Cs:

- competition;
- change management;
- complexity of business;
- control needs; and
- creativity.

The business has to try to cope with the changes by establishing:

- new organisational structures;
- re-engineering;
- re-structuring;
- alliances;
- mergers;
- hostile take-overs;
- acquisitions;
- joint ventures; and
- other potential arrangements.

As has been seen above, changes in the organisation and its priorities are becoming an integral part of the global business scenario. The business has to be alive to all changes to meet them effectively. Management has to anticipate and meet effectively the challenges in order to align their business with changing markets and customer needs and implement this proactive approach into its culture. In this context it is helpful to consider the role of what are often described as strategic alliances.

Strategic alliances 8.4

Bearing in mind the issues of brand and culture that have developed over the last few decades, the business world has witnessed many strategic alliances as a way to:

- secure competitive advantage;
- share costs;
- leapfrog into new markets; or
- protect existing ones.

Such strategic alliances can appear in a variety of forms, from full-blown mergers and acquisitions or joint ventures to co-operation agreements in areas such as:

- licensing;
- technology and research, technology and development agreements;
- long-term buyer-seller agreements; and
- market alliances.

They can provide positive advantages, by helping companies to:

- gain competitive advantage with less risk and expense than going it alone;
- gain economies of scale from partners in the same sector;
- share development costs;
- swap technical know-how;
- gain a wider presence – this being useful when takeover targets are scarce;
- overcome cultural and language barriers;
- enter complementary product lines; and
- enlarge distribution reach.

In this respect too change management is another significant area that requires ongoing business focus. Not only should business therefore develop a proactive culture as part of its ongoing cultural due diligence, but also it should seek a similar approach in the area of potential business partners and business transactions.

Culture clash in mergers and acquisitions: risk mitigation 8.5

In CHAPTER 1's consideration of transactional and operational concerns some comment was made regarding the high failure rate of merger and acquisition that occurs post-completion. This is the case in spite of the traditional due diligence processes. The research data on why some 55–77 per cent of mergers and acquisitions fail in meeting their intended results is absolutely clear. This is the case in many parts of the world. Many experts have found that the failures are overwhelmingly attributable to 'culture clash' that occurs as attempts to bring the two organisations together are made. The problem is that this makes merging the two organisational cultures or establishing a new culture for the merged organisation extremely difficult, if not impossible. The inevitable result is the expensive demerger. In this connection it is important to understand how to avoid the culture clash and what to do post-merger when expected results have not occurred.

In those instances where the organisations are merged, the ongoing direct and indirect costs of unresolved 'culture clash' issues are high, and require the merged organisation to focus on internal issues and problems rather than on the market place, the customers and the competition. In this connection a full and appropriate cultural due diligence exercise can make a huge difference.

The Cultural Due Diligence process in mergers and acquisitions 8.6

As has been emphasised above the cultural direction of managers and their decisions are noted by stakeholders, both direct and indirect. It is important to bear in mind that organisational culture is critical to organisational effectiveness. Also, managing the culture is vital to successful business operation. This is particularly important in times of large-scale change, such as an acquisition or alliance. Fiduciary responsibility and due diligence require careful examination of the cultural aspects of any acquisition as a major component of the ability to actually run the operation and achieve the potential synergies.

The Cultural Due Diligence (CDD) process can be described as a systemic, systematic and research-based methodology for significantly increasing the odds of success of mergers, acquisitions and alliances. Until-recently it has been an overlooked parallel process to the traditional financial and legal due diligence that is considered absolutely essential to any merger or acquisition (see comments on environmental due diligence by way of comparison in chapter **16.1**). As with other forms of due diligence the CDD process is proactive and problem solving in advance of the transaction, the merger and acquisition or deal.

By assessing the characteristics of both organisations' cultures as soon as possible in the merger process, potential culture clash problems can be predicted, prioritised and focused on in a comprehensive Cultural Integration Plan. Such a plan will guide the integration of the two cultures, or the building of a new culture for the merged organisation in full consideration of the cultural issues and 'landmines' that are a part of the terrain. Bearing in mind the failure rate and the costs involved in every way, CDD is at least as vital and necessary as traditional legal and financial due diligence in providing an informed basis for executive decision making and planning, and perhaps more so in increasing the odds of success of the merger or acquisition.

CDD involves sound leadership and management. It offers decision makers in both organisations:

- comprehensive, data-based predictions of culture clash problems that will occur within the merger process;
- the relative priority of those problems; and
- recommendations on how to eliminate their cause or minimise their impact before they occur.

The results of the research in this regard are clear: CDD is overlooked at the peril of the success of the merger or acquisition. It is therefore also important that a business develops a positive proactive culture as part of its ongoing due diligence and corporate governance practice before any transactional issues are raised.

As in the case of any sound organisational research the CDD process employs both qualitative and quantitative data collection, and includes:

- interviews;
- focus groups;
- workplace observations;
- documentation reviews; and
- web-based CDD surveys.

One approach to analysing and organising CDD data is to group the findings within twelve domains of the CDD process, although the data can be organised around the key elements of the business plan, or in a manner that will be of greatest value to the two organisations.

The CDD model 8.7

When performing CDD during, for example, a merger or acquisition, it is necessary to gather operational and behavioural data on the relevant domains in both organisations; the one acquiring and the one acquired. Once data is collected from both organisations, this can be contrasted and compared, having regard to potential areas of conflict and/or misunderstanding and of synergy and leverage.

The twelve domains of the CDD process 8.8

A brief description of each of the twelve domains follows. These descriptions provide a general sense of each area and are not meant to be definitive. They enable an understanding of the key issues that fulfil the broad requirements of good corporate governance.

The twelve domains generally cover the relevant issues of corporate culture. However, at least two areas commonly mentioned in discussions of corporate culture may appear to be overlooked:

- values and beliefs; and
- myths, legends and heroes.

In actuality, data on these issues are embedded in the twelve domains. By analysing each domain, underlying values and beliefs are uncovered. This is far more effective than simply asking a representative 'what are the values and beliefs around here?'

The same is true of myths, legends and heroes. These are simply the anecdotal versions that give more direct and immediate meaning to the belief systems operating in a company. Myths, legends and heroes will present themselves when the twelve domains are considered, but only if qualitative data gathering techniques are used.

1. Intended direction/results **8.9**

It is important to ascertain, from the top of the organisation to the bottom, what the company intends to accomplish. Key questions that should be raised are:

- what is the business plan about?
- what is the intent and purpose of the organisation?
- what results are expected from the business activity of the organisation? and;
- how are these issues discussed and communicated?

2. Key measures **8.10**

This domain encompasses what the company measures, why and what happens as a result. The key measures demonstrate a great deal about the manner in which the company and its executives and staff are driven, particularly when the consequences for each measure are also considered. A comparison of key measures across the two companies is an important consideration and cultural indicator.

3. Key business drivers **8.11**

This demonstrates how the company views its industry and its subsequent efforts within the industry. If one company defines success in terms of total market share while another defines it as net profit margin, there is considerable room for disagreement around such matters as what actions are appropriate to correct unacceptable results, or decisions on appropriate new product offerings. Questions to be raised include:

- what are the primary issues driving the business strategy?
- is the focus on competitive edge? and if so, how is that defined, eg by price differentiation, quality, market share, service, reliability, etc

4. Infrastructure 8.12

Under this heading relevant questions are:

1. How is the company organised?
2. What is the nature of the reporting relationships?
3. How do the staff systems interface with the line systems? and
4. What is the nature of the relationship between groups and units in the organisation?

5. Organisational policies and practices 8.13

This domain raises such issues as:

1. What formal and informal systems are in place?
2. What part do they play in the daily operations of the business?
3. How much flexibility is allowed – at what levels and in which systems?
4. What is the relationship between political reality and business reality?

6. Leadership/management practices 8.14

While there are clear behavioural differences between management and leadership functions, both are clearly important in running a successful business. The issue is around which approach is predominant in each area/ department of each company. This domain relates primarily to the middle management group but has obvious impact on the next area. Questions to consider are listed below.

1. What is the balance between leadership and management approaches with staff?
2. What basic employee value systems, such as the degree of supervision or delegation, are in place?
3. How are employees treated and why?
4. How is the business plan implemented through the management system?

5. How are decisions made?
6. Who is involved in what, and when?

7. Supervisory practices **8.15**

Supervisory practices have a major impact on employees and their feelings about the company and the work they do. The nature of the interaction between the employee and their immediate supervisor is one of the primary tone-setters for the culture of the company. The main issue is what is the degree of delegation, level of trust and responsibility between the supervisor and the employee?

8. Work practices **8.16**

The manner of work practices is also a vital area of concern, especially considering overseas cultures and their differing practices. Sample questions are:

- how is the actual work performed?
- is the emphasis on individual responsibility or group responsibility?
- what degree of control, if any, does the individual worker have on the workflow, quality, rate, tools utilised and supplies needed?

9. Technology utilisation **8.17**

This is an important domain both in relation to internal systems and equipment, as well as the services and products provided to customers. (see also **CHAPTER 9**). Key questions are:

- how current is the technology being utilised?
- what are people used for in relation to technological support/resources?

10. Physical environment **8.18**

Many aspects of the physical environment can have a bearing on how people feel about work and the company. Changes in these areas, particularly if it is perceived as arbitrary, can result in bad feelings for years. Useful questions are:

- how do the workplace settings differ?
- what are the differences between:
  - open work spaces versus private offices;
  - high security versus open access, buildings, furniture, grounds?

11. Perceptions/expectations **8.19**

This as another crucial domain affecting governance of the business. Questions for consideration are:

- how do people expect things to happen?
- what do they think is important?
- what do they *think* should be important, versus what they believe the company feels is important?

12. Cultural indicators/artifacts **8.20**

This domain covers the following key issues:

- how do people dress and address each other?
- what is the match between formal work hours and actual hours spent working?
- what company-sponsored activities exist and what are they like?

CDD deliverables **8.21**

All of the data collected by means of the CDD process is carefully analysed and organised into a number of extremely valuable management tools to be used by executive and senior management in planning the integration of the two organisational cultures into a desired new culture of the merged organisation. These tools include:

- detailed cultural profiles of both organisations;
- perceptions of various departments of both organisations;
- assessments concerning the current culture and the merger;
- specification of cultural similarities within the twelve cultural domains;
- specification of cultural differences within the twelve cultural domains;
- prediction, specification and prioritisation of 'culture clash' problems and their impact on the merger;
- specific recommendations on avoidance and/or minimisation of culture clash problems;
- integrated road map for implementation of recommendations.

Given this information and these management tools, key decisions can be made early in the merger process that will:

- minimise culture clash problems;

- facilitate the optimum integration of the two cultures; and
- greatly increase the probability of success of the merger.

CDD and other activities 8.22

It should be noted that the points raised above in the discussion of the CDD process in conjunction with mergers and acquisitions are also relevant to a discussion of cultural due diligence relating to an organisation's ongoing operations. Also the points raised are discussed in other chapters in relation to best practice and corporate governance.

Appendix I

Outsourcing and offshoring – the Indian case study

Introduction

Today, the world is talking about the 'Indian BPO (business process outsourcing) success story' – a story of an industry that is growing at a phenomenal pace of 59 per cent year-on-year and contributing nearly a quarter to the total IT exports revenue. Many strategy experts attribute this growth to the inherent advantages that India possesses, for example low cost base, vast English speaking population and a highly skilled manpower. Undoubtedly, cost savings due to labour arbitrage continues to be one of the most compelling factors for offshoring processes to India. In addition to this, rising vendor sophistication, improved process and project management skills and quality standards have given India a pre-eminent position in the $US 275bn global outsourcing market. However, the real challenge before India lies in its ability to maintain its strengths and build newer capabilities to tackle increasing competition from countries like Ireland, China and Israel, for example.

With the objective to chart a strategic road map for unhindered growth of this industry and to gauge the perception of Indian BPO service providers on various issues, ASSOCHAM (The Associated Chambers of Commerce and Industry of India – one of the leading Indian corporate representative bodies) conducted the 'First BPO Industry Confidence Survey' in the months of May–June, 2003. Coincidentally, this was also the period when domestic and international media highlighted the outsourcing backlash in the US, Australia and European countries and the possible impact it could have on the Indian BPO industry. This nation-wide survey evinced a huge response, especially from the BPO small and medium sized enterprises (SMEs), and gives some view from the India perspective while so much media coverage has dealt with the issues form the developed world. In India, leading companies like eFunds International, Bhilwara Infotech, 24/7 Customer Access, Transworks, HCL

BPO, ITI Limited and Indigo Lever also participated in this opinion poll. Overall, 160 Chief Executive Officers (CEOs) responded to this survey.

Research methodology

In planning this survey, the ASSOCHAM BPO research team implicitly accepted the potential, opportunity and drivers of the trend to outsource processes to India. The questionnaire survey was specifically modelled to study the underlying issues that could become future threats for this industry. Accordingly, the questions covered the following aspects:

- short and long-term impact of:
  - slow pace of regulatory reforms;
  - infrastructure bottlenecks;
  - lack of streamlined approach for branding Indian BPO services;
  - increasing competition from other countries with similar capabilities;
  - increasing service level requirements;
  - geo-political situation in sub-continent;
  - rising attrition rates/other HR issues; and
  - US and European backlash against outsourcing;
- level of satisfaction on the HR and telecom infrastructure in India;
- acceptability of the government policies/initiatives in this field;
- future direction of the government's BPO agenda; and
- overall confidence on the future prospects of this industry.

Findings of the survey

Short and long term impact analysis

Slow pace of regulatory reforms: the members of ASSOCHAM BPO Steering Committee have recognised regulatory issues as one of the major concerns for Indian companies in this area. While the government has increased support to the IT/ITES (information technology enabled services) industry, many industry experts point to the tardy pace of these reforms.

Industry opinion:

- short to medium term (0–2 years) – moderate impact (50 per cent); and
- long term (2–5 years) – serious impact (54 per cent).

The survey clearly showed that industry is concerned over the long-term implications of slow pace of regulatory reforms. Broadly, the policies/acts pertinent for this industry are:

- *Information Technology Act 2000;*
- Software Technology Park policy;
- Foreign investment policy;
- Venture Capital investment policy;
- overseas investment policies (mergers and acquisitions, American depository receipts/global depository receipts (ADR/GDR), remittance of profit, for example); and
- other fiscal incentives;

The government in consultation with industry associations should chalk out a time-bound plan that covers the various loopholes in the above-mentioned policies.

Infrastructure bottlenecks

Availability and reliability of Infrastructure facilities is still cause of serious concern. However, the government is attempting to strengthen telecom infrastructure and build fiberoptic networks in city centres of software activity, as well as providing uninterrupted power supply. Much of these efforts have drastically improved power availability and telecom density.

Industry opinion:

- short to medium term (0–2 years) – serious impact (86 per cent); and
- long term (2–5 years) – very serious impact (47 per cent).

An overwhelming 86 per cent of the respondents view infrastructure as a serious cause of concern in the short term. In the long term, majority felt that poor infrastructure would have a very serious impact.

Lack of streamlined approach for branding Indian BPO services

Industry experts have often suggested that India should have an 'umbrella brand' for IT and ITES industry. The marketing of India as a BPO destination until now has been mostly at the behest of service providers themselves. The industry-government partnership in this field needs to be strengthened. Currently there is lack of a streamlined approach on branding Indian BPO abroad.

Data protection laws in India

One of the key aspects of outsourcing relate to the resolution of any concern that business partners should ensure that the culture is honoured. By way of example the date protection laws of India must be understood. Currently there is no statutory legislation in relation to data protection in India. Unlike the UK/EU countries, India has no codified data protection act setting out the rules for processing personal information applicable to paper records as well as those held electronically.

The issue of privacy has, in a very limited way, been addressed under the *Information Technology Act of 2000* (*IT Act*).The *IT Act* provides protection from unauthorised disclosure of information by person who has secured access to such information in pursuance of powers conferred under the *IT Act*. The provision is also limited to information accessed and passed on 'without the consent of the person' who the information relates to. The provision is thus extremely narrow in its application.

There is, therefore, a growing concern over the absence of data protection laws in India and the impact inadequate legal protection will have on personal data being transferred to India from the western world. Such concern (together with a host of other reasons) is also hindering Indian BPO companies from gaining lucrative contracts in certain key segments such as government tenders and contracts. The Indian government has, therefore, been facing tremendous pressure to enact an appropriate data protection law to ward off any adverse impact on the Indian BPO industry.

Today, the largest portion of BPO work coming to India is low-end call centre and data processing work. If India has to exploit the full potential of the outsourcing opportunity, then it has to move up the value chain. Outsourced work in intellectual property rights intensive areas such as clinical research, engineering design and legal research is the way ahead for Indian BPO companies. In the absence of adequate data protection laws, Indian BPO outfits are finding it difficult to move up the value chain and risk being relegated to doing low end work like billing, insurance claims processing and of course transcription. Accordingly, the move up the value chain for Indian BPO entities will greatly be facilitated by enactment of an appropriate data protection legislation.

Further, with the globalisation of business and ease of transfer of information, the requirement to protect the personal/sensitive data of Indians as a diaspora is also being felt. Today Indians who use mobile phones are regularly being inundated with unwanted advertisements, their email boxes are being flooded with unsolicited data and their personal information is being freely traded/ exploited by persons who possess it. Given the lack of an adequate data protection regime there is very little that an ordinary Indian citizen can do to prevent such exploitation.

Methods adopted to satisfy concerns over data transfer

While the absence of adequate data protection laws in India is a serious deterrent, for data flowing into the country in connection with the offshoring activities being carried out, Indian BPO outfits have alleviated concerns of their customers by attempting to adhere to major US and European regulations either by contracting to adhere to such regulations or by obtaining appropriate certifications.

Most Tier I BPO companies today have certifications that comply with regulations like the *Sarbanes Oxley Act of 2002* (*SOX*), *Safe Harbor Principles of 2000*, Financial Modernisation Act 1999 (Gramm-Leach-Bliley Act (GLBA)), the *Fair Debt Collection Practices Act* (*FDCPA*), OCC regulations for banking and the *Health Insurance Portability and Accountability Act of 1996* (*HIPAA 1996*) for healthcare. While most laws and certifications are oriented around verticals, there are laws like the UK *Data Protection Act 1998* (*DPA 1998*) and the US *SOX*, which are laws for data security across different industries. The requirements of such horizontally oriented legislations are overcome by contractual arrangements between the BPO vendor and their customers.

Forthcoming changes

A recent study has found that more than 40 countries around the world have enacted, or are preparing to enact, laws that protect the privacy and integrity of personal consumer data. India is not, however, one of them. Given the above stated concerns, enactment of appropriate legislation in other countries and need for an adequate data protection regime being felt, it is becoming extremely important for India enact the same if it desires to continue its leading position as the preferred offshoring destination.

Recognising this need, since the turn of the century there has been talk of India enacting some sort of data protection regime. The first such action was initiated by NASSCOM (National Association of Software and Service Companies – the apex technology body of India) when the trade body tried to push through a drafting exercise but it appears that the exercise has not been pursued further.

Recently, there has been talk of the Indian government recognising the urgency of the situation and to hasten the process, is considering an amendment to the *IT Act*. It is still unclear what the exact shape of the final legislation will be, however, it is expected that such legislation will be comprehensive. The amendment is expected to come into force later this year or early 2005. However, whether or not such a legislation is introduced in the Indian parliament or the shape such legislation will take will depend on the result of the on going elections in India, result of the pending legislations against outsourcing in the US and pressure by the EU to enact the same.

It is very important to understand the fundamental purpose of any data protection law and existing regulations in other countries in order to comment on the shape that any proposed Indian legislation would take. The driving force, worldwide, for the enactment of data protection laws has been the protection of personal data and information of citizens and providing a framework that facilitates trade and commerce between countries, while not compromising an individual's privacy. There is no reason to conclude that in the Indian context it is likely to be any different – except that the overriding consideration of enactment of a data protection regime is to ensure India maintains its pole position as the preferred destination for offshoring activities.

According to the EU guidelines, EU countries may transfer personal data only after determining that

> 'the third country in question ensures an adequate level of protection'.

It further goes on to provide that the EU shall consider the 'rules of law..in the third country' to make this determination.

The EU guidelines are an outcome of the Organisation for Economic Co-operation and Development (OECD) guidelines of 1980 which has listed eight broad principles to be adhered in protecting personal information of the citizens of the country. They are covered below.

Collection Limitation Principle

There should be limits to the collection of personal data and any such data should be obtained by lawful and fair means and, where appropriate, with the knowledge or consent of the data subject.

Data Quality Principle

Personal data should be relevant to the purposes for which they are to be used, and, to the extent necessary for those purposes, should be accurate, complete and kept up-to-date.

Purpose Specification Principle

The purposes for which personal data are collected should be specified not later than at the time of data collection and the subsequent use limited to the fulfilment of those purposes or such others as are not incompatible with those purposes and as are specified on each occasion of change of purpose.

Use Limitation Principle

Personal data should not be disclosed, made available or otherwise used except:

- with the consent of the data subject; or
- by the authority of law.

Security Safeguards Principle

Personal data should be protected by reasonable security safeguards against such risks as loss or unauthorised access, destruction, use, modification or disclosure of data.

Openness Principle

There should be a general policy of openness about developments, practices and policies with respect to personal data. Means should be readily available of establishing the existence and nature of personal data, and the main purposes of their use, as well as the identity and usual residence of the data controller.

Individual Participation Principle

An individual should have the right:

(i) to obtain from a data controller, or otherwise, confirmation of whether or not the data controller has data relating to him;

(ii) to have communicated to him, data relating to him within a reasonable time; at a charge, if any, that is not excessive; in a reasonable manner and in a form that is readily intelligible to him;

(iii) to be given reasons if a request made under subparagraphs (i) and (ii) is denied, and to be able to challenge such denial; and

(iv) to challenge data relating to him and, if the challenge is successful to have the data erased, rectified, completed or amended.

Accountability Principle

A data controller should be accountable for complying with measures which give effect to the principles stated above.

Given that the overriding concern for the proposed Indian legislation is to facilitate the growth of the nascent BPO industry, the Indian government can be expected to consider the above principles in enacting any legislation for data protection. Further, the predominant issue that the Indian legislation is likely to address is the protection of the foreign nationals' data, if transferred to the Indian sub-continent. Whether or not the proposed amendment to the *IT Act* will also cover data of Indian citizens is yet to be seen.

In view of the above, any Indian legislation is likely put in place a mechanism that will protect against the misuse of personal data that is in one's possession rather than misuse of data that is in one's ownership and the various measures of adequacy that the proposed legislation is likely to address include:

- widening the scope of data protection to include information collected from foreign citizens;
- putting in place an obligation on both the Indian government and others to maintain the integrity of that data and providing for appropriate procedures (eg enforcement authorities) to ensure compliance;
- restricting access to data of foreign citizens by statutory and regulatory authorities; and
- making data theft and misappropriation a criminal offence.

Conclusion

As regards the issue of corporate culture, due diligence and corporate governance there is no doubt that India provides a useful case study – the requirements of good corporate governance largely follow the English pattern. Yet the regulatory framework, and its need for change, must be considered. Having regard to the outsourcing work that is being transferred to India and the growing importance of the BPO industry to the Indian economy there is no doubt that appropriate legislation is required.

The question that remains to be seen is what shape the regulations take and how soon consensus is reached to actually enact the laws that are undoubtedly needed in India.

One way forward would be for business to deal with the issues in a proactive manner that demonstrates a positive approach to CSR in a practical manner as is set out below.

Appendix 2

Bridging the digital divide in India through corporate and social responsibility and the reuse of redundant IT equipment

Corporate and social responsibility (CSR) is fast becoming a mainstream business issue, particularly within the larger companies, as managers and stockholders alike become increasingly concerned about potential risks to a company's reputation, brand and customer-clientele from the mismanagement of environmental and social issues. Businesses are looking for opportunities to demonstrate good corporate citizenship to their stakeholders and to the outside world.

Within the EU, new Directives [2002/96/EC and 2003/108/EC] on waste electrical and electronic equipment (WEEE) are set to require the recovery, reuse and recycling of end-of-life IT products through the imposition of mandatory product take back schemes for these kinds of goods.

Large IT-user organisations, such as major companies and governmental departments, can replace IT equipment in as little as 18 months to two and a half years during times of economic prosperity. The present economic downturn has considerably lengthened this replacement cycle and has led to businesses postponing the decision to replace IT assets; however this will lead to pent up demand for replacement equipment as existing equipment ages and moves ever closer towards obsolescence in terms of meeting current corporate needs. This will result in large volumes of used IT equipment becoming available during the coming years as businesses seek to dispose of and replace outdated equipment. Nonetheless, although such equipment may become obsolescent in terms of corporate usage, if refurbished it can have a useful second life. So the resale of such IT equipment in developing countries is becoming a growing trade.

Original equipment manufacturers (OEMs), as well as retailers and large-scale users, might well be persuaded to view the export of refurbished IT equipment to India as a delivery channel for fulfilment of their product take-back and waste management responsibilities, and for demonstrating both awareness and application of CSR by management and employees within the businesses.

The acquisition of second user IT equipment can serve to help meet the increasing demand for useful IT products in developing nations such as India where average users may not require the latest high specification equipment. On the contrary, the functions which most users need, such as word processing, spreadsheet calculus and use of email and Internet, require relatively low specification equipment.

Extending the useful life of such equipment via resale of IT equipment discarded in Europe to eager users in India can promote sustainability by providing individuals in India with access to the information and know-how available from it, thereby helping to bridge the so-called 'digital divide'. In practice, however, there have been serious problems in instances where equipment supposedly destined for use or recycling abroad is merely being dumped there or sent to dubious recycling facilities with few environmental and health and safety protection measures in place, if any. There is also the challenge of dealing with such equipment when it eventually breaks down again.

Therefore, what is needed is a sustainable second use 'resale supply chain' helping to ensure that second user IT equipment is not merely dumped on arrival and that when it reaches the absolute end of its useful life it can be appropriately recycled (ie reuse of its constituent parts) or safely disposed of.

There is already some existing momentum in the arena of the sustainable use of IT and electronics in India. For example, the Indian Ministry of Information Technology (MIT) partnered by the United Nations Development Programme (UNDP) has already initiated a new programme to:

'disseminate the environmental measures adopted by various agencies in the developed countries to the Indian electronics industry'.

The aim of a CSR project would be to promote this further. It would consist of two components.

1. It should be possible to work with western companies (manufacturers/ large IT users), western IT equipment recyclers, Indian importers/IT resellers and Indian recyclers to establish a sustainable supply chain for the provision of good quality second user IT equipment in India to eager users, particularly educational establishment, with a strong CSR element being retained.
2. Also, a programme to set up one or more centres in India for training workers in handling, refurbishing and recycling used IT equipment in a safe and responsible fashion, whilst learning ancillary skills (eg computer repairs/servicing and the use of software programmes). By training individuals in this way it may also help them obtain jobs in the longer term and even set up new micro businesses (eg in computer servicing and repairs).

A preliminary feasibility study would need to be carried out to first, in particular to ascertain regulatory and policy incentives and obstacles (eg legitimate compliance routes for the WEEE Directives, export/import duties and restriction, local (Indian) environmental, social and health and safety laws), as well as for identifying and securing the participation of appropriate European and Indian businesses, NGOs and other partners needed to make the sustainable supply chain work as intended.

Chapter 9

Information Technology and E-Commerce: Issues of Due Diligence, Risk Management and Corporate Governance

Introduction 9.1

As the business world continues to reel from the corporate scandals on both sides of the Atlantic, companies also have to cope with a stringent new regulatory framework. Compliance with a whole range of new legislation and standards has become a major concern and, increasingly, companies are looking to technologies such as business intelligence to offer some solution. As the discussion in **CHAPTER 11** demonstrates, the *Sarbanes-Oxley Act of 2002* (*SOX*) legislation has provided much of the pressure in the US in that it requires companies to strengthen their internal controls and increase levels of financial disclosure. As noted, while the US framework is primarily aimed at US business SOX will also apply to companies from other countries that are listed on the US stock exchanges. As a matter of corporate governance US companies will expect suppliers to conform to the same standards regarding the retrieval and retention of records. Meanwhile in Europe the new International Financial Reporting Standards (IFRS) come into force on 1 January 2005. IFRS is intended to make it easier to compare the performance of listed companies in different EU countries. Its adoption will require greater transparency from companies' financial reports. For instance, it will become compulsory to account for certain financial instruments at market level. Also the UK's operating and financial review (OFR) requirements, referred to in **CHAPTER 4**, **7** and **10**, must be considered.

While all of this legislation may seem overwhelming the overall intention should be recalled: to enable stakeholders to see financial statements clearly and accurately. The regulations require businesses to be able to produce, when requested, both financial information and records of how decisions were made.

In practice data must be accurate and accessible rapidly so that:

- auditors are able to see a company's financial statements; and
- auditors are able to look in detail at the data that was input into those statements.

The presence of a good business intelligence solution system is becoming increasingly important. It should have a data warehouse that ensures that all reports are based on the same information. It should provide a solution that allows users to follow an audit trail by 'drilling down' from the high level into more detail. For instance commentators have remarked that had Shell been compliant with *SOX* they would not have been able to state that they thought that they had one third more oil reserves than they actually had. One of the requirements of *SOX* is for real-time reporting so that if a major event takes place that has an impact on a company's financial statements then the company must respond quickly by making updated financial reports available. This requires:

- having a very agile infrastructure;
- being able to capture information in relative real time;
- having the tools to assess it very quickly; and
- assessing the implication for the profit-and-loss account or balance sheet.

Some experts warn that many firms implementing solutions are losing sight of the need for a holistic view by considering, for example, the regulations separately. For instance initially the Basel II Convention (which revised the international capital framework and caused enterprise risk management in banking to align adequacy assessment with underlying credit risk, market risk and operational risk) and IFRS requirements on reporting losses were not compatible. Although the issue has been resolved, such conflict demonstrates the need to take a high level view. In addition, by taking a short term view that solves current problems businesses fail to protect themselves against possible merger and acquisition issues. To deal with the constant increase in new demands solutions should provide a coherent view across the organisation and the functions that are in place within it.

Therefore, it is clear that in today's complex business world companies that use the right business intelligence should be more successful in handling all of the issues in a more holistic way. However, it also means that many of the complex concerns related to IT and e-commerce should also be considered as a priority. As a result of the growing reliance upon technology more risks must be managed both as regards the internal use of technology in monitoring business concerns and as regards the handling of external relationships and matters. Many smaller businesses, for example, cannot afford to protect themselves in an ideal way. Therefore in this chapter an overview is provided of some of the developments that have occurred – and current issues – in the context of the IT debate. This is such a broad subject that spans both macro and micro issues (from handling the above requirements to dealing with everyday email nuisance such as spam) that it requires a manual in itself.

Data handling risks

Intercepting communications 9.2

For some time, monitoring communications (such as email, telephone calls and internet use) has been possible without any legal restriction. However, the UK's legal requirements are included in the *Telecommunications (Lawful Business Practice) (Interception of Communications) Regulations 2000 (SI 2000/2699)*. Further guidance is contained in the publication 'Electronic Communications at Work: What you need to know' published by the e-Government Unit and available from www.e-envoy.gov.uk. Further guidance, albeit contradicting some of the advice above, is available in the publication 'The Use of Personal Data in Employer/Employee Relationships', which was issued by the office of the Information Commissioner in October 2000 (available from www.dataprotection.gov.uk).

Whilst this is a controversial area and discrepancies have appeared between the statutory instrument and the guidance published by the Information Commissioner, business must ensure monitoring continues for one of the reasons set out in the statutory instrument:

- to establish the existence of facts;
- to conform with regulatory compliance;
- to determine quality control; and
- to provide training.

Interception has been authorised only if it is for the purpose of monitoring or recording communications that are relevant to the business, and the business has made all reasonable efforts to inform all users that communications may be intercepted. Failure, for instance, to inform or make aware to users other than employees that their communications may be intercepted may cause serious problems in the future. Additionally, getting the agreement of employees to monitor communications may also be difficult. There is a dividing line between an employee accepting that their communications will be monitored and the company informing the employee that they cannot remain an employee (or be employed) unless they agree to the monitoring of their communications.

If a company decides to monitor communications, the board must take an active role in deciding whether to monitor and how to implement the policy, otherwise the company could face action from several sides: employees, customers and suppliers.

Protecting data 9.3

Any failure to deal with personal data under the terms of the *Data Protection Act 1998* (*DPA 1998*) may damage the reputation of many companies in the future. The Information Commissioner has greater powers to enforce the provisions of the *DPA 1998* than previously granted under the *Data Protection Act 1984*. Boards can expect to be put under pressure to treat this matter seriously, especially with the advent of e-business. For instance, in the summer of 2000, PowerGen was required to respond to an alert by a customer who stumbled upon the names, addresses and bank card details of approximately 7,000 customers who paid their electricity, gas and telephone bills online. Several points emerge from this event:

- the company clearly breached their duty of confidentiality to each customer;
- various sources have estimated that this incident cost PowerGen about £35,000 in payments to customers, amongst other expenses;
- the company demonstrated that it was not looking after data properly under the terms of the *DPA 1998*, and
- adverse publicity resulted because of the breach itself, unfounded threats that suggested that the person finding the breach did not do so innocently, and because this incident will be used time and again to remind people that such incidents will occur unless you have a proper security policy in place.

Protecting personal data is part of the wider problems relating to risk management. The risks are both internal (PowerGen seemed to have failed to put adequate security measures in place) and external, which includes hacking, computer viruses and actions of ex-employees (CD Universe, a US company, was hacked by a teenager in 2000 and the personal details of thousands of customers and their credit card details was released on the Internet after the company refused to pay a sum of money). Directors are expected to ensure that personal data is managed properly, which means both technical and organisational measures must be implemented to prevent the unauthorised or unlawful processing of data or the transfer of data outside the European Union.

Online transactions 9.4

Security breaches on the Internet, such as the breach at CD Universe (see **9.3**) are not the only sources of potential embarrassment and liability to a company. The electronic environment has a wide range of risks that directors are only just beginning to be made aware of. The types of attacks by electronic means that have occurred or can occur include:

- the interception of confidential information such as user names or passwords by the use of what has been referred to as a 'packet sniffer';

- hackers obtaining rights of access to a system by hiding behind, for instance, email addresses that the system is configured to trust, this is known as 'IP spoofing';
- attacks on a website by bombarding the server with queries and messages until the server stops working because of the sheer volume, known as 'denial of services attacks';
- attempts to gain access to the system by 'brute force', where the hacker submits repeated login information in an attempt to guess a correct password to gain entry;
- successful access to a server by exploiting known weaknesses in application level software; and
- the use of viruses, worms and 'Trojan horses' to wreak havoc to the system in a variety of other ways.

For some time it has been clear that external threats are not the only risks that can cause extensive damage, as John Leaver of PrestigeCars.com reported in the *Financial Times* in an article published in the 23 October 2001 edition. A rogue piece of code designed to manage the deletion of advertisements caused thousands of new adverts to simultaneously disappear from the website two days before the launch of a marketing campaign. With the help of a dedicated team and extra help, the problem was resolved, but it was a close call for a fledgling company.

A similar internal problem occurred to Ikea's website in 2000. Customers who requested catalogues on the company's website received an error message containing the name of a database. If the customer used the name of the database as a web address, they would obtain access to the names, mailing addresses and telephone numbers of people who had previously requested Ikea catalogues.

These two examples indicate how the protection of personal data and the security and technical framework of the computer system interact. Many commentators cite the perceived lack of security on the Internet as a reason why people do not buy goods and services online. Clearly buying online can be more secure if personal details are transferred by means of encryption. Indeed it does not matter which method is used by the customer to provide personal information, whether by telephone, post, facsimile transmission or by e-mail. Invariably the information is put on a database, which in turn is held on a computer. The key issue is the security that is in place to prevent unauthorised people from obtaining access to the database. If that database can be hacked into from the outside, no matter how the data was obtained, it was never placed in a secure environment from the day it was received. As a result, the risk is within the company as a consequence of poor procedures and inadequate security. In the UK, a hacker can be liable under the terms of the *Computer Misuse Act 1990* for committing a criminal act, but, of course, they cannot be held liable for the poor risk control mechanisms of the organisation.

As has been emphasised, directors of boards are custodians of the company they lead. They have responsibilities for the reputation of the company, and duties to shareholders and stakeholders. In today's business climate it is foolhardy to ignore many of the obvious risks that should be identified, analysed and dealt with in accordance with the adverse affect such a risk could have on the business. Directors should be aware of the benefits of managing risk. If other members of the board refuse to consider the issue seriously, or the culture of the board and senior management does not allow for the problem to be treated with the gravity it deserves, the individual director might, perhaps, look to their own future with the company. There is only so much an individual director can do to counteract a culture or environment that does not acknowledge the responsibilities of the board (see also **CHAPTER 7** and **8**).

There are benefits to managing risk, and any process introduced to the company that seeks to bring risk to the attention of every employee can be regarded as a healthy procedure that can reap positive rewards through better performance and higher sales. Consider the process of risk management as an asset which, when analysed, may well produce higher profits and better internal procedures, thereby helping to bring about a virtuous circle.

Contract risk in e-commerce 9.5

In an e-commerce environment, the survival of a business should include an analysis of contractual obligations and liabilities. By examining the legal issues and understanding its contractual duties and rights, a business will be better placed to ensure valuable legal relationships are not put at risk. This discussion of the issues relating to e-commerce illustrates how such an analysis can be beneficial.

In the e-commerce environment a considerable number of problems can occur if something goes wrong. For instance, if a business operates a web site selling goods or services, the following types of problem can occur.

1. The entire plan for the development of an e-commerce website can be seriously jeopardised if the web designer does not deliver the website on time, or the Internet service provider fails to keep the website up and running when the e-commerce site is generating a substantial and sustained income from regular sales every day.
2. A hacker can cause serious disruption. For instance, the content of the front page could be replaced with racist or pornographic images, sensitive files could be opened and customer credit card details downloaded, or the prices posted on the website altered. Whichever action is adopted by a hacker, the affect on the business would be similar: not only will the company have to devote valuable resources to dealing with the police, national press and the concerns of customers, but there may also be a breach of the provisions of the *DPA 1998*, which could lead to action by the Data Protection Commissioner.

3. Internal failure can occur, as it did for the retailer Argos, when television sets were offered at £2.99 each in the summer of 1999, rather than the full price of £299.99.

When assessing risk, the failure to focus on contractual liabilities can be construed as poor risk management. This is because the business needs to consider what, if any, action it is contractually obliged to take if it faces a disruption. As has been noted in **CHAPTER 5** any disruption to the smooth running of the business can be caused due to factors that are both internal and external. In both instances, where there are written contracts in place or the use of standard trading terms, the contract will invariably include a term relating to *force majeure* or the occurrence of unforeseen circumstances. These terms are drafted to incorporate events that might occur which are beyond the control of either party. The inclusion of such clauses absolves the party affected of carrying out their obligations under the terms of the contract – especially if they are not able to fulfil their contractual responsibilities.

Any business should aim to assess what, if any, potential legal challenges it might face and consider what action to take to avoid the expense and uncertainty of litigation (see also **CHAPTER 5**). In the process of dealing with disruption, the action that it takes and how it deals with the situation may not only affect its ability to trade, but whether its reputation is damaged or enhanced. If the business faces a serious disruption, it is crucial that all trading partners and customers are kept informed. By keeping suppliers and customers up-to-date with a regular flow of relevant news, they will be far more likely to assist by providing technical or financial support.

By looking at its contracts, a business will be able to assess its most important legal relationships. In so doing, it can determine which are crucial to it and, as a result, where relationships are affected by a problem, for whatever reason, it will be in a better position to deal with the crisis.

The information gathering exercise **9.6**

When looking at its contractual obligations, the business will need to consider:

- the rights, obligations and liabilities relating to it; and
- any future inability to fulfil contractual obligations.

In carrying out this exercise, it will inevitably want to focus on the principal areas of its e-commerce operation. However, it is also worth considering the full range of business activities. Identify the areas of of the business that will suffer if something goes wrong. A business only needs one vital supplier to fail to deliver, for instance, and it could find itself in a very difficult position.

The following list, which is not exhaustive, provides some idea of the breadth of contracts that should be taken into account: contracts with customers (if any), suppliers, distributors and wholesalers' utilities Internet service providers and web designers.

Questions relating to suppliers

Parties to the contract 9.7

For each contract the legal identity of the other party should be established. In some instances, a business may have agreed with a web designer to provide it with a website, and may not be aware that they are a sole trader or operate within the identity of a limited company. By establishing the precise identity of the other party, it will be able to assess the risk to the business if, for instance, the web designer fails to produce the website on time or at all (see also **CHAPTER 4**).

When planning for the first website, or for a re-design of the present website, a business will invariably accept that the process will probably take longer than originally agreed. No doubt it will provide for a flexible timetable for the development of its e-commerce site. For instance, it may have established a flexible timetable with suppliers, the warehouse operator and the internet service provider. Although it can resist entering into legally binding agreements for a period, it will have to commit itself at some stage. It may well be that it has to enter such contracts before the website is finished. As a result, any failure to have the website ready on time may lead to rising costs, because of the commitments necessarily entered into for the successful implementation of the website.

In such circumstances, not only should the business ensure that it knows with whom it has entered a contract, but it should also set out to establish a timetable and the consequences of failure to adhere to the timetable. The higher the value of the project, the greater its concern should be in identifying who its future partners are and the greater degree of attention it should give to any trading terms that might apply to the contract.

Pre-contract representations and issues 9.8

Quite often, when a person tries to effect a sale, they make comments to assist a decision to buy. Such comments are usually made orally. They can also be made in writing, for instance on literature provided about the product. In both cases, representations will have been made that might cause the purchaser to enter into the contract.

In the context of IT, it is wise to discuss some of the issues set out below with a number of Internet service providers before deciding which company to place one's business with. Therefore some or all of the following practical issues should be incorporated into a contract with an Internet service provider and steps taken:

- establish the power of the servers;
- ask whether the temperature of the room in which the server is located is temperature controlled;

- seek to negotiate a service level agreement that suits the needs of the business;
- ask the following questions:
  - does the company offer hardware maintenance?
  - does the company provide software that can balance the load?
  - how much spare network capacity can be relied on?
  - if the business website suddenly becomes busy, will the Internet service provider allocate more bandwidth?

While each Internet service provider will discuss the issues raised above (the list is not meant to be exhaustive) and careful attention must be paid to the points below about such representations. The Internet service provider should set out their answers in writing their responses should be included in the terms of the contract.

Although a service provider may try to prevent any representation, whether made orally or in writing, becoming part of the contract, the representation made, whether by the person selling something, or if written in company literature, can still form part of the contract. This is a general statement, and the position will depend on the precise circumstances relating to the contract. Any comments noted down at the time a sale was effected might be important in the future. If comments were made prior to the contract, they should be noted down at the time they were made. They can then be confirmed, together with the affect the comments have on the agreement, by setting them out in writing to the other party prior to the contract.

If a company has a set of trading terms, one clause might refer to representations. The clause usually seeks to avoid responsibility for any oral representations made by employees or agents. In addition, an attempt might be made to refuse liability for any written representations. Such clauses might, depending on the circumstances, be effective. Whether a clause can prevent the purchaser from relying on a representation will depend on a number of factors that are beyond the scope of this chapter. In any event a business is strongly advised to consider whether any representations were made, whether oral or written, for all contracts. By doing so, it will heighten the awareness of this issue. Further, its contractual rights might be strengthened if the representations, if any were made, were incorporated into the contract. All of these issues need to be considered in the context of IT and e-commerce: some practical tips are set out further below.

Analysis of agreements **9.9**

Evidently the paper trail can determine the method by which a business entered the contract. How it entered the agreement will have a bearing on rights, and therefore the strength of any negotiating position. For example, if it uses its own purchase order, the probability is that its terms of trading, if any, will apply to the contract. If this is the case, in the event of a dispute, the other party to the contract is likely to argue that their terms do not apply to the contract if their terms are generous. This is an issue that should be considered carefully. It is better not to assume that one's own trading terms are advantageous. Consider trading terms carefully because they seek to apportion risk and attempt to negotiate changes if necessary. It is not always possible to negotiate changes with a bigger and more powerful company. However, if a dispute occurs, it does not necessarily follow that the bigger company will succeed in having all their limitation clauses accepted by a judge. This is a complicated area and a business should be aware that onerous clauses limiting liability can be challenged.

The purpose of a purchase order is usually to provide evidence of the intention to enter a legally binding contract. If a business is willing to accept the trading terms of the organisation it is buying from, it is important to ensure, when it provides a purchase order number, that the provision of the purchase order number is merely evidence of the acceptance of the offer to enter into a legally binding contract on their terms of trading.

If goods or services are purchased without a purchase number, the probability is that the terms of the supplier will apply to the contract and a copy should be requested. It is in the interest of good practice to establish whose terms apply to the contract. Failure to do so might mean the need to litigate to establish the position, which is costly and time consuming (see **CHAPTER 5**). Moreover, whether the supplier can enforce the terms of the agreement might also depend on answers to the following questions.

1. Did the purchaser know the terms before it entered into the contract?
2. Were the terms on the invoice?
3. Were the terms, or mention of them, on any literature or other documents sent to it?

It is important to know whether the literature or other documents were sent before or after buying the goods or service. The terms of trading accompanying goods, for instance, will only apply to the contract if a business knew they existed before it entered into a contract.

Many companies erroneously print their terms of trading on the back of invoices. This is a waste of money. Terms that are presented to the customer after the contract has been concluded do not have any affect.

If the contract is not governed by any terms – whether that of the purchasing business or that of the supplier – then in the UK it can rely on the statutory rights provided by Parliament. When planning for continuity, therefore, it should try to establish an answer to this question. By being aware of its legal position, a business is better informed. As a result, it can more readily determine the strengths and weaknesses of its predicament, should an issue arise.

On-line practical contractual issues

Administering the website effectively **9.10**

Issues relating to copyright and trademarks must be carefully considered. Ensure the web designer (or employee, if the website is administered in-house) does not use the trade marks of competitors in meta-tags, for instance. Meta-tags are the hidden indexing system that help search engines to find websites. For example in the UK, Road Tech Computer Systems successfully applied to the High Court for a summary judgement against Mandata, where Mandata wrote trademarks belonging to Road Tech into their meta-tags 27 times. Mandata was ordered to pay £15,000 damages and the costs of both sides, which amounted to £65,000.

Consider having an administration process in place to ensure that all changes to the content of the website are checked before the replacement goes live. Paying a person to undertake a quality check can be cheaper than resolving the adverse publicity and increased costs associated with mistakes on websites. Consider having trading terms drawn up: an added contractual problem for Argos (see **9.5**) related to the legal position of the price. Clearly, customers may have thought the price somewhat low, but they could legitimately argue that it is possible that Argos could have been offering a short, one-off sale as a marketing gimmick. Alternatively, Argos would say that the price was so low as to be a clear mistake.

An additional complexity to add to the legal analysis is whether the website is termed an invitation to let a customer think about buying from the business, or an offer to sell an item at the price noted on the website. If the website can be construed as an offer, it may find itself liable for every order made by each customer electronically. This is a very important point. If a business fails to establish the precise nature of its website, and if demand outstrips the ability to supply goods ordered, it could be embarrassed by the national press as yet another online trader that fails to live up to the promises that e-commerce is supposed to bring to traders and customers alike.

It is important to try to minimise the risks when trading on-line, as the Argos example illustrates. It is important to avoid unlimited liability to deliver thousands of items at very low prices. The contractual position with a potential customer should be considered very carefully, and provision should be made in a specifically designed set of trading terms to ensure the business can quickly resolve such a problem, if it were to occur.

Business interruption **9.11**

It is probable that most businesses with an online presence will not suffer from the attacks of hackers. However, the activities of hackers are not the only problem businesses face in relation to e-commerce. The proliferation of viruses and similar unpleasant electronic attempts to make computer systems inoperable can all affect the smooth operation of the business. In such circumstances, it is imperative that the business scrutinises any contract it has with its Internet service provider and any other relevant service provider, to ensure that it understands the provision, if any, in relation to the security system. If it is not given sufficient assurances in writing about the ability of a service provider to keep the e-commerce site running 24 hours a day, seven days a week, the business should either consider an alternative provider or the services of a good quality electronic security company.

Analysing the contractual obligations when running an e-commerce website can pay dividends. By assessing the strength of the contractual position, the business can plan more effectively. If it has good relations with customers and suppliers, it may consider incorporating clauses into contracts relating to risk and continuity. It should not allow tight deadlines to force it to gloss over the conditions of important contracts. By establishing an accurate picture of its contracts, the business is in a better position to take steps to minimise those risks that it has taken the care to identify. (see also **CHAPTER 5** regarding business continuity and operational risk management).

Technology due diligence: managing legal risk exposure 9.12

In earlier sections comment has been made about the handling of IT and data as part of the ongoing due diligence and operations of the business. Another issue concerning the use of IT relates to technology due diligence in so far as it is often one key aspect of the overall due diligence process performed by those wanting to acquire or invest in companies. In this context it is primarily an exercise about managing both commercial and legal risks. It is about managing risks in a way that will enable the parties involved in the business transactions ultimately to maximise their respective commercial returns or to minimise bottom line losses. Technology due diligence is about assessing the quality of technology assets and any attendant risks that an acquirer would assume in acquiring such assets. In this spectrum of risks, legal risk exposures probably rank among the top concerns of senior management.

In the era of Internet-based business, one of the key assets of corporations is its intellectual asset. Acquiring intellectual assets is often one of the key objectives in mergers and acquisitions and they are also increasingly exploited for strategic advantage. In this discussion 'intellectual assets' covers those generic cluster of intangible assets which, when narrowly defined within a precise legal context, emerge as rights which are protected or can be protected by law. Such intellectual property (IP) rights typically include patents, copyrights, trademarks, design and trade secrets. They are often intangible and come in the form of technology or know-how which can be software or hardware driven or simply business processes driven by technology.

Companies typically want to acquire technology from others for any one of the following reasons:

- the technology is useful in itself;
- the technology may enhance the acquirer's own product range, service offerings or technology development cycle;
- the acquisition of the technology will enhance the acquirer's corporate branding;
- the acquirer sees long-term investment value by acquiring the technology at what the acquirer regards as fair market value; and
- the acquirer sees benefits in terms of positive public perception of the acquisition that would result in better market price for its shares.

Investors typically insist on having a very clear view of the strength and weaknesses of the technology in the overall corporate and industry context. Research and development houses may also use technology due diligence to assess or validate their performances while technology start-ups may use the result of technology due diligence for their own self-assessment or evaluation for the purpose of fund raising. The ultimate aim of any technology due diligence exercise is to use it as a management tool that helps increase the probability of a successful investment, a partnership or a merger and acquisition. As with all due diligence exercises, a risk assessment has to be made.

In a technology due diligence process, the purchaser or investor typically would want to achieve one or more of the following objectives:

- to identify technology asset strength and weaknesses that would help in closing the deal successfully;
- to remedy any identifiable flaws in a way that will be conducive to further negotiations but often in the hope that it will reduce the acquisition cost for the acquirer or purchaser and secure better terms in general.
- to ensure the eradication or minimisation of all if not most legal risk exposures from the acquisition of that technology.

The main assessment that an adviser has to make typically arises from the following scenarios:

- an acquirer intending to purchase a technology company; or
- an investor taking up an equity stake in a technology company; or
- parties to any transaction with a technology component.

The questions to be asked cover:

- whether the technology that is the primary asset of the company being acquired does in fact do what it is supposed to do: its capabilities must be evaluated and verified;

- whether the acquirer can verify that it would be able to extract business value from acquiring the technology; in particular whether the management team responsible for the creation of the technology can in fact deliver what the technology purports to be able to do for businesses, in short, the commercial prospect of the technology;
- whether the owner of the technology does in fact legally own all the rights to the technology;
- the nature and magnitude of legal risks that the acquirer would be assuming if the acquisition is made; and
- what the true measure of the value of the technology being acquired is, that is, its strengths, weaknesses and therefore its real worth.

Given the primacy of intellectual assets in the Internet-based economy and the essence of such assets being the technology itself, the issue of protection of such assets is of paramount importance. In a technology due diligence, managing legal risk exposures in relation to the intellectual property rights of the company is always one of the primary concerns of parties. Therefore this central aspect of technology due diligence, that is, the evaluation of and the risk assessment of intellectual property rights issues, both internal and external to the company or individual owning the technology is considered below.

Management of intellectual property issues 9.13

The proper management of the IP issues, in particular, the management of any legal risk exposures to the acquirer may often make or break a deal. Even if the technology is valuable, the danger of legal risk exposure over a mismanagement of IP issues could unravel business deals. In any technology due diligence, the identification of the nature of assets or more specifically intellectual property rights is often the first step. In most jurisdictions around the world, the class of intellectual property rights would typically include:

- patents;
- copyrights;
- trademarks;
- designs; and
- trade secrets.

Some rights arise automatically, such as copyrights, in most jurisdictions. Other types of rights, for example patents, must be applied for and the examination process is rigorous. One major issue that arises in cross-border technology due diligence is the question of the applicability of different laws which affects the creation of intellectual property rights. Most intellectual property laws are territorial in nature and any sound technology due diligence should adequately address this issue to ascertain with some precision the existence or non-existence of intellectual property rights. For example in most countries patent protection is secured from the date of first filing with the exception of the US which uses a mix of first-to-file and first to invent system. Also the test for patentability of technology varies from country to country and this would have a major impact on the value of technology which is to be used across multiple jurisdictions.

A particular technology being acquired in turn may have multiple intellectual property rights. In a typical e-commerce site, the text, the layout and the 'look and feel' of the site, accompanying music and the software codes are the subject of copyrights while logos, prominent marks or signs are subject of trademark. The process of electronic payment which could be software driven and which has a demonstrable technical effect could be patented. An investigation into the nature of the property would therefore require a thorough analysis of the nature of legal rights of that class of assets to ensure that any overlap could be positively exploited for the benefit of the purchaser or investor.

Another aspect in addressing the nature of the technology asset is the question of legal compliance. The development of any technology must comply with the prevailing laws and regulations in which the operating company is domiciled. Thus if a particular country prohibits a particular type of technological endeavour for example cloning of human cells, any patenting of such technology is likely to be against public policy and this sort of issues should be uncovered during the technology due diligence process.

Ownership issues **9.14**

The next issue relates to the ownership of assets. Rights which are registrable such as trademarks, patents and in which there is some kind of searchable governmental registry should be the first line of enquiry in any technology due diligence. This can be done comprehensively. Nevertheless, issues of the territorial nature of IP laws must be borne in mind. So if a technology-based company uses Internet technology across multiple jurisdictions for its business operations, searches in different jurisdictions must be made.

However, ownership issues become murkier when such intellectual property rights are not yet registered or are not capable of being registered. Thus an employee in an research and development department may have obtained source codes for software development (that is copyrighted materials) that rightfully belong to the employee's former employer. Any technology due diligence must therefore uncover such legal risk exposures for the acquirers of such technologies.

Assets can be owned or be verified to be owned by a particular entity but the legal protection itself is weak owing to gaps in legal drafting. While warranties and other forms of representation complete with indemnity provisions would clearly help the legal position of a purchaser or an investor of companies, there is nothing more reassuring than a thorough background check including physical examination to ensure that nothing – whether in hard or soft copies – of any form of property that belongs to someone else is being used at all by the staff or management of companies in which the technology is purportedly owned.

Other ownership issues that should be addressed during technology due diligence include:

- identities of all parties involved in the development of that technology;
- when the technology was developed – for example it could have a prior existence in a different legal entity;
- whether there are any third party rights that attach to the property;
- whether the registration of rights (eg granted patents) can be challenged and subsequently revoked;
- whether the documentation evidencing ownership (if not registrable in governmental search agencies) exist or clear enough for avoidance of doubt;
- whether all legal compliance requirements have been fulfilled for purposes of evidencing ownership;
- whether all renewals and payments have been made (even if the governmental authorities have not issued any reminders or warnings of expiry)

Duration of protection 9.15

Different types of intellectual property rights have different life spans. For example, a trade secret can last forever while trademarks can last forever subject to fee renewals. Patents in turn typically last for 20 years or so depending on the jurisdiction. Therefore the technology due diligence process should elicit a very clear position on the duration of such rights as this will have an impact on the commercial value of such technologies.

Rights involving other parties 9.16

A sound technology due diligence should also be able to ascertain whether the key features and functionalities of the technology are dependent on any licensed component. Are there, for example, third party relationships with contractors and other service providers in which the successful exploitation of the technology being acquired are dependent? These issues would determine the quality of technology assets.

Ascertaining the risk of infringement of the intellectual property rights of others is probably one of the more difficult part of any technology due diligence. Again tightly drafted representations and warranties to cover all forms of contingencies are necessary.

Sufficiency of legal documentation 9.17

The technology due diligence must also review all available legal documentation to ensure the eradication or minimisation of legal risk exposures. So in a high risk transactions, for example, where the ownership issue is in question, there has to be legal provisions typically in the form of legal rights that must be included in the agreement to protect the interests of the purchaser or investor. Representations, warranties and indemnity provisions in legal documentation must be carefully scrutinised in this part of the technology due diligence process.

Technology due diligence is but one critical aspect of the whole due diligence process (as discussed in **CHAPTER 1**). Investors or purchasers interested to acquire enterprises in which the intellectual assets are one of its quality assets are well-advised to evolve a systematic way to manage its technology due diligence process and focus on intellectual property rights issues and the minimisation of any attendant legal risk exposures. This would maximise the chance of a successful acquisition or investment.

Legal risk issues in Internet commerce 9.18

There are several characteristics of internet commerce that require legal risk issues to be assessed in a different light.

Digital and other information assets 9.19

Internet commerce deals with hitherto new types of digital and information assets. Such assets in a way define what Internet commerce is all about for the traditional brick and mortar enterprises. In cases where the enterprise itself is a 'pure' Internet company, that is, one without a physical presence, the Internet-based business model is actually the very business itself. These digital and information assets are particularly vulnerable to attacks which can threaten the commercial viability of the business. The interception of data and breach of confidentiality can create specific legal risks that have focused the minds of many organisations in terms of risk management, especially since the Turnbull Report requirements and the importance of embedding risk management into core corporate policy at board level (see further **CHAPTER 10**). The legal risks are considered in more detail below.

Borderless and global activities 9.20

Internet commerce is by definition a borderless and global activity. The Internet is a global network of networks. Internet connectivity itself crosses political boundaries with no hindrances so long as the networks in two different jurisdiction are connected. Business methods that are effective and in compliance with the laws and regulations in one enterprise's home market may not work in markets that operate in a totally different legal environment, and might even expose the enterprises to unexpected legal liability. An example would be a US online bank trying to offer its services to citizens of other countries located in different legal jurisdictions. This kind a business model would probably be affected by the laws affecting such citizens in their respective home markets.

Timing for product and service roll-outs 9.21

In internet commerce, the 'go to market' time for a new project is much shorter compared to brick and mortar commerce. This reduced time frame means that legal issues must be addressed much earlier than it is traditionally expected to.

Managing legal risk issues in Internet commerce 9.22

As a result of the more Internet-intensive commercial environment, technology-related legal risk management is now becoming an increasingly familiar concept to the board and senior management of all enterprises. If it is not, it should be.

If the legal risks that flow from technology risks are serious enough to threaten the legal and commercial interests of the enterprise, senior management needs to ensure the establishment of a legal risk management framework to identify these risks and take adequate measures to address them. The company's board of directors, for instance, have a fiduciary duty to protect the organisation from security attacks and other forms of cybercrime and security risks which may have a critically negative impact on the organisation's reputation, assets and commercial viability.

Enterprises should ensure that adequate steps are taken to protect themselves legally. Apart from liabilities for breaches of contractual obligations, the failure to take reasonable and adequate steps to provide security measures may possibly lead to an enterprise being liable for negligence, either in not taking sufficient steps to protect data and information where it has a duty of care to protect, or in being used as a platform or a channel to mount an attack against another party. Preparatory steps should therefore be taken in advance in planning the procedures to handle security breaches.

The board and senior management should therefore review and approve the organisation's legal risk management policies taking into account technology risks and the capacity of the organisation to deal with such problems. Legal risk management in this new technology-intensive environment cannot be a task that is merely carried out periodically, ie yearly or half yearly. In today's accentuated security risk environment, legal risk management has to be regarded as an oversight process undertaken by senior management on a continuous basis. This process involves legal risk identification, assessment, control and mitigation. Also the scope of legal risk management should embrace a broader horizon which incorporates proactive legal risk management. A key component in this legal risk management framework is the protection of digital assets.

Protecting digital assets 9.23

To protect its Internet-based business, enterprises should first begin by identifying the assets to be protected before it begins to do its business. This will avoid any potential loss of time, resources and when the enterprise finds itself losing control of these assets. Potential assets at risk are described below.

Data 9.24

This includes customer information, financial data, equity and market index data online and other proprietary data.

Applications or software 9.25

Such applications or software includes those which run corporate IT systems and its workflow (for example an Internet commerce software or an enterprise resource planning software which may cost millions of pounds).

Digital products and services 9.26

Information products sold by the enterprise such as financial planning software, e-toolkits or e-guides and business information. Legal advisers should help ensure that the enterprise has the right to sell these assets and can help improve the chances of successful litigation against digital asset violators and pirates.

Intellectual property rights 9.27

Such intellectual property rights could include those that are in digitised form (for example copyrights in e-commerce software or trade secrets stored in a digital format). The enterprise's business identity in turn can be embodied in its trademarks, logos, and its domain name. These assets should be protected by registration in commercially important jurisdictions, to ensure the highest level of protection for the enterprise.

Websites 9.28

As has been touched on above (see **9.5** to **9.8**), the enterprise's information website and its transactional site or portal itself needs to be protected through effective contracts governing its formation and enforcement. Such websites should also be monitored and controlled as well through effective contracts involving users of the site such as the enterprise's customers and other third parties. Pre-emptive action should be taken against users who violate the enterprise's intellectual property rights and other digital assets.

Contractual obligations 9.29

From the business perspective, the legal risk exposures that result from major service disruptions are to be given greater priority. Such legal risk exposures usually arise out of contractual obligations in the following two situations:

- where there is service disruption affecting their customers, which, if not clearly regulated in legal terms, may expose the enterprise to potential legal suits for non-performance of its contractual obligations;
- in the event of service disruption to the enterprise's partners or other third parties who rely on the enterprise technology infrastructure to fulfil other transactional requirements.

Compliance relating to business continuity 9.30

Another legal issue that enterprises have to address in the provision of Internet commerce services relates to compliance requirements in relation to business continuity planning. Enterprises such as banks and financial institutions typically operate in a legal environment that is very tightly regulated. The regulatory authorities may require legal compliance in terms of having a sound business continuity plan or disaster recovery that is subject to regulatory review and penalties for non-compliance. Such regulatory non-compliance is one form of legal risk exposure that the enterprise's legal advisers must address.

A business recovery and continuity plan is essential for every business that owns any mission critical application or system. To ensure adequate availability, enterprises typically provide for contingency back-up systems to mitigate denial of service attacks or other events that potentially cause business disruptions. As has been mentioned in **CHAPTER 5** a business continuity plan or disaster recovery plan is an essential part of the overall risk management framework of the business. Such a risk management framework typically also includes issues pertaining to data confidentiality, system and data integrity and security practices in general. The board of directors has a fiduciary duty to ensure that in the event of system failure for whatever reason, there is continuity of service for the enterprise's clients and partners.

Relationship with technology providers 9.31

Most commercial enterprises are not in the business of providing technology solutions and they rely greatly on external parties such as Internet commerce technology service providers to provide the technology infrastructure to enable them to provide Internet commercial services. This is another dimension in the legal portfolio that senior management must handle.

As has been discussed above (see **9.7** and **9.8**) a vitally important, aspect of the legal protection framework in internet commerce is the use of effectively drafted contracts with third party vendors and solution providers to ensure the enterprise's potential legal liabilities are adequately managed. These are contracts that typically manage the relationships that enable the enterprise to provide secure and continuous services, covering such matters as:

- web hosting;
- development of applications (for example Internet commerce software);
- access services provided typically by infrastructure providers such as telecommunication and Internet service provider companies;
- security services including the supply of security products such as firewalls and encryption software.

Since the provision of technology services are typically not part of a commercial enterprise's core competencies, such services are typically outsourced to external providers. However, the enterprise's primary responsibility to its customers is to provide an accessible, secure service direct. In the event of the failure of the enterprise's service provider, the enterprise itself would still be accountable to their customers. There is therefore a need for enterprises to ensure sufficient counter-indemnity arrangements are entered into between themselves and the third party technology providers.

Therefore when there is a major service disruption which is caused by technology or system failure, the issue that often arises is the extent to which the enterprise is able to pass on or share any legal risks to the technology service providers. This typically takes the form of indemnity provisions which require the technology service providers to indemnify the enterprise for losses that result from the service provider for failure to ensure business continuity.

Managing liability issues **9.32**

The task of a business and its legal advisers in an Internet commerce business is to ensure that once the types of technology risks have been identified, the legal ramifications are clearly understood and analysed. Any potential economic loss should be quantified wherever possible. With this information, the enterprise would then be able to prioritise the legal risks and make legal risk mitigation decisions. As has been indicated above in **9.5** to **9.8**, enterprises can minimise, if not eradicate, such legal risk exposures by designing terms and conditions in their service agreement that exclude or limit their liability in the event of system failure that causes non-delivery of essential services.

As is also seen above, by the very nature of enterprise being in 'big businesses', it is not uncommon to see 'pro-company' terms being imposed on the customers. While customers might simply accept such terms that exclude or limit the liability of the enterprise, particularly when they are not in a strong negotiating position, it makes a lot of sense for enterprises to focus on managing their relations with their customers in other more productive ways such as in the form of client education.

Consumer interests 9.33

For most consumers transacting over the internet, the primary concern when a transaction fails is usually whether he suffers a pecuniary loss, for example payment made but the goods or services are not received, or when the wrong or unsatisfactory goods or services are delivered. From the perspective of the customers, confidence is about knowing what the customers can expect from the enterprise when there is a disaster or an attack that affects their commercial transactions. Individuals and consumers also need to understand the available remedies of a failed transaction over the Internet, regardless of whether it is attributed to a merchant that was a target of hacking or the action of fraudulent third parties. In general customers think in legal terms only when there is a major economic loss on their part (although the trend in society is towards more litigation). In any event, the way to manage possible legal risk exposures that might result from contractual obligations is an assurance programme that is sound, well-publicised and that engages the clients of the enterprise in times when there is no disaster.

While taking the legalistic approach of protecting one's interests by defining and controlling legal risk through 'small print' might serve its purpose, a better strategy is therefore to focus on assurance and effective communication to parties that may potentially sue the enterprise in the event of major service disruptions.

Evidential issues 9.34

In any cases involving breaches of security, companies must have in place work policies and procedures to ensure that evidence can be properly presented to the prosecuting agencies and the courts. If proper steps are not taken in relation to digital evidence, the chances of proving one's case or to disprove the other side's case will be much less. Given the fragility of digital evidence and the need to collect, preserve and present evidence to the prosecuting agencies in a criminal legal proceedings, enterprises should ensure that digital evidence can be properly detected, preserved and presented in a manner that legally complies with the local laws of the country. Also, given the transient nature of digital evidence, time is of the essence in all cases involving information security breaches. In general, companies should have in place policies and procedures to include the following:

- steps to isolate or quarantine the evidence;
- recovery of evidence;
- reproduction of evidence;
- processing and analysis of evidence; and
- preparation of report by an expert for use in the courts

In the event that digital evidence and data are not properly secured or preserved, such evidence may subsequently be found inadmissible in court for the purposes of criminal or civil proceedings. Therefore, as part of the enterprise's post incident operation procedure in areas of disaster recovery and business continuity planning, there is a need to ensure that legally-compliant procedures be pre-established so that they can be activated expeditiously when the incident happens.

Businesses should also seek legal advice on how to determine whether a crime has been committed and the possible courses of action that can be taken based on the evidence available. Digital forensics work will invariably have to be undertaken together with legal personnel to identify the crime, the offender, and to collect and reconstruct the necessary evidence which are typically found on disks, logs and other media. Legal advice should be sought on issues such as preservation of evidence, issues of admissibility and the overall presentment of such evidence to the prosecuting agencies in a manner that not only comply with the law but also are managed in a manner that would make a strong case for the prosecution. Aside from criminal proceedings that the Public Prosecutor may take against the perpetrator for the offences committed, the victim of the crime may also consider filing civil claims for damages and other losses that may have been suffered as a result of the attack.

Designing a risk management framework 9.35

In designing the overall legal risk management framework, businesses should, as a general rule, have a proactive and structured programme of action involving the following elements:

- an overall system to identify, classify, measure, prioritise and assess legal risks that are relevant to the enterprise's operations;
- a plan that is documented in the form of an operation manual (both hard copy and embedded into the system in the form of web-based documents containing policies, practices and procedures that addresses and controls these risks). Such a plan must specify the responsibilities of all parties involved in the whole risk management process from the operational level right up to the CEO;
- a regular test plan that when implemented approximates all possible worst-case scenarios for the purpose of testing the system to its fullest potential;
- a monitoring programme to assess all types of technology and other operational risks and the evaluation of the effectiveness of such programmes;
- the regular updating of such plans in the light of developments in the technology, law and business practices;
- post-incident recovery procedures which must incorporate digital evidence collection, preservation and presentment techniques which are legally compliant;

- the fulfilment of legal compliance requirements as specified by the regulatory bodies;
- a security awareness programme that will help nurture a more security conscious environment.

Legal audit 9.36

In designing this legal risk management framework, it is best to begin with an audit. This phase involves the senior management in the enterprise and the legal team conducting an audit on the adequacy of legal strategies, legal documentation and work procedures and guidelines that affect the day to day legal management of Internet commerce services. Examples of issues that are usually addressed during the audit include:

- the overall legal strategy to handle legal risks that are technology-driven and the objectives and plans currently guiding the business in the area of technology risk management;
- an assessment of the legal compliance environment as well as the developments in the legal standard of care in providing security services to protect the business from intrusions or such other forms of attack;
- the form and effectiveness of the legal standard operating procedures and guidelines of the business, as well as the general organisation and administration of legal matters;
- legal cost and economics, for example the cost benefit analysis of getting external lawyers to advise on the legal issues that the enterprise is currently addressing;
- legal department resources and capabilities, that is, strengths and weaknesses relating to resources, reputation, services and legal market position. Issues that need to be addressed include whether existing lawyers who are competent in the traditional type of commercial activities such as loan document preparation are competent to handle IT-related legal liability issues for example in the area of security breaches in Internet commerce.

A forecast of the technology and security risk environment and the legal consequences that flow from it that will affect the legal position of the enterprise and its clients.

Enterprises involved in Internet commerce should address and manage legal issues in a manner that is structured and proactive. In Internet commerce it is imperative that not only is the physical security be assured but a sound legal protection regime that protects and secures the enterprise's other commercial interests should also be in place. If planned and executed in such a structured and proactive manner, such a legal protection regime would bolster the enterprise's overall corporate governance framework.

Summary **9.37**

Bearing in mind all of the above, a business should be aware that the value of its digital assets could surpass that of its physical assets. Moreover, given the unique nature of digital assets, legal risk exposures can be particularly accentuated. A structured and proactive legal risk management approach has to be one of the central pillars of an organisation's management of its digital assets. The elements of such a management framework can be dealt with by a corporate digital asset management policy (see the Appendix below). This may help organisations better manage their legal risk exposures when they create, protect and exploit digital assets and assist them to fulfill the need for ongoing due diligence and good corporate governance as considered in this handbook.

Appendix

Checklist for implementing a corporate digital asset management policy

Developing a corporate digital asset management policy

- ☐ Have you identified all digital assets in your workplace?
- ☐ Has an audit of your company's digital asset been conducted? Is there a digital asset inventory?
- ☐ Have you registered all of your digital assets that can be registered as intellectual property rights?
- ☐ Do you have a corporate digital asset policy?
- ☐ Is there a separate operating procedure to complement the company digital asset policy?
- ☐ Is your company's digital asset policy regularly reviewed or updated? How often?
- ☐ Do you have in-house expertise to review your digital asset policy and procedures?
- ☐ Is your company's digital asset policy documented (manual and in soft copy)?
- ☐ Is there a digital asset management procedure that your employees must comply with?

Communicating digital asset policies

- ☐ Have you communicated the company's digital asset policy to all employees?

- ☐ Are employees required to acknowledge in writing that they have read and understood the company's digital asset policy?
- ☐ Is your workplace designed in a manner such that signs or notices about the need to protect digital assets are clearly visible?
- ☐ Do you regularly remind your employees about your company digital asset policy?
- ☐ Do you have a hire and exit procedure that takes into account digital asset policy matters and are these adequately communicated to all employees?
- ☐ Do you have a digital asset compliance officer that ensures that all employees abide by the company's digital asset policies?

Implementation of company digital asset policy

- ☐ How is the company's digital asset policy implemented?
- ☐ Is there a company officer in charge of implementation or execution?
- ☐ Do you have a pre-agreed criteria to assess the success of your company's digital asset management programme?
- ☐ Are there both quantitative or qualitative aspects in the assessment criteria?

Documentation

- ☐ Is your legal documentation for all your digital asset rights complete?
- ☐ Have all contracts and agreements been reviewed to ensure compliance with the company's digital asset policies?
- ☐ Do you have a procedure for clearing publication of professional papers for presentations at industry or trade shows that meets the requirements of your company's digital asset management policy?

Violation of digital asset management policy

- ☐ Do you monitor your digital asset rights for possible violation?
- ☐ Have you ensured that your employees have not violated the digital assets of their previous employers?
- ☐ Have you ensured that employees who leave the company have not violated the company's digital asset policy and procedures?
- ☐ Are there measures to restrain key personnel from violating digital assets for example restrictive covenants?

Chapter 10

Corporate Governance Issues

Corporate governance definitions and drivers

Definitions 10.1

There is considerable and ongoing discussion regarding the meaning of the concept of corporate governance. On one hand it can be regarded in a limited sense that covers financial controls and on the other hand it can be taken to extend to all of the responsibilities and policies of the business that include such matters as environmental concerns (see further in **CHAPTER 16**. Between these two approaches there are, of course, many variations. What is clear is that the trend is in favour of widening the scope of the term. As with previous debates over such issues as sustainability, it is recognised that the role of business in today's global economy is profound and that accountability should be extensive. The responsible partners are the public and private sectors. The stakeholders, including members of the public, and not only those with a direct interest in the company, are demanding more transparency and evidence of responsible behaviour.

Another important matter concerns the repercussions of the corporate governance debate for smaller business. As with the demands of other regulatory frameworks including product liability, the environment and health and safety, the corporate governance framework affects small business. In addition to the supply chain pressures there is the consideration that today's small growing business can become a big business very quickly.

This chapter is intended to give an overview of key points that have come together to demonstrate where we are today. In the earlier chapters, such as **CHAPTER 4** some comment has already been made as to the meaning of corporate governance and in the chapters that follow international comparisons are made (see **CHAPTERS 11** to **15**). Moreover, the term governance is considered in various contexts in **CHAPTER 18.** In order to reduce the potential for duplication it should be emphasised that in this chapter the more recognised definitions are referred to. In addition, the main developments towards today's understanding of what is meant by corporate governance issues are reviewed bearing in mind that it is an evolving area. Since the management of risk is very relevant to the development of corporate governance, risk management is also considered later in the chapter (see **10.8** to **10.15**). Clearly the way in which a business manages risk is closely interlinked with its corporate governance.

As has been noted throughout this handbook the concept of corporate governance has been attracting public attention for quite some time in many parts of the world and the practical developments and responses have been gaining momentum. Therefore some attention is given to selected jurisdictions, such as India, Hong Kong, China and Australia (see further in **CHAPTERS 12** to **14**). The topic is no longer confined to the halls of academia and is increasingly finding acceptance for its relevance and underlying importance in the industry and capital markets. Progressive firms in many places have voluntarily put in place systems of good corporate governance. The focus on corporate governance and related issues is an inevitable outcome of a process, which leads firms to increasingly shift to financial markets as the pre-eminent source for capital. In the process, more and more people are recognising that corporate governance is indispensable to effective market discipline. This growing consensus is both an enlightened and a realistic view. Corporate governance can enable a company to have its systems in place and give it sufficient freedom to operate within a framework of accountability. It helps to enhance transparency and responsibility while maximising long-term wealth of investors.

However, the concept of corporate governance is defined in several ways because it potentially covers the entire array of activities having direct or indirect influence on the financial health of the corporate entities. As a result, different people have recommended different definitions, which often reflect their special interests in the field.

To start with it would be useful to recall the earliest definition of corporate governance by the economist and Nobel laureate Milton Friedman. According to him, corporate governance is to conduct the business in accordance with owner or shareholders' desires, which generally will be to make as much money as possible, while conforming to the basic rules of the society embodied in law and local customs. This definition is based on the economic concept of market value maximisation that underpins shareholder capitalism. At first sight Friedman's definition might be assumed to be narrow in its scope, but as the 'basic rules of the society embodied in law and local customs' are dynamic and have evolved rapidly over recent times to embrace the well-being of nearly all, then Friedman's definition holds good today and is likely to in the foreseeable future. It is now generally accepted that a company is not only responsible to its shareholders but to, amongst other things, its employees, customers, suppliers, the public at large and the environment. The great challenge for a company now is in aligning and managing the complexities of the various stakeholders' interests, values and expectations.

According to some experts corporate governance means doing everything better, to:

- improve relations between companies and their shareholders;

- improve the quality of outside directors;
- encourage people to think long-term;
- ensure that the information needs of all stakeholders are met; and
- ensure that executive management is monitored properly in the interest of shareholders.

Experts at the Organisation for Economic Co-operation and Development (OECD) have defined corporate governance as the system by which business corporations are directed and controlled. According to them, the corporate governance structure specifies the distribution of rights and responsibilities among different participants in the corporation, such as the board, managers, shareholders and other stakeholders, and spells out the rules and procedures for making decisions on corporate affairs. By doing this, it provides the structure through which the company objectives are set, and also provides the means of attaining those objectives and monitoring performance.

It is helpful to compare definitions in other jurisdictions. For example, according to the relevant report in India – the Report of the Kumar Mangalam Birla Committee on Corporate Governance:

> '...in an age where capital flows worldwide, just as quickly as information, a company that does not promote a culture of strong, independent oversight, risks its very stability and future health. As a result, the link between a company's management, directors and its financial reporting system has never been more crucial. As the boards provide stewardship of companies, they play a significant role in their efficient functioning.'

This report goes on to state that:

> 'Studies of firms in India and abroad have shown that markets and investors take notice of well-managed companies, respond positively to them, and reward such companies, with higher valuations. A common feature of such companies is that they have systems in place, which allow sufficient freedom to the boards and management to take decisions towards the progress of their companies and to innovate, while remaining within a framework of effective accountability. In other words they have a system of good corporate governance.
>
> Strong corporate governance is thus indispensable to resilient and vibrant capital markets and is an important instrument of investor protection. It is the blood that fills the veins of transparent corporate disclosure and high-quality accounting practices. It is the muscle that moves a viable and accessible financial reporting structure. Without financial reporting premised on sound, honest numbers, capital markets will collapse upon themselves.'

For further comment on corporate governance in India see **CHAPTER 12**.

According to some economists, corporate governance is a field in economics that investigates how corporations can be made more efficient by the use of institutional structures such as contracts, organisational designs and legislation. This is often limited to the question of shareholder value – that is how the corporate owners can motivate and/or secure that the corporate managers will deliver a competitive rate of return. In brief, therefore, corporate governance calls for three factors:

- transparency in decision-making;
- accountability which follows from transparency because responsibilities could be easily established for actions taken or not taken; and
- the accountability to safeguard the interests of the stakeholders and the investors in the organisation.

For the purposes of this discussion it is also useful to cite the important definitions of corporate governance found in the Cadbury Report[1] (see **10.3**) and the Hampel Combined Code as follows:

- 'The system by which companies are directed and controlled';
- 'The greatest practicable enhancement over time of their shareholders' investment. All boards have this responsibility and their policies, structure, composition and governing processes should reflect this';
- 'Corporate governance is the system by which business corporations are directed and controlled. The corporate governance structure specifies the distribution of rights and responsibilities among different participants, such as board, managers, shareholders and other stakeholders, and spells out the rules and procedures for making decisions on corporate affairs. By doing this, it also provides the structure through which the company objectives are set, and the means of attaining those objectives and monitoring performance.' See also **CHAPTER 7.**

Summary of corporate governance drivers 10.2

Bearing in mind the matters referred to throughout the handbook, particularly in **CHAPTERS 1** and **4** there are clear drivers that have resulted in:

- the establishment of committees to assess best practice;
- the main reports that have been published; and
- the regulatory and voluntary frameworks that now exist.

It may be said that we live in a climate of corporate governance that has largely come about through the series of high profile corporate disasters in the 1990s for example BCCI, major recent financing scandals such as at Enron, the ongoing controversy over directors' pay and an overall trend towards risk awareness and management.

Whereas many of the developments are referred to elsewhere such as **CHAPTER 7** it is useful to set out and understand the chronology of events.

The Cadbury Report **10.3**

In 1991 the Cadbury Committee was set up by the Financial Reporting Council, the London Stock Exchange and the accountancy profession. Following the consultation period the Cadbury Report[1] was published in December 1992. The main terms of reference of the report were:

- the structure and responsibilities of boards of directors;
- the role of auditors;
- the rights and responsibilities of shareholders; and
- a code of best practice.

The relevant highlights of the report were:

- boards should have a formal schedule of matters specifically reserved to them including..risk management policies (paragraph 4.23);
- directors need to maintain a system..designed to minimise the risk of fraud (paragraph 4.31); and
- the prime responsibility for the prevention and detection of frauds is that of the board (paragraph 5.23).

The stated objective of the Cadbury Committee was 'to help raise the standards of corporate governance and the level of confidence in financial reporting and auditing by setting out clearly what it sees as the respective responsibilities of those involved and what it believes is expected of them'.

The Committee investigated accountability of the board of directors to shareholders and to society. It submitted its report and associated 'Code of Best Practices' in December 1992. In its findings it spelt out the methods of governance needed to achieve a balance between the essential powers of the board of directors and their proper accountability.

The resulting report, and associated 'Code of Best Practices', published in December 1992, was generally well received. Whilst the recommendations themselves were not mandatory, the companies listed on the London Stock Exchange were required to clearly state in their accounts whether or not the Code had been followed. The companies who did not comply were required to explain the reasons for that.

The Cadbury Code of Best Practices had 19 recommendations. Since it may be described as a pioneering report on corporate governance, it would be in order to make a brief reference to them. The recommendations are in the nature of guidelines relating to the board of directors, non-executive directors, executive directors and those on reporting and control.

As regards the board of directors, the Cadbury Code of Best Practices made the recommendations that are summarised below:

- the board should meet regularly, retain full and effective control over the company and monitor the executive management;
- there should be clearly accepted division of responsibilities at the head of a company, which will ensure balance of power and authority, such that no individual has unfettered powers of decision. In companies where the chairman is also the chief executive, it is essential that there should be a strong and independent element on the board, with a recognised senior member;
- the board should include non-executive directors of sufficient calibre and number for their views to carry significant weight in the board's decisions;
- the board should have a formal schedule of matters specifically reserved to it for decisions to ensure that the direction and control of the company is firmly in its hands;
- there should be an agreed procedure for directors in the furtherance of their duties to take independent professional advice if necessary, at the company's expense;
- all directors should have access to the advice and services of the company secretary, who is responsible to the board for ensuring that board procedures are followed and that applicable rules and regulations are complied with. Any question of the removal of company secretary should be a matter for the board as a whole.

Regarding the non-executive directors the recommendations are:

- non-executive directors should bring an independent judgement to bear on issues of strategy, performance, resources, including key appointments, and standards of conduct;
- the majority should be independent of the management and free from any business or other relationship, which could materially interfere with the exercise of their independent judgement, apart from their fees and shareholding. Their fees should reflect the time, which they commit to the company;
- non-executive directors should be appointed for specified terms and reappointment should not be automatic; and

- non-executive directors should be selected through a formal process and both, this process and their appointment, should be a matter for the board as a whole.

For the executive directors the recommendations in the Cadbury Code of Best Practices are:

- directors' service contracts should not exceed three years without shareholders' approval;
- there should be full and clear disclosure of their total emoluments, those of the chairman and the highest-paid UK directors, including pension contributions and stock options. Separate figures should be given for salary and performance-related elements and the basis on which performance is measured should be explained; and
- executive directors' pay should be subject to the recommendations of a remuneration committee made up wholly or mainly of non-executive directors.

As regards reporting and controls the Cadbury Code of Best Practice stipulates that:

- it is the board's duty to present a balanced and understandable assessment of the company's position;
- the board should ensure that an objective and professional relationship is maintained with the auditors;
- the board should establish an audit committee of at least three non-executive directors with written terms of reference, which deal clearly with its authority and duties;
- the directors should explain their responsibility for preparing the accounts next to a statement by the auditors about their reporting responsibilities;
- the directors should report on the effectiveness of the company's system of internal control; and
- the directors should report that the business is a going concern, with supporting assumptions or qualifications as necessary.

Corporate governance – the Cadbury Report and beyond **10.4**

The emphasis in the Cadbury Report is on the crucial role of the board and the need for it to observe a code of best practices. Its important recommendations include the setting up of an audit committee with independent members. The Cadbury model is one of self-regulation. It was recognised that in the event British companies failed to comply with the voluntary code, legislation and external regulation would follow.

It is interesting to note how the corporate world reacted to the Cadbury Report. The Report in fact shocked many by its boldness, particularly by the code of practices recommended by it. The most controversial and revolutionary requirement – and the one that had the potential of significantly impact on the internal auditing – was the requirement that 'the directors should report on the effectiveness of a company's system of internal control'. It was the extension of control beyond the financial matters that caused the controversy.

Accordingly, the Paul Ruthmann Committee was constituted later to deal with this controversy and watered down the proposal on the grounds of practicality. It restricted the reporting requirement to internal financial controls only as against the 'the effectiveness of the company's system of internal control' as stipulated by the Code of Best Practices contained in the Cadbury Report.

It took another five years to get the original Cadbury recommendations on internal control reporting reinstated. Meanwhile public confidence in UK continued to be shaken by further scandals and Ron Hampel was given the task of chairing the Committee on Corporate Governance with a brief to keep up the momentum by assessing the impact of Cadbury and developing further guidance (see **10.6**).

The Greenbury Report **10.5**

The Greenbury Report[2], which was submitted in 1995, addressed the issue of directors' remuneration. This Report was largely in response to increasing concerns over huge pay increases for directors, share option gains and directors' contracts. It was published in 1995 and made these issues the main focus of the report since directors' pay issues had been of growing interest and concern to stakeholders and the media as a matter of public interest. The Greenbury Report set out a Code of Best Practice that covered:

- remuneration committees;
- reporting to shareholders; and
- service contracts and compensation.

The Report made it clear that remuneration for directors should be aligned with corporate objectives and bonuses should be paid for the achievement of challenging targets.

The Hampel Review and the Combined Code **10.6**

The Final Report submitted by the Committee chaired by Ron Hampel[3] had some important and progressive elements, notably the extension of directors' responsibilities to 'all relevant control objectives including business risk assessment and minimising the risk of fraud...'

The Combined Code was subsequently derived from Ron Hampel Committee's Final Report, from the Cadbury Report and the Greenbury Report. The Combined Code is appended to the listing rules of the London Stock Exchange. As such, compliance is mandatory for all listed companies in the UK.

The stipulations contained in the Combined Code require, among other things:

- that the boards should maintain a sound system of internal control to safeguard shareholders' investment and the company's assets;
- that the directors should, at least annually, conduct a review of the effectiveness of the group's system of internal control and should report to shareholders that they have done so; and
- that the review should cover all controls, including financial, operational and compliance controls and risk management.

Therefore the Hampel Review and the Combined Code demonstrated a change of emphasis and the positive contribution that corporate governance can make. Principle AIV, for example, stated that the board of a company should be supplied with information to enable it to discharge its duties. This was bearing in mind that the board consists of an active or dynamic group of people charged with running the company as opposed to a passive body. In addition:

- Principle DII stated that the board should maintain a sound system of internal control to safeguard shareholders' investment and the company's assets; and
- Hampel recommended that Guidance Notes should be prepared to expand upon and clarify what is meant by 'internal controls'.

The system of internal control was intended to cover not only financial controls but also operational and compliance controls and risk management.

The culmination of the above was the publication of the Combined Code – published in June 1998. It was a result of the work and recommendations of Cadbury, Greenbury and Hampel and covered:

- principles of good governance;
- the role of the board;
- the role of shareholders;
- directors' remuneration;
- accountability and audit; and
- the public reporting of risk

The Combined Code 2003 superseded the previous Combined Code issued by the Hampel Committee in 1998.

The revised Code derived from a review of the role and effectiveness of the non-executive director by Derek Higgs and a review of audit committees by a group led by Sir Robert Smith (both reviews published January 2003). They had reported simultaneously on 20 January 2003. The final Higgs report contained many recommendations relating to:

- the structure of the board;
- the role and other commitments of the chair;
- the role of the non-executive director;
- the recruitment and appointment procedures to the board;
- induction and professional development of directors;
- board tenure and time commitment;
- remuneration;
- resignation procedures;
- audit and remuneration committees;
- board liability; and
- relationships with shareholders.

Higgs recommended that the Financial Reporting Council (FRC) and the Financial Services Authority (FSA) should process the review's proposals rapidly. The FRC announced that it would progress the recommendations of both the Higgs and the Smith reports for changes to the Combined Code by 1 July 2003[4]. The FRC was the body charged with incorporating the Higgs recommendations into the Combined Code – its consultation on the Higgs recommendations resulted in 181 submissions[5]. After much controversy, particularly regarding the prescriptive nature of the recommendations, the FRC concluded that:

- the chairman should be allowed to chair the nomination committee;
- no limit should be placed upon the re-election of non-executive directors;
- that companies outside the FTSE 350 should not have to have at least half independent directors; and
- that some of the provisions in the draft were more like principles,

These changes were included in the new Combined Code published in 23 July 2003, by the FRC, due for implementation in company reports from 1 November 2003 and were at last accepted with widespread support.

The Financial Services Authority (see **10.7**) has also replaced the 1998 Code with the revised Code. The 2003 Code applies to reporting years beginning on or after 1 November 2003.

The 2003 Code included guidance on how to comply with particular parts of the Code:

- 'Internal Control: Guidance for directors on the Combined Code', produced by the Turnbull Committee, which relates to Code provisions on internal control (C.2 and part C.3 in the Code); and
- 'Audit Committees: Combined Code Guidance', produced by the Smith Group, which relates to the provisions on audit committees and auditors (C.3 of the Code).

In both cases the guidance suggests ways of applying the relevant code principles and of complying with the relevant Code provisions. In addition, it includes suggestions for good practice from the Higgs Report.

The Combined Code includes the corporate governance principles which the London Stock Exchange listed companies are expected to adhere to as part of their listing rules.

> 'Listed companies..have to describe how they apply the Code's main and supporting principles and either confirm that they comply with the Code's provisions or provide an explanation to shareholders.'
>
> (*FRC Press Notice 75, 23 July 2003*) (*see also* **10.7**)

The Combined Code reflects the principles developed in and following the Cadbury Report (see **10.3**).

The 2003 Code does not include material in the earlier Code on the disclosure of directors' remuneration. This is because the *Directors' Remuneration Report Regulations 2002* (*SI 2002 No 1986*)have entered into force and supersede earlier provisions in the 1998 Code.

London Stock Exchange Listing Rules **10.7**

Mention has already been made of what are referred to as listed companies (see **Chapter 4**). In order to apply to be a member of, and remain listed on, the London Stock Exchange companies must comply with the Financial Services Authority (FSA) Listing Rules. The FSA Listing Rules require listed companies to supply the following information in their annual report and accounts under section 12.43A:

- a narrative statement of how it has applied the principles set out in Section 1 of the Combined Code, providing an explanation that enables its shareholders to evaluate how the principles have been applied;
- a statement as to whether or not it has complied throughout the accounting period with the Code provisions set out in Section 1 of the Combined Code. A company that has not complied with the Code provisions, or complied only with some of the Code provisions or, in the

case of provisions whose requirements are of a continuing nature, complied for only part of the period, must specify the Code provisions with which is has not complied and, where relevant or applicable, for what part of the period such non-compliance continued. The company must give reasons for any non-compliance; and

- a report to the shareholders by the board that must contain a statement of the company's policy on executive directors' remuneration.

The rules require detailed discussion of financial issues such as:

- share options;
- long term incentive plans; and
- service agreements.

The Turnbull Report 10.8

Subsequent developments with regard to corporate governance in the UK led to the guidance of the Turnbull working Party on Internal Control, known as the Turnbull Report[6], in September 1999.This required the board of directors to confirm that there was an on-going process for identifying, evaluating and managing the key business risks. Shareholders, after all, are entitled to ask if all the significant risks have been reviewed and whether appropriate actions have been taken to mitigate them. They are also entitled to question why any event causing financial damage was not anticipated and acted upon.

In this context, it was observed that the one common denominator behind the past failures in the corporate world was the lack of effective risk management. As a result, risk management subsequently grew in importance and is now seen as highly crucial to the achievement of business objectives by the corporate world.

It was clear, moreover, that boards of directors were not only responsible for, but also needed guidance on, reviewing the effectiveness of internal controls and for providing assurance that all the significant risks had been reviewed. Furthermore, assurance was also required that the risks had been managed and an embedded risk management process was in place. In many companies this challenge was being passed on to the Internal audit function.

Therefore, in terms of the framework for corporate governance in the UK the Turnbull Report represents a major key document to enable an understanding of the key steps in the background to corporate governance. It provides guidance on all of the controls as opposed to only internal financial controls. The final date for compliance with the Turnbull Report was 23 December 2000. In setting out internal control guidance notes it has regard to the following:

- in Principle D2 of the Combined Code it is stated that the board 'should maintain a sound system of internal control to safeguard shareholders' investment and the company's assets'; and
- guidance was sought to enable companies to establish what was required to comply with Principle D2.

In accordance with the above, the Working Party stated that in determining its policies with regard to internal control, and thereby assessing what constitutes a sound system of internal control, the board should consider:

- the nature and extent of risk facing the company;
- the extent and categories of risk which it regards as acceptable for the company to bear;
- the likelihood of the risks concerned materialising;
- the company's ability to reduce the incidence and impact on the business of risks that do materialise; and
- the costs of operating particular controls relative to the benefit thereby received.

It is helpful to summarise the Turnbull Report for the purpose of the discussion on risk that follows. Regarding the review of the effectiveness of internal control it states that:

- the board should receive and review reports; and
- the board should undertake an annual assessment to the effect that it has considered significant risks.

As regards the board's statement on internal controls:

- evaluating and managing significant risks; and
- the disclosure should provide meaningful, high level information.

A series of questions to assess the effectiveness of the company's risk and control processes include:

- a risk assessment;
- a control environment and the control of activities;
- information and communication; and
- monitoring.

Developments post-Turnbull 10.9

Since the guidance was published in 1999 and as a result of the Report, there have been many media reports, conferences and workshops to highlight the growing importance of risk management. Since then there has been a growing awareness of risk management and, together with the ongoing scandals that have been the subject of media headlines globally, it has become true to say that risk management is the key issue for business in terms of its internal due diligence and corporate governance.

Relevant matters that have been under discussion have included:

- how far companies have changed their approach to risk in practice;
- which issues have been singled out as priority concerns;
- to what extent listed companies simply pay lip service to the review of internal processes in their annual report; and
- how far risk management is embedded in the management of the business in accordance with the Turnbull requirements.

This has also meant that businesses have had to change their approach to risk management in order to:

- align their risk management structure to follow the Turnbull requirements;
- define and disclose the optimal structure for their risk management framework;
- assess and explain their risk controls to promote shareholder confidence;
- assess responsibilities with regard to their risk management function;
- identify risk management tools; and
- ensure that their organisation is able to respond quickly to the ever changing nature of risks.

Optimising the impact of new requirements **10.10**

Developments concerning the role of the audit committee and the risk manager in the business, together with their relationship with the board, have led to the optimising of corporate governance, business risk assessment and the internal control framework. Companies have had to consider:

- linking their corporate governance framework with their internal control framework to achieve the best value;
- whether Internal audit and risk management should be merged or form an alliance;
- using their governance and control framework to embed risk management into their organisation;
- ensuring that they implement a dynamic internal control framework; and

- determining and embedding responsibilities and accountabilities for governance risk management and internal control.

The commitment of the board 10.11

In order to obtain the involvement and commitment from the board of directors to an improved risk control framework, risk management and control advice has been important for all relevant staff across the organisation. In addition the role of the audit has had to be considered to fulfil the Turnbull requirements. Companies have needed to provide independent and objective assurance through the audit to the board about the adequacy and effectiveness of key internal controls and other risk management activities across the organisation. The auditors can act effectively as risk and control educators to all relevant staff in the business. In order to do so steps should be taken to:

- educate the board to the benefits and opportunities of the Turnbull recommendations for improving the management of risk, providing transparency and clarity of controls and improving business performance and shareholder value;
- determine the priorities of the board and identify the risk areas that affect shareholder value;
- communicate the internal control failures and necessary actions, using business – rather than risk – language to enable improvements to the existing internal control framework; and
- gain sponsorship from the board.

Risk management and shareholder value 10.12

When considering the impact of the Turnbull Report on risk management and shareholder value over the last few years companies have worked to:

- analyse the impact of greater disclosure of risk management processes on shareholder value;
- manage all aspects of risk effectively;
- address the possibility of insurance as a relevant strategy to identify whether risk transfer addresses the correct risks and is desirable in a modern economic model;
- assess the level of technical expertise among the directors and identify the new issues that they need to address such as the efficacy of the information supply lines, the need to appoint new directors and the expectation of shareholders; and
- understand how the failure to provide the right people with the right skills can lead to executive errors.

When reviewing board issues, companies have also had to address:

- how the board should be structured to compensate for shortcomings;
- the extent to which directors need effective training;
- the way in which the risk of material mis-statement can be mitigated effectively;
- the identifying of the role of directors and auditors in providing assurance; and
- complying with the requirement to separate critical boardroom roles and establish clear responsibilities between the chairman, chief executive and the independent senior non-executive director.

Embedding the risk management processes 10.13

The business world has improved risk awareness in order to:

- embed risk management processes in the organisation to meet the requirements;
- improve business performance; and
- gain competitive advantage.

The goals are to:

- establish best practice, using the Turnbull Report to create an embedded risk management process;
- establish a business risk exploitation programme;
- identify and evaluate the significant risks;
- deal with exposures and exploit opportunities;
- integrate the process within any existing risk management framework;
- monitor key risks;
- co-ordinate assurance activities across the organisation;
- refocus the function of internal audit;
- create a continuous embedded process; and
- measure benefits.

Legal risks 10.14

As has been indicated in other chapters (such as **CHAPTERS 6** and **8**) he management of risks has evident implications for legal risk management and internal due diligence.

In order to minimise legal risk a company must:

- understand the legal challenges if directors exercise judgement in areas in which they have not done before and things go wrong;
- identify the specific types of proceedings to be concerned about, such as potential exposure to negligence claims or the potential for litigation of US shareholders;
- review the role of the in-house lawyer to enable a dynamic growing business;
- grasp the connection between risk management and legal relationships with customers, suppliers, regulators, employees and consumers;
- adopt a legal risk management approach that is flexible enough to identify the changes in risk; and
- achieve an active embedded cultural concept of legal risk management.

It is also important to relate the above to the company's corporate governance approach.

Reporting Turnbull compliance 10.15

There is no doubt that the reporting trend is here to stay. Companies that have become compliant with the Turnbull requirements have had to consider:

- what to write in their disclosure in the annual report in order to optimise the wording disclosure and reporting of their risk management structure;
- identifying which information is 'meaningful' or 'high level' and should therefore be disclosed;
- ensuring that transparency is promoted and misleading statements are avoided;
- implementing a risk management strategy to assess the probability and impacts of key risks on the company's finances and reputation; and
- providing reasonable assurance to executive management and the board about the adequacy and effectiveness of the risk management and control framework.

Recent developments 10.16

As has been indicated earlier the reporting of major corporate scandals has led to a rapid regulatory response in many countries that has in turn impacted hugely on the way companies do business. One of the main examples of such regulatory response has been the enactment of the *Sarbanes Oxley Act of 2002* in the US. This legislation, known as the *SOX* legislation, has had such ramifications for business that the legislation is assessed in depth in **CHAPTER 11.**

For the present purposes it is important to bear in mind that the *SOX* legislation:

- relates to all US listed companies, including overseas companies with US listing;
- introduced a huge change as regards the accountability of the board; and
- brought about a 'sea change' in audit practice.

Other major developments have been the review in the UK of the role and effectiveness of non-executive directors known as 'the Higgs Report' and the audit committees Combined Code Guidance known as 'the Smith Report'. The former considered the role of non-executive directors in depth and proposes changes to the Combined Code and the latter deals with the role of the audit committee.

Practical corporate governance 10.17

Having regard to the definitions referred to in **10.1** and the interaction with risk management that has been highlighted, certain steps may be identified to establish a system and structure of corporate governance. The key steps of a system that has regard to the requirements referred to in **10.8** to **10.15** can be proposed. These key steps are:

- establish and implement a risk management framework;
- ensure the full commitment and support of the board and the chief executive officer (CEO) to best practice as regards corporate governance standards and activities;
- establish an audit committee;
- establish a remuneration committee that fulfils the Greenbury requirements;
- establish a nomination committee;
- organise the board agenda to cover standing items that deal properly with corporate governance;
- devise a programme of shareholder communication;
- prepare an annual timetable to ensure effective preparation for the AGM;
- monitor changes in the context of corporate governance and plan appropriate responses; and
- ensure that there are adequate resources for corporate governance and risk management functions.

All of these steps can be approached following some of the tips below.

Risk management framework **10.18**

Bearing in mind the Turnbull Guidelines in particular, in today's business world it is now imperative for companies to establish:

- an internal risk management structure using a formal risk management methodology, including:
  - risk identification techniques;
  - quantitative and qualitative risk assessment; and
  - cost effective methods for risk reduction and transfer;
- a process such as the UK Risk Management Process contained in the Standard (see **10.33**); and
- a review process by internal audit to ensure the effective operation of the risk management framework.

Board commitment **10.19**

Without top to bottom commitment the introduction of most policies will not succeed. This is certainly true of corporate governance where, as has been seen, many of the issues are directed at the board in particular. Therefore the CEO and the board must be fully committed to good standards of corporate governance and must support and promote all related corporate governance activities. Without the full commitment of the board any system will fail. To assist the implementation of a practical system the board should agree and sign off the risk management activities through, for example:

- the engagement of a risk manager and support staff;
- the financial resources to enable the undertaking of risk control activities; and
- proper disaster recovery planning.

It is vital that the board creates the culture of commitment to good corporate governance by demonstrating an example and expecting effective risk management and corporate governance from senior management and staff.

The audit committee **10.20**

It is important that the company establishes an audit committee as part of the corporate governance system and structure. The Smith Report, which was a major report entitled 'Audit Committees – Combined Guidance: A report and proposed guidance by an FRC-appointed group', should be examined carefully by the chairman of the audit committee. In any event:

- the company should aim to have an audit committee or explain in the annual report why it does not have one and review the need for one from time to time:
- the audit committee should consist if at least three non-executive directors and have clear terms of reference and authority and duties; and
- the audit committee should keep the work of the internal audit function and the results of the external auditors findings under review.

The remuneration committee 10.21

In view of the sensitivity of the remuneration of directors and the manifold problems that have occurred in this regard, strict and clear guidance has emerged for companies, as follows:

- the company should establish a remuneration committee;
- the remuneration committee should consist of non-executive directors who are free from any business or other relationship that could materially affect or interfere with their judgement;
- all bonuses that are payable to directors should be based on performance and designed to enhance the business;
- the remuneration committee should establish levels of remuneration for directors that are sufficient to attract and retain the directors needed to run the company successfully, having regard to the external market and relevant benchmarks;
- a stated policy on remuneration should be prepared for communication to the shareholders; and
- the statement of actual salaries of directors and the remuneration report should be prepared for the annual report.

The nomination committee 10.22

The establishment of a nomination committee also helps to endorse the commitment to corporate governance. Certain points should be borne in mind:

- a nomination committee should be set up to make recommendations to the board in connection with all appointments to the board unless the board is small;
- the nomination committee should mainly consist of those non-executive directors that are most suitable for the task; and

- the board should be ready, willing and prepared to respond to shareholders' questions regarding why a particular director has been nominated or has been approved.

The board agenda 10.23

It cannot be overemphasised that the board must be positively engaged in risk management in an ongoing manner. It is also important for the board to be guided on these issues with an appropriate agenda. Therefore the company should consider that:

- the company secretary should draft board agendas and set aside appropriate time in order that meetings regarding corporate governance matters receive full and proper attention;
- the meetings should occur on a regular basis, generally monthly, so that standing items on the agenda can be dealt with including:
  - the risk management report, with details of any breaches of internal financial controls;
  - proposed new business activities; and
  - any new risks that are material;
- the items that are periodic would generally cover:
  - reports – usually quarterly – from the audit committee;
  - the remuneration report, usually annually or as required; and
  - board nominations, at least annually or as required.

Shareholder communication 10.24

Many of the recent concerns have shaken stakeholder confidence in companies and can be addressed only by appropriate communication. Key points on appropriate communication are listed below.

1. The Combined Code gives a clear expectation that institutional investors should use their voting powers with discretion.
2. Boards of companies should be ready to enter into a dialogue with institutional investors.
3. In accordance with the Code boards should use the AGM to communicate with private investors and encourage their participation, by voting on resolutions, expressing opinions and asking questions.
4. The board should consider the best way forward for a programme of communication.

The annual timetable 10.25

As part of the commitment to demonstrate best practice and good corporate governance, the company should have an appropriate annual timetable that fulfils the requirements of the regulatory framework. Such a schedule ensures proper preparation for the AGM. With this in mind the schedule should:

- ensure that all board meetings are set a year in advance;
- confirm regular dates for reports from the risk management team, audit committee, nomination committee and remuneration committee; and
- ensure that directors are provided with all relevant information in good time to enable them to conduct a review of the effectiveness of the system of financial control and sign-off of the report to shareholders that they have done so.

Monitor changes and plans 10.26

It should always be borne in mind that this is a dynamic subject and that it has practical repercussions on the day to day operation of the company. Therefore any changes in corporate governance should be followed up and appropriate responses developed. In particular it should be recalled that:

- since 1992 the development of the corporate requirements of listing authorities has taken place;
- meanwhile the expectations of shareholders have been developing and higher standards of management are required now;
- the Higgs Report and the Smith Report have endorsed the expansion of corporate governance; and
- the new Combined Code is showing the way forward.

Resources for risk management and corporate governance 10.27

The allocation of resources is always a sensitive issue. There is no doubt that, as with other policies, the success of a strategy for corporate governance depends upon the resources allocated to it. Therefore:

- successful risk management and corporate governance strategies depend upon the qualifications, experience and commitment of the organisation's staff;
- as this may be a new, and fast growing, area of corporate activity there should be an ongoing assessment of the numbers of staff required; and

- as a result of the link to the audit function it should be noted that there is a trend for accounting personnel to be selected into risk management positions even though this is not always appropriate.

The role of risk management 10.28

In many contexts, particularly in events following the Turnbull Report in 1999 risk and risk management was regarded as a threat to business. Risk management should not only be considered to be a threat – it should also be recognised as an opportunity. With this in mind a company should consider that:

- the role of risk management is a function to optimise outcomes such as profit objectives, return on investment and performance measures, that is value creation;
- there are strong loss prevention implications;
- effective and comprehensive risk management facilitates improved risk taking analysis to improve decision taking;
- this is a role that has grown from insurance and health and safety, as well as accounting; and
- it is very closely related to corporate governance.

The development of risk management is a subject that requires extensive comment. While it is not the place to comment in great detail in this handbook, some of the developments that are relevant to corporate governance should be touched upon in the interests of clarity. A brief discussion is set out below in the following paragraphs.

The risk manager 10.29

In order to fulfil the corporate governance requirements it is proposed that there should be a risk management round table. The members of the round table would carry out their tasks while the risk manager would undertake facilitation and administration. The round table would consist of representatives from the following departments:

- legal/secretarial;
- finance;
- treasury;
- IT;
- corporate communications and branding;
- health and safety;
- human resources;
- security;
- quality; and
- procurement.

Senior management 10.30

The matter of commitment has already been pointed out on several occasions. In order to ensure change it is vital that senior risk management members of staff support the risk management strategy. Senior management can ensure:

- authorisation for the risk manager's function by the senior management team;
- top level sponsorship, which usually provides the most effective way to open doors and build networks;
- identify and prioritise necessary changes through risk analysis;
- enable best practice reports for the board as they bear ultimate responsibility for decision making; and
- recognition and value for the function.

Periodically the corporate governance and risk management activities of the organisation should be assessed externally. Advice from external auditors should be taken by the board on their performance to understand whether they are effective in what they do. It has been recognised that peer group reviews, as well as major investors' views should be sought.

The future 10.31

Bearing in mind the evolving approach to corporate governance and risk management in the future, the focus of the role of the risk manager could include:

- risk financing as a vital balance sheet management tool;
- a strong interest in risk reduction rather than avoidance;
- an emphasis on profiling and mapping;
- stewardship of the risk management and corporate governance framework;
- selling the agenda for the risk management round table;
- undertaking research and keeping or storing risk information; and
- assisting completion reports to senior management;

It seems clear that the demand for the traditional risk manager will continue to expand in today's demanding business world, as well as the role or function of systems and methodologies.

Systems and methodologies for risk management **10.32**

In order to address risk management in practice and having regard to the Cadbury Report's definition of corporate governance (see **10.1**) the meaning of methodologies and systems should be also considered.

According to the Concise Oxford Dictionary a system is 'a complex whole, a set of connected things or parts'. A method is defined as 'a special form of procedure; the orderly arrangement of ideas'. Whereas risk management can be generally understood as the management of inherent risk, thereby leaving 'residual risk' for the business, risk management methodology can be defined as 'a set of connected ideas that are applied in practice as an orderly arrangement of actions'. Risk managers and practitioners, however, often have very different views on the order of actions and the terminology that is applied in the area of risk management can vary considerably. Nevertheless for the purpose of this discussion a practical approach to risk strategy is helpful. As can be seen elsewhere in this handbook, (such as **CHAPTERS 4**, **6**, **7** and **8**) as well as in other legal and business publications, a risk strategy can, for example, cover:

- ethical, social and reputational risks;
- new and emerging risks;
- developments in directors' liabilities;
- Turnbull and its practical implications; and
- exploiting risks as opportunities.

Therefore a risk management framework might cover:

- types of product – liability/recall;
- intellectual property; licensing;
- reputation;
- environment;
- health and safety;
- business interruption; and
- ethical issues.

As is mentioned in the discussion ofTurnbull in particular, it is most important to achieve the commitment of the board and buy-in for business risk management. It is vital that the board is convinced that risk management is good practice and that it:

- can be used to demonstrate a level of corporate governance that complies with the requirements of the Combined Code and associated Turnbull Report;

- will avoid a poor audit report;
- will help mitigate health and safety issues;
- will combat directors' criminal liabilities (including corporate manslaughter charges); and
- is a value-adding, profit related exercise.

With this in mind, a risk manager has to apply a risk management methodology that maximises efficient use of the board's time when addressing risk control measures. They should build the confidence of the board that all significant risks have been identified and are being controlled so that the impact of embedding risk management into the framework of the business reaps clear results.

Accordingly, risk management standards have been introduced to provide a methodology for risk management that can be referred to and be adapted to the needs of the business.

Below **10.33** highlights the main aspects of such standards – set out as a practical overview. One useful approach is to base the organisation's methodology on the risk management standard (below IN **10.34**).

Risk management standards **10.33**

In 1995 the first comprehensive risk management standard was published in Australia after a consultation process administered by academics. This was updated in 1999. This Standard was published by the Australia and New Zealand Standards Authorities (for further discussion of corporate governance in Australia see **CHAPTER 14** and www.standards.com.au). In 2002 the UK Risk Management Standard was published after a joint initiative was undertaken by the Institute of Risk Management (IRM), the Association of Insurers and Risk Managers (AIRMIC) and ALARM (see www.irm.org). This received input from the Association of British Insurers (ABI) and the Institute of Internal Auditors (IIA) in the UK but is yet to be adopted by the British Standards Institute (BSI).

The main features of the Australasian and British Standards are:

- the concept of 'flow-process' risk is promoted by both;
- both contain definitions of risk terminology;
- the UK Standard incorporates the International Organization for Standardization (ISO) Guide 73[7] risk management vocabulary; and
- neither Standard is compulsory or prescriptive.

The UK Risk Management Standard **10.34**

The UK Risk Management Standard sets out a risk management process. The key characteristics of the process that form part of the audit process and modification are:

- setting the organisation's strategic objectives;
- risk assessment;
- risk analysis;
- risk identification;
- risk description;
- risk estimation;
- risk evaluation;
- risk reporting;
- threats and opportunities;
- decision;
- risk treatment;
- residual risk reporting; and
- monitoring.

For more detail regarding the elements of the Standard see www.theirm.org and the ISO Risk Vocabulary[7]. In accordance with the Standard, the approach an organisation has towards risk management will be guided by its strategic objectives such as:

- the risk environment including financial and political issues;
- the nature of the business, that is the sector, for example chemicals or healthcare, and the high risk activities as in the airline, shipping or construction business;
- the effect of external influences such as regulatory framework and listing requirements.

As a matter of best practice, risk management should take into account the legal, regulatory, moral and contractual limitations or constraints.

Following is a summary of a selection of the key characteristics of the process.

Risk assessment **10.35**

Risk assessment is the first stage of the process and is defined by the ISO Guide as 'the overall process of risk analysis and risk evaluation'. Since the identification of risks is vital for the treatment of risk, risk evaluation is part of the risk analysis. The risks should be set out or be noted down. The estimates of the potential likelihood and impact of the risk event must be made.

Risk analysis 10.36

Risk analysis forms the second stage, in the following manner:

- risk identification;
- risk description; and
- risk estimation.

Risk identification 10.37

Risk identification is the 'process to find, list and characterise [describe] elements of risk according to the Risk Management Standard'. While it is possible to identify risks without formal processes this is unlikely to be so effective. Therefore it should be approached in a structured way in order to ensure that all significant activities within the organisation have been captured and the risks flowing from these activities are identified. The results must be co-ordinated and centralised. There are very well established formal risk identification techniques and these include:

- the brainstorming of risks through workshops at different organisational levels including the board, management and operational staff, as well as across different departments, subsidiaries and geographic regions;
- business process and scenario analysis;
- flow process analysis;
- incident investigation;
- auditing and inspection;
- hazard and operability studies (HAZOPs);
- fault trees; and
- statistical analysis.

In terms of quantitative and qualitative risk analysis each risk that is identified and described must be measured, either using quantified techniques or qualitative views. There are various known quantitative techniques such as Monte Carlo simulation, which randomly generates values for uncertain variables over and over to simulate a model; commercial software is also available, for example @Risk and loss trend forecasting.

The brainstorming period is crucial since it enables input from across the organisation. During the brainstorming and risk identification workshops subjective views and financial impact can be gathered. The brainstorming session should be facilitated by someone who is familiar with risk management. This will help direct the debate and identify relevant risk issues. Nevertheless all comments are valid and all ideas noted for later discussion. When the risks have been captured they can be grouped according to type, such as:

- strategic – long-term objectives;
- operational – day-to-day activities;
- financial – internal and external financial risk factors;
- knowledge management – control and protection of knowledge resources; and
- compliance – legal or regulatory.

Risk description and estimation **10.38**

The purpose of risk description is to display the identified risks in a structured format, for instance through a risk map (see **10.40**). As regards the estimation of risk, this can be quantitative, semi-quantitative or qualitative in terms of the probability of occurrence and potential impact. The estimate of impact may relate to threats or downside risks and opportunities or upside risks. The estimated probability and impact is over-laid on the risk map and can be finite or grouped by consequence and risk. By way of example as regards fire risk:

- probability of fire: low, medium or high or one in five years, ten years or 25 years etc; and
- potential impact of fire: low, medium or high: up to £1m, £1–5m or over £5m.

Risk prioritisation and risk mapping **10.39**

The advantages of risk prioritisation are that an agenda for action is produced and cost benefit judgements can be made. There are some tips that can be referred to by the business such as:

- each risk will have some probability and frequency rating attached to it;
- the current risk control status will reduce a 'gross' risk to a 'net' or 'current' risk;
- the risks can be listed according to the greatest financial impact; and
- the listing will never be final as it is so hard to accurately quantify exposures.

Risk mapping is the identification, quantification and prioritisation of risks that enables a clear risk profile or risk map to be constructed. A risk map enables a central focal point for risk treatment activities and to record gross and net risk. Based upon the risk map summarised reports to the board can be prepared. Risk mapping is therefore a process that is a key part of monitoring and review of the risk management process.

Risk treatment **10.40**

Corporate governance practices require directors to have good risk management information. According to the Cadbury Report[1] 'Boards should have a formal schedule of matters specifically reserved to them including..risk management policies' (Cadbury, paragraph 4.23). In order to comply with this requirement the organisation's upward reporting structure should be such that the right people are informed of the risk profile. In addition risk treatment activities should be properly assigned so that 'risk owners' are noted in the risk profile schedules. Risk treatment is the process of selecting and implementing measures to modify the risk. The key elements of risk treatment are:

(*a*) risk control – it is quite clear that risk control is the core function of risk management; its objective is the reduction or removal of the underlying risks of the business. Risk controls can of course cover a wide range of activities, such as:

- internal financial controls;
- good contract management;
- product efficacy and safety;
- health and safety of employees and the public;
- sprinkler and other fire protection;
- construction standards; and
- security.

(*b*) risk transfer other than insurance – can be achieved by several methods, other than insurance, the most usual being by contract conditions and outplacement or joint venture structures:

- contractual risk transfer includes statements in the contract regarding who has responsibility for any loss events, known as indemnity and 'hold harmless' clauses (see also **CHAPTER 1**); and
- risk transfer by outplacement has the effect of moving responsibility for risk control to another party.

Despite such methods the risk transfer alone does not control or reduce the 'gross' risk.

(*c*) risk financing including insurance – also involves several methods that do include insurance. The various sources of finance include:

- insurance;
- cash flow;
- bank lines of credit;
- reserves in the bank; and
- shareholders funds.

Generally insurance is acquired when there is a legal requirement and when the organisation decides that the premium payable – the price – is worth the transfer of the risk.

(*d*) risk avoidance – can occur through strategic decisions such as withdrawing from a market sector or region such as the manufacture of chemicals and being in the volatile film industry or the sale of products in the US. Apart from such withdrawal other risk avoidance strategies include the purchase of forward contracts for the supply of raw materials or for the sale of the organisation's product. One caution is that there may be very limited scope in practice to avoid risks since the avoidance of one risk can often mean exposure to other risks.

Residual risk **10.41**

It is important to recall that the above exercise is intended here to enable a risk management strategy to fulfil the requirements of corporate governance. Part of a good risk management structure is the presence of contingency planning to address the effects of adverse events as and when they occur. Since effective risk management involves the continuous assessment of both the risk environment and the operation of the identification, analysis and treatment process monitoring should take place. Risks are always changing and the response should be effective and rapid. Therefore an external review or audit of the overall process is necessary to enable the senior management to be confident that good risk management is in place and that the residual risk really reflects good corporate governance practice.

The Operating and Financial Review **10.42**

The 'Company Law Review' (Department for Trade and Industry (DTI), July 2002) identified the fact that company accounting and reporting was backward-looking and based on financial indicators. There are few statutory requirements to report on the main qualitative factors which underlie past and future performance, in particular on intangible, and so-called 'soft' assets (which may contribute significantly to success but are not well captured in traditional financial statements); and on key business and wider relationships. A good reputation is considered central to every company in its acquittal of its responsibilities (see also **CHAPTER 7**).

The legal requirement to produce an Operating and Financial Review (OFR) will apply to 1290 companies in the UK (see also **CHAPTERS 4** and **7**). The Accounting Standard Board (ASB) will issue an exposure draft of the first OFR standard in 2004, and finalise it by 1 January 2005. Companies will be expected to prepare OFRs for the first time for financial years beginning on or after 1 January 2005. The process is linked to the annual return and accounts, so the first OFRs could appear soon after.

Subsequently, the OFR, under principles, advises directors to explore and understand the agendas not only of shareholders but also of other stakeholders – employees, customers, suppliers, non governmental organisations (NGOs), local communities, society – that are likely, directly or indirectly, to influence the performance of the business and its value. A further requirement is to include appropriate performance indicators, where directors will want to include quantitative data wherever they can. The OFR may need to address anticipated events, and include an assessment of non-financial issues that can have significant consequences for future performance and value. Directors may wish to address the link between their strategy and the key attributes upon which the company's reputation is built. Directors should consider the approaches adopted by others in the same industry to see whether any consistent approaches might be appropriate in their own case.

In terms of process, the OFR recommends mapping, ie audit trails, and references to best practice. Account needs to be taken of actual and potential sources of value creation in the business as well as the factors that may prejudice value creation. There may be an incentive for special work to be done or advice obtained on a particular issue or topic that the board feels may be important but that does not fit readily into existing management structures, in the social and environmental areas for example. Businesses with significant groups of stakeholders are likely to benefit from the results of consultations with these groups to ascertain their view of the key issues, and the potential effect on business reputation. Boards may need to decide on an appropriate approach for 'new' information.

Conclusion 10.43

Today there are several issues relevant to corporate governance such as, for instance, company law reforms, director's compensation, self-regulation and the role of audit committees. However, several business imperatives will drive the move towards high levels of corporate governance in the twenty-first century.

1. What are the pressures or fears that force business leaders to make decisions detrimental to the company?
2. How can they be helped to be more integral to their own organisations?
3. How can the board of directors play a more useful role?

The owners have to bring about an attitudinal change in them and identify with the aspirations of each stakeholder. One way in which organisations can improve their standards of corporate governance in the future is by focusing on how they choose external directors – to be selected mainly on the basis of their experience and expertise.

Modern corporate governance also emphasises the role of shareholders and financial institutions. Shareholders should be involved in all major decisions and financial institutions should appoint functional experts to represent them on the boards of companies in which they have substantial holdings.

Effective corporate governance requires a clear understanding of the respective roles of the board, of senior management and their relationships with others in the corporate structure. The relationships of the board and management with shareholders should be characterised by candour; their relationships with employees should be characterised by fairness; their relationships with the communities in which they operate should be characterised by good citizenship and their relationships with government should be characterised by a commitment to compliance.

The corporate environment elsewhere, such as India and Australia, could not remain indifferent to the developments that were taking place in the UK. In fact, as indicated earlier in **10.1**, the developments in the UK have had tremendous influence on the developments in other countries. In places such as India they triggered the thinking process that has led them to laying down their own ground rules on corporate governance (see **CHAPTER 12**). For example, as a result of the interest generated in the corporate sector by the Cadbury Committee's report[1], the issue of corporate governance was studied in depth and dealt with by the Confederation of Indian Industry (CII), the Associated Chamber of Commerce and Industry of India and the Securities and Exchange Board of India (SEBI) through Kumar Mangalam Birla Committee Report (cited In **10.1**). Though some of the studies did touch upon shareholders' right to vote by ballot and a few other issues of general nature, none can claim to be wider than the Cadbury report in scope. With this in mind it is interesting to look upon the corporate governance model in other countries (see **CHAPTERS 12** to **15**).

References

1. Committee on the Financial Aspects of Corporate Governance. *Report with Code of Best Practice,* (Cadbury Report), London: Gee Publishing. 1992.

2. Confederation of British Industry. The Greenbury Committee. *Directors' Remuneration – Report of a Study Group* (chaired by Sir Richard Greenbury). July 1995.

3. Committee on Corporate Governance – The Hampel Committee (January 1998), Final Report, chaired by Sir Ronald Hampel. Gee Publishing, London, ISBN 1 86089 034 2.

4. Financial Reporting Council press Notice 71, 20 January 2003.
5. Financial Reporting Council Press Notice 74, 14 May 2003.
6. Institute of Chartered Accountants. *Internal Control: Guidance for Directors on the Combined Code.* (The Turnbull Report.) 1999.
7. ISO/IEC Guide 73: *Risk management. Vocabulary. Guidelines for use in standards* – available from ISO national member institutes (see www.iso.org).

Chapter 11

The Sarbanes-Oxley Act of 2002

> 'The Sarbanes-Oxley Act is arguably the most sweeping and important US federal securities legislation affecting public companies and other market participants since the SEC was created in 1934.'
>
> *Ethiopis Tafara, Acting Director, Office of International Affairs, US Securities and Exchange Commission (June 10, 2003)*

Introduction **11.1**

The last five years have seen an unprecedented upswing in reports of accounting and other irregularities at public companies. There was a time, a few years ago, when every day seemed to bring with it a new report that a public company was in the process of re-stating its earnings, seeking bankruptcy protection, fighting criminal indictments, or sometimes all of the above. In the wake of these developments, it was a hardly a surprise when the United States legislature took action. On July 30 2002, President Bush signed into law the *Corporate Accounting Reform and Investor Protection Act of 2002*, which is now almost universally referred to as the *Sarbanes-Oxley Act of 2002* (*SOX*) (Pub. L. No 107–204 116 Stat. 745). Despite its swift drafting and passage, the *SOX* will likely have profound implications not just for public companies, but for all companies doing business in the United States.

It is important to first acknowledge the scope of *SOX*. While the scope of the Act may seem, at least on its face, to be quite limited, the more accurate answer may be that only time will tell the true reach of the Act. By its own terms, the *SOX* applies only to public companies and the accounting firms that audit them. In the long-run, however, the changes to corporate governance and accounting imposed by the *SOX* may gradually become institutionalised as a series of best practices that are adopted by non-public companies as an appropriate baseline. Furthermore, it is certainly possible that states, and indeed the United States itself, may ultimately decide that the principles enunciated in the *SOX* should be equally applicable to private companies.

At its core, the *SOX* has five primary areas of focus: the Act:

- establishes a Public Company Accounting Oversight Board (sometimes referred to as the PCAOB or the Board);

- sets forth new and revised standards for auditor independence, many of which are carryovers from the auditor independence rules adopted by the Securities and Exchange Commission (SEC) in 2002;
- promotes revised corporate governance guidelines addressing audit committees, conflicts of interest for officers and directors and enhanced financial disclosure;
- authorises numerous studies and reports to assess various concerns that gave rise to the Act; and
- imposes new and strengthens existing criminal penalties for violations of the Act and the securities laws.

See generally *Summary of SEC Actions and SEC Related Provisions Pursuant to the Sarbanes-Oxley Act of 2002*, issued by the SEC (July 30, 2003) (summarising SEC-related aspects of the Act).

In so far as this US legislation has had such extensive ramifications globally it is important to examine the details of this regulatory framework.

Basic goals of act

Public Company Accounting Oversight Board 11.2

The *SOX* established the Public Company Account Oversight Board (PCAOB) as an oversight body for accounting firms (*SOX § 101, 15 USC § 7211*). Given the historical focus on self-regulation in the accounting industry, the concept of any type of external regulatory body was tremendously contentious. Ultimately, the compromise reached by the drafters of the Act was that the PCAOB, unlike the SEC and other regulatory bodies, would be a non-profit corporation, not an agency or establishment of the government. Accordingly, the PCAOB embodied a compromise and something of a hybrid – neither a part of the government, nor of the accounting industry.

> 'The Board shall be a body corporate, operate as a nonprofit corporation, and have succession until dissolved by an Act of Congress … The Board shall not be an agency or establishment of the United States Government, and, except as otherwise provided in this Act, shall be subject to, and have all the powers conferred upon a nonprofit corporation by, the District of Columbia Nonprofit Corporation Act.' (*SOX § 101(a)-(b), 15 USC § 7211(a)-(b)*)

The PCAOB is not, however, entirely autonomous and is subject to a certain degree of SEC oversight (*SOX § 107(a), 15 USC § 7217(a)*).

The PCAOB has three primary responsibilities which may be generalised as registration, promulgation, and regulation. Under the Act, all accounting firms that prepare or issue, or participate in the preparation or issuance, of audit reports for public companies were required to register with the PCAOB within 180 days of its creation (*SOX § 102(a), 15 USC § 7212(a)*). The registration fee ranges, depending upon how many public company clients the registrant has, from $250 to $390,000, (*SOX § 102(f), 15 USC § 7212(f)*) (authorisation of PCAOB to charge fees), PCAOB Rule 2103 (stating that the PCAOB will, from time to time, announce the amount of the registration fee). As of the date of this chapter, all of the accounting firms listed in the foregoing sentences should have registered with the PCAOB.

The PCAOB also has the duty of establishing auditing, quality control, ethics, independence, and other standards relating to the preparation of audit reports, (*SOX § 103, 15 USC § 7213*). Effective from 16 April 2003, the PCAOB adopted temporary, transitional standards addressing auditing (PCAOB Rule 3200T), attestation (PCAOB Rule 3300T), quality control (PCAOB Rule 3400T), ethics (PCAOB Rule 3500T), and independence (PCAOB Rule 3600T). The temporary rules were approved by the SEC on 25 April 2003, SEC Release No 33–8222. On 14 May 2004, the SEC approved the PCAOB's first auditing standard, *Auditing Standard No 1, References in Auditors' Reports to the Standards of the Public Company Accounting Oversight Board. SEC Release No 34–49707*. All of the PCAOB's rules are available in a compiled form on the PCAOB's website at http://www.pcaobus.org/rules/RulesOfTheBoard.pdf.

Finally, in order to 'police' auditing companies, the Act authorised and empowered the PCAOB to investigate registered accounting firms and to impose sanctions on them (everything from censure to monetary penalties), *SOX §§ 104–105, 15 USC § 7214–15*. The PCAOB has submitted rules to the SEC for approval which would allow the PCAOB to:

> 'conduct investigations concerning any acts or practices, or omissions to act, by registered public accounting firms and persons associated with such firms, or both, that may violate any provision of the Act, the rules of the Board, the provisions of the securities laws relating to the preparation and issuance of audit reports and the obligations and liabilities of accountants with respect thereto, including the rules of the Commission issued under the Act, or professional standards.' http://www.pcaobus.org/pcaob_enforcement.asp.

The PCAOB's rules would also:

> 'require registered public accounting firms and their associated persons to cooperate with PCAOB investigations, including producing documents and providing testimony ... [and would] permit the PCAOB to seek information from other persons, including clients of registered firms.' http://www.pcaobus.org/pcaob_enforcement.asp..

In order to close a potential loophole for foreign public accounting firms, the Act provides that any such firm that prepares an audit for an issuer must be subject to the Act and the jurisdiction of the PCAOB, *SOX § 106(a)(1), 15 USC § 7216(a)(1)*. Moreover, the PCAOB is empowered to subject foreign firms to registration and oversight requirements if the firm plays a 'substantial role' in the preparation of audits, *SOX § 106(a)(2), 15 USC § 7216(a)(2)*. By the same token, the SEC or the PCAOB (with SEC approval) may exempt foreign firms from some or all of the requirements of the Act, *SOX § 106(c), 15 USC § 7216(c)*. Note that a foreign firm's registration with the PCAOB does not, by itself, subject that firm to unlimited Federal and state court jurisdiction, *SOX § 106(a)(1), 15 USC § 7216(a)(1)*.

The PCAOB is governed by five PCAOB members, *SOX § 101(e)(1), 15 USC § 7211(e)(1)*, two of whom must be certified public accountants, *SOX § 101(e)(2), 15 USC § 7211(e)(2)*. PCAOB members may serve up to two five-year terms, provided that the first terms of the initial PCAOB members are staggered over a four-year period, SOX § 101(e)(5), 15 USC § 7211(e)(5). The current members of the PCAOB are: William J McDonough (Chairman), Kayla J Gillan, Daniel L Goelzer, Bill Gradison, and Charles D Niemeier. The PCAOB's website, http://www.pcaobus.org, contains information about various aspects of the PCAOB's activities, including registration information and the PCAOB's Standards Documentation.

Auditor independence 11.3

One of the most significant perceived problems addressed by the Act was the inherent conflicts in the relationship between auditors and their clients. In an effort to prevent future conflicts, the Act set forth new and revised standards for auditor independence, many of which are carryovers from the auditor independence rules adopted by the SEC in 2002, but are even more restrictive.

Under the Act, the following non-audit services, if provided to an audit client, would impair an accounting firm's independence and are therefore prohibited:

- bookkeeping or services related to accounting or financial statements;
- financial information systems design and implementation;
- appraisal or valuation services, fairness opinions or contribution in-kind reports;
- actuarial services;
- internal audit outsourcing services;
- management functions or human resources;
- broker or dealer, investment adviser or investment banking services; and
- legal services and expert services related to audit. (*SOX § 201(g), 15 USC § 78j-1(g)*).

In addition, an auditor may only provide tax services to a client if those services were pre-approved in advance by company's audit committee, *SOX § 201(h), 15 USC § 78j-1(h); SOX § 202, 15 USC § 78j-1(i)* (describing pre-approval process). In addition to audit committee pre-approval of tax services, the SEC adopted final rules requiring disclosure of the amount of fees paid to the accounting firm for tax services, *17 CFR § 240.14a-101*. As with many provisions of the Act, the PCAOB may grant exemptions to the above described rules, *SOX § 201(b), 15 USC § 7231(b)*.

The Act also acknowledged that the relationship between an audit partner and the corporation her or she audits may, over time, become too close and thus inhibit the auditor's necessary objectivity. Accordingly, an audit partner can only serve in that capacity for five fiscal years; after that, a new audit partner must be appointed, *SOX § 203(j), 15 USC § 78j-1(j)*.

Revised corporate governance guidelines 11.4

The Act promoted revised corporate governance guidelines addressing:

- audit committees;
- conflicts of interest for officers and directors; and
- enhanced financial disclosure.

For example, the Act makes it unlawful for an accounting firm to serve as the company's independent auditor if any of the following officers of the company were employed by that firm and was involved in the company's audit during the one-year period prior to commencement of the audit: CEO, controller, CFO, or chief accounting officer, *SOX § 206, 15 USC § 78j-1(l)*.

Studies and reports 11.5

In order to determine what additional legislative and regulatory measures might be necessary and appropriate, the Act authorised the undertaking of numerous studies and reports. Among the studies to be undertaken pursuant to the Act are studies on:

- special purpose entities to determine the extent of off-balance sheet transactions, and whether Generally Accepted Accounting Principles (GAAP) result in financial statements of public companies reflecting the economics of such transactions, *SOX § 401, 15 USC § 78m and 15 USC § 7261*;
- the consolidation of public accounting firms to identify factors that led to consolidation and resulting reduction in number of firms capable of providing audit services to large businesses; affect on capital formation; ways to increase competition and number of firms capable of providing services, *SOX § 701*;
- the role and function of credit rating agencies in the operation of the securities markets, *SOX § 702*;

- the number of securities lawyers practicing before the SEC who have aided and abetted a violation of the federal securities laws but not been sanctioned, or who have been primary violators, *SOX § 703*;
- enforcement actions by the SEC involving violations of reporting requirements and restatements of financial statements, *SOX § 704*;
- investment banks and financial advisers, *SOX § 705*; and
- the potential effects of requiring mandatory rotation of accounting firms, *SOX § 207, 15 USC § 7232.*

Many of the studies are available on the SEC website at www.sec.gov.

New and enhanced criminal penalties 11.6

An integral element of the Act is forcing high-ranking corporate officers to take greater responsibility for financial irregularities of their companies. In order to accomplish this, the Act created a number of new criminal penalties and strengthened various existing penalties for violations of the Act and the securities law. *SOX §§ 801–807*, codified at *18 USC § 1348, 1514A, 1519, 1520, 11 USC § 523(a), 28 USC § 994, 1658, SOX §§ 901–906*, codified at *18 USC § 1341, 1343, 1349, 1350, 29 USC § 1131, SOX § 1106, 15 USC § 78ff(a)*, and *SOX § 1107, 18 USC § 1513(e)*.

Beyond the addition of new criminal penalties and the enhancement of existing criminal penalties, the Act also attempts to make such actions more difficult to avoid. For example, the Act expressly provides that debts of a company are non-dischargeable in bankruptcy if those debts were incurred in violation of the securities fraud laws, *SOX § 803, 11 USC § 523(a)*. Similarly, the Act amended existing statute of limitations law by allowing suits for, among other things, violations of securities laws, to survive for the earlier of:

- two years after the discovery of the facts constituting the violation; or
- five years after such violation, *SOX § 804, 28 USC § 1658.*

Specific aspects of the act

CEO/CFO certificates 11.7

Following the Enron hearings, there was a general trend towards requiring CEOs and CFOs to certify their periodic filings. Prior to the adoption of the Act, on 14 June 2002, the SEC had issued a Proposed Rule on the Certification of Disclosure in Companies' Quarterly and Annual Reports, which was subject to public comment until 19 August 2002, SEC Release No 34–46079. On June 27, 2002, the SEC issued an order requiring that CEOs and CFOs file a sworn statement with the SEC that, to each executive's knowledge, no 'covered reports' contain an untrue statement of a material fact, or omit to state a material fact necessary to make the statements contained in the report not misleading, SEC File No 4–460. Sworn statements were required to be filed by 14 August 2002 for fiscal year-end filers.

The Act contains its own unique certification requirements found in *§ 302* and in *§ 906* of the Act, *SOX § 302, 15 USC § 7241; SOX § 906, 18 USC § 1350*. Initially, there was much consternation over whether these were two separate requirements, or whether *§ 906* simply contained the penalty provisions for a violation of *§ 302*. Section 906 Certification was effective immediately, and therefore required on the Form 10-Qs that were being filed for the period ending 30 June 2002 (due by 14 August 2002 in most instances). The CEO and the CFO were required to certify that:

- the periodic report containing financial statements fully complies with the requirements of *Section 13(a)* and *15(d)* of the *Securities Exchange Act of 1934*; and
- the information contained in the periodic report fairly presents, in all material respects, the financial condition and results of operations of the issuer, *SOX § 906(b), 18 USC § 1350(b)*.

The Section 906 Certification contains no knowledge qualifier and signatories are subject to criminal penalties for knowing or willful false certifications up to $5m and 20 years in prison, *SOX § 906(c); 18 USC § 1350(c)*. In implementing the Section 906 Certification, the SEC has not inserted a knowledge qualifier, and also has not specified the exact text of the certification.

Effective 29 August 2002, the *§ 302* certification was required following the issuance of final rules by the SEC on 28 August 2002. This certification differed slightly from the Section 906 certification in that it did contain a 'knowledge' qualifier ('Based on my knowledge, this report does not contain any untrue statement..'), *SOX § 302(a)(2) and (3), 15 USC § 7241(a)(2)* and *(3)*. This certification also required a more extensive certification about the company's internal controls and disclosure controls and procedures. Notably, the *§ 302* certification makes no reference to the financial information being presented in accordance with GAAP, in light of the view that such disclosure should meet a standard of overall material accuracy and completeness that is broader than financial reporting requirements under GAAP. Additionally, senior officers are certifying that required material non-financial information, as well as financial information, is included in the periodic reports through the certification of the company's 'disclosure controls and procedures'. The SEC prohibited modification of the language required in the *§ 302* certification other than deletion of certain language with respect to internal control over financial reporting during in the interim period before effectiveness of *section 404* of the *SOX*. The text of the *§ 302* certification is codified at *17 CFR § 229.601(b)(31)*.

Congress anticipated that the Act could encourage some public companies to 'seek greener pastures' and attempt to shift their primary domicile outside of the United States to evade the reach of the Act. To forestall such corporate domicile shopping, the Act made clear that a company could not avoid the *§ 302* reporting requirements by simply reincorporating or changing the domicile of the company to a country other than the United States, *SOX § 302(b), 15 USC § 7241(b)*.

On 21 March 2003, the SEC issued additional proposed rules on the *§ 302* and *§ 906* certifications, requiring that they be included as exhibits to periodic reports. The *§ 302* certification is 'filed' with the report, while the *§ 906* certification is simply 'furnished' with the report.

A final note on the subject of CEO reporting and certificates – although *section 1001* of the *SOX* is not binding, Congress used it as an opportunity to opine that 'Federal income tax return[s] of a corporation should be signed by the chief executive officer of such corporation.' *SOX § 1101*. It goes without saying that the enhanced focus on reporting violations has increased significantly the workload for CEOs and forced them to take a hard look at corporate disclosures, with a view towards protecting themselves against personal liability as well as protecting the corporation. As of the date of this chapter, it was still too early to tell who would take up the Congressional 'free advice' provided in *section 1101* of the Act.

Prohibition on personal loans to officers and directors 11.8

Section 402 of the Act severely limits an issuer's ability to make personal loans to executive officers and directors, *SOX § 402, 15 USC § 78m(k)*. Personal loans maintained on 30 July 2002 are grandfathered so long as there are no material modifications to the terms of the loan or any renewal of the loan after 30 July 2002, *SOX § 402(k)(1), 15 USC § 78m(k)(1)*. The Act also permits an issuer to make loans to executive officers and directors to the extent that such loans are:

- made in the ordinary course of the issuer's consumer credit business;
- generally made available by the issuer to the public; and
- made on market terms that are no more favorable than those offered by the issuer to the general public, *SOX § 402(k)(2)(A)-(C), 15 USC § 78m(k)(2)(A)-(C)*.

The prohibition also does not apply to insured depository institution loans where the lender is subject to the *Federal Depository Insurance Act* (*12 USC § 1813*) insider lending restrictions of the *Federal Reserve Act* (*12 USC § 375b*), *SOX § 402(k)(3), 15 USC § 78m(k)(3)*.

Until guidance is issued under the Act with respect to personal loans, issuers face many unanswered questions concerning common practices with respect to officers and directors. Issuers should act to examine benefit plans and programs to determine where there are potential extensions of credit in violation of the Act. Commonly used practices include:

- split-dollar life insurance arrangements;
- cashless option exercises;
- personal loans from 401(k) plans;
- loans in connection with recruiting new executives (including loans to purchase a residence pending sale of an old residence or forgivable loans to replace unvested benefits);
- company assistance with third party lenders which may include company credit cards; and
- expense and other advances to be reimbursed by the officer or director.

In the absence of further guidance, the grandfather provisions for extensions of credit maintained on 30 July 2002 do not address loans to an employee who subsequently becomes an executive officer or binding arrangements in effect on 30 July 2002 that provide for future draw-downs. In particular, split-dollar life insurance arrangements typically contain a binding agreement for future advances for premium payments. Under recent IRS guidance, split-dollar insurance established with a collateral assignment is taxed under the theory of a loan from the employer to the employee. However, a loan theory is not applied to endorsement split-dollar insurance arrangements. Many public companies have chosen to terminate such arrangements in light of this uncertainty.

Forfeiture of certain bonuses and profits 11.9

If an issuer is required to restate its financial statements 'as a result of misconduct', the CEO and the CFO must reimburse the issuer for any of the following amounts received during the twelve-month period following the first public issuance of the document containing the financial statements which are restated:

- any bonus or other incentive-based or equity-based compensation received by the CEO or CFO from the issuer; and
- any profits realised from the sale of securities of the issuer, *SOX § 304(a), 15 USC § 7243(a)*.

The SEC may, in its discretion, exempt an officer from the *§ 304* forfeiture provisions, *SOX § 304(b), 15 USC § 7243(b)*.

Although this section was effective 30 July 30 2002, the Act provided little guidance as to what constitutes 'misconduct'. The courts appear to be the venue in which this term shall be defined, with much debate as to its meaning in the interim. As with the prohibition on loans, the SEC is not required to engage in any additional rulemaking with respect to this provision although it is permitted to do so.

Audit Committee Independence 11.10

Under the Act, all audit committee members of a public company must be 'independent', *SOX § 301(m)(3), 15 USC § 78f(m)(3)*. The Act defines 'independence' as follows: a board member may not, except in his or her capacity as a member of the audit committee, the board of directors, or any other board committee:

- accept any consulting, advisory or other compensatory fees from the issuer; or
- be an affiliated person of the issuer or any subsidiary thereof.

The Act allows for national securities exchanges to adopt more rigorous definitions of independence, *SOX § 301(m)(3)(B), 15 USC § 78f(m)(3)(B)*. The SEC has the ability to exempt a particular audit committee member from the general independence requirements. *SOX § 301(m)(3)(C), 15 USC § 78f(m)(3)(C)*.

The SEC adopted a final rule directing the exchanges to prohibit listing of the securities of any issuer that does not have an independent audit committee by the earlier of:

- their first annual shareholders meeting after 15 January 2004; or
- 31 October 2004, 17 *CFR § 240.10A-3*. However, the provisions of the Act requiring audit committee independence are already effective.

Audit Committee Financial Expert 11.11

Section 407 of the Act requires the SEC to issue rules requiring the disclosure of:

- whether or not the issuer's audit committee had a member who was a financial expert, as such term was defined by the SEC; and
- if not, the reasons for not having such a financial expert, SOX *§ 407, 15 USC § 7265*.

A 'financial expert' must have:

- an understanding of generally acceptable accounting principles;
- experience in the preparation or auditing of financial statements of generally comparable issuers and the application of such principles in connection with the accounting of estimates, accruals, and reserves;
- experience with internal accounting controls; and
- an understanding of the audit committee functions, *SOX § 407(b)(1)–(4), 15 USC § 7265(b)(1)–(4)*.

The company's board of directors must make the determination of whether a director qualifies as an audit committee financial expert.

On 22 October 2002, the SEC issued proposed rules requiring each issuer to disclose: the number and names of persons that the board of directors has determined to be the financial experts serving on the company's audit committee; whether the financial expert or experts are 'independent', as that term is used in *section 10A(m)(3)* of the *Exchange Act*, and if not, an explanation of why they are not; and if the Company does not have a 'financial expert' serving on its audit committee, the Company must disclose that fact and explain why it has no financial expert, SEC Release No 33–8138.

On 15 January 2003, the SEC voted to adopt rules implementing the new financial expert rules, SEC Release 2003–6 (15 January 2003). The final rules significantly expand the types of experience and background that will satisfy the definition of financial expert from the definition previously proposed.

The rules define 'audit committee financial expert' to mean a person who has the following attributes:

- an understanding of financial statements and generally accepted accounting principles;
- an ability to assess the general application of such principles in connection with the accounting for estimates, accruals and reserves;
- experience preparing, auditing, analysing or evaluating financial statements that present a breadth and level of complexity of accounting issues that are generally comparable to the breadth and complexity of issues that can reasonably be expected to be raised by the registrant's financial statements, or experience actively supervising one or more persons engaged in such activities;
- an understanding of internal controls and procedures for financial reporting; and
- an understanding of audit committee functions, *17 CFR § 229.401(h)(2)*.

Under the final rules, a person can acquire such attributes through any one or more of the following means:

- education and experience as a principal financial officer, principal accounting officer, controller, public accountant or auditor or experience in one or more positions that involve the performance of similar functions;
- experience actively supervising a principal financial officer, principal accounting officer, controller, public accountant, auditor or person performing similar functions;

- experience overseeing or assessing the performance of companies or public accountants with respect to the preparation, auditing or evaluation of financial statements; or
- other relevant experience, *17 CFR § 229.401(h)(3)*.

The final rule also provides a 'safe harbor' to make clear that an audit committee financial expert will not be deemed an 'expert' for any purpose, including for purposes of *section 11* of the *Securities Act of 1933*, and that the designation of a person as an audit committee financial expert does not impose any duties, obligations or liability on the person that are greater than those imposed on such person as a member of the audit committee in the absence of such designation, nor does it affect the duties, obligations or liability of any other member of the audit committee or board of directors, *17 CFR § 229.401(h)(4)*.

The SEC's final rule requires that companies include disclosure on whether they have at least one audit committee financial expert serving on the audit committee and, if so, the name of the audit committee financial expert and whether the expert is independent of management in their annual reports on Form 10-K or 10-KSB, *17 CFR § 229.401(h)* and *17 CFR § 249.310, Item 10*. Since the information would be included in Part III of Form 10-K, issuers would have the option of including the disclosure on financial experts in their proxy statements, as long as the proxy statement was filed with the SEC no later than 120 days after the end of the fiscal year covered by the Form 10-K or 10-KSB. A company that does not have such a financial expert will be required to disclose this fact and explain why it has no such expert.

Code of ethics **11.12**

Section 406(a) of *SOX* requires the SEC to issue rules requiring each issuer to disclose whether or not, and if not, the reason for the issuer's lack of one, such issuer has adopted a code of ethics for senior financial officers, applicable to the issuer's principal financial officer and comptroller or principal accounting officer, or persons performing similar functions, *SOX § 406(a), 15 USC § 7264(a)*. The SEC was also required to adopt rules requiring the prompt disclosure of any change in or waiver of the code of ethics, *SOX § 406(b), 15 USC § 7264(b)*. The Act defined 'code of ethics' as such standards as are reasonably necessary to promote:

- Honest and ethical conduct, including the ethical handling of actual or apparent conflicts of interest between personal and professional relationships;
- Full, fair, accurate, timely and understandable disclosure in the periodic reports required to be filed by the issuer; and
- Compliance with applicable governmental rules and regulations, *SOX § 406(c), 15 USC § 7264(c)*.

On 23 January 23 2003, the SEC issued final rules which required issuers to disclose:

- whether they have adopted a code of ethics applicable to its principal executive officer, principal financial officer, principal accounting officer or controller, or persons performing similar functions; and
- if the issuer has not adopted such a code of ethics, the reasons it has not done so, SEC Release No 33–8177; *17 CFR § 229.406.*

The SEC utilised a broader definition of it as 'code of ethics', defining written standards that are reasonably necessary to deter wrongdoing and to promote:

- honest and ethical conduct, including the ethical handling of actual or apparent conflicts of interest between personal and professional relationships;
- full, fair, accurate, timely, and understandable disclosure in reports and documents that a company files with, or submits to, the Commission and in other public communications made by the company;
- compliance with applicable governmental laws, rules and regulations;
- the prompt internal reporting of violations of the code to an appropriate person or persons identified in the code; and
- accountability for adherence to the code, *17 CFR § 229.406(b)*.

Under the SEC's final rules, companies are required to disclose in their annual reports whether they have a code of ethics that applies to the company's principal executive officer, principal financial officer, principal accounting officer or controller, or persons performing similar functions, *17 CFR § 229.406* and *17 CFR § 249.310, Item 10.* Companies which have not adopted such a code of ethics must explain why they have not done so.

A company will be required to make available to the public a copy of its code of ethics, or portion of the code which applies to the company's principal executive officer, principal financial officer, principal accounting officer or controller, or persons performing similar functions, *17 CFR § 229.406(c)*. The company may do so by: filing the Code as an exhibit to the annual report; providing the Code on the company's internet web site (provided that a company choosing this option also must disclose its internet address and its intention to provide disclosure in this manner in its annual report on Form 10-K); or undertaking in its annual report to provide a copy of its Code to any person without charge upon request, *17 CFR § 229.406(c)*.

A company is also required to disclose in a Form 8-K filing or through posting the disclosure on its Internet website (but only if the company disclosed in its most recent Form 10-K: (a) its intention to disclose these events on its Internet website; and (b) its Internet website address):

- the nature of any amendment to the company's Code that applies to its principal executive officer, principal financial officer, principal accounting officer or controller, or persons performing similar functions; and
- the nature of any waiver, including an implicit waiver, from a provision of the Code granted by the company to one of these specified officers, the name of the person to whom the company granted the waiver and the date of the waiver, *17 CFR § 229.406(d)* and *17 CFR § 249.308, Item 10.*

In each case, the Form 8-K must be filed or the disclosure posted on the company's website within five (5) business days of the amendment or waiver, *17 CFR § 249.308, General Instruction B.1.*

Assessment of internal controls and procedures **11.13**

As part of the *§ 302* certification, an issuer's principal executive officer and principal financial officer are required to certify that they are responsible for establishing and maintaining "disclosure controls and procedures" and that such disclosure controls and procedures have been designed to ensure that material information is made known to them, *SOX § 302(a)(4)(B)*, *15 USC § 7241(a)(4)(B)*. Furthermore, they must certify that they have evaluated the effectiveness of the company's disclosure controls and procedures as of the end of the period covered by the periodic report, and have presented their conclusions about the effectiveness of such controls and procedures in the periodic report, *SOX § 302(a)(4)(C)-(D)*, *15 USC § 7241(a)(4)(C)-(D)*. Finally, the certifying officers must confirm that they have disclosed any deficiencies in the design or operation of internal control over financial reporting and any fraud involving management or employees who have a significant role in the company's internal control over financing reporting to the company's auditors and the audit committee of the board of directors, *SOX § 302(a)(5)*, *15 USC § 7241(a)(5)*.

'Disclosure Controls and Procedures' is a newly-defined term meaning controls and other procedures of an issuer that are designed to ensure that information required to be disclosed by the issuer in the reports filed or submitted by it under the *Securities and Exchange Act of 1934* is recorded, processed, summarised and reported within the time required by the SEC, so that appropriate information is disclosed in a company's periodic filings, SEC Release No 33–8124 (August 29, 2002). The SEC has not required any particular procedures for conducting the required review and evaluation, but instead expects each company to develop a process that is consistent with its business and internal management and supervisory practices. The SEC did recommend the creation of a 'disclosure committee' with the responsibility for consideration of the materiality of information and determining disclosure obligations on a timely basis. Internal Controls' is a pre-existing term relating to a company's internal control over financial reporting, *15 USC § 78m(b)*; AICPA Professional Standards AU Section 319.06.

Section 404 of the Act furthers the increased focus on internal control mechanisms, *SOX § 404, 15 USC § 7262.* In accordance with *section 404*, each annual report filed by an issuer must now contain an 'internal control report' (or 'internal control assessment') containing certain information specified by the Act, *SOX § 404(a), 15 USC § 7262(a)*. The auditor must attest to and report on the internal control report, *SOX § 404(b), 15 USC § 7262(b)*.

On 5 June 2003, the SEC adopted final rules implementing *section 404* of the Act, SEC Release No 33–8238 (5 June 5 2003). The final rules define 'internal control over financial reporting' and require management to:

- annually assess the effectiveness of the company's internal control over financial reporting; and
- to include a report on the effectiveness of the company's internal control over financial reporting in the company's annual report, *17 CFR § 229.308.*

Additionally, the company's independent auditor must attest to management's assessment of the company's internal control over financial reporting, and the attestation must be included in the company's annual report. Companies which are 'accelerated filers' as defined by the SEC must comply with the report and attestation requirements in their annual reports for their first fiscal year ending on or after 15 November 2004, and all other companies must comply for their first fiscal year ending on or after 15 July 2005.

Audit committee responsibilities **11.14**

Section 301 of *SOX* requires the audit committee to have certain responsibilities:

- the audit committee shall be directly responsible for the appointment, compensation, and oversight of the work of any registered public accounting firm employed by the issuer (including resolution of disagreements between management and the auditor regarding financial reporting) for the purpose of preparing or issuing an audit report or related work, and each such public accounting firm shall report directly to the audit committee;
- the audit committee must establish procedures for: (a) the receipt, retention and treatment of complaints received by the issuer regarding accounting, internal accounting controls, or auditing matters; and (b) the confidential, anonymous submission by employees of the issuer of concerns regarding questionable accounting or auditing matters;

- the audit committee must have the authority to engage independent counsel and other advisers as it deems necessary to carry out its duties; and
- each issuer must provide for appropriate funding, as determined by the audit committee, in its capacity as a committee of the board of directors, for payment of compensation (a) to the registered public accounting firm employed by the issuer for the purpose of rendering or issuing an audit report; and (b) to any advisers employed by the audit committee, *SOX § 301, 15 USC § 78f(m)*.

On 8 January 2003, the SEC issued proposed rules which would direct national securities exchanges and national securities associations to prohibit the listing of any security of an issuer that is not in compliance with the audit committee independence and other responsibility requirements contained in *section 301* of the *SOX*. SEC Release No 33–8173. The proposed rules clarified that an audit committee's oversight responsibilities would include the power to retain and terminate the outside auditor, as well as having ultimate authority to approve all audit engagement fees and terms, and all significant non-audit engagement fees and terms. The proposed rules did not, however, include any mandated procedures for handling complaints and concerns..

On 9 April 2003, the SEC issued final rules which require national securities exchanges and national securities associations to prohibit the listing of any security of an issuer that is not in compliance with the audit committee independence and other responsibility requirements contained in *section 301* of *SOX*, SEC Release No 33–8220. Under the final rules:

- each member of the audit committee must be independent according to specified criteria;
- the audit committee of each issuer must be directly responsible for the appointment, compensation, retention and oversight of the work of the issuer's outside auditor, and the outside auditor must report directly to the audit committee;
- the audit committee must establish procedures for the receipt, retention and treatment of complaints regarding accounting, internal controls or auditing matters, including procedures for the confidential, anonymous submission by employees of the issuer of concerns regarding questionable accounting or auditing matters;
- the audit committee must have authority to engage independent counsel and other advisers as it deems necessary;
- each issuer must provide appropriate funding for the audit committee; and
- listed issuers must be in compliance with the new listing rules by the earlier of (1) their first annual shareholders meeting after 15 January 2004, or (2) 31 October 2004, *17 CFR § 240.10A-3(a)*. Certain other issuers must be in compliance by 31 July 2005, *17 CFR § 240.10A-3(a)(5)(i)(A)*.

Insider trading during pension fund blackout periods **11.15**

One of the most disturbing tales that was told after the collapse of Enron and other high profile companies concerned the company employees who had poured their savings into the purchase of company stock only to see the value of that stock fall to virtually nothing. During those same periods, the story went, corporate 'insiders' were able to quickly liquidate their stock to avoid a similar cataclysmic decline in its value. To prevent the recurrence of such perceived inequities, the Act established 'blackout periods' so that during times when employee stock plans could not sell company stock, officers and directors would also be unable to sell their stock in the company, *SOX § 306, 15 USC § 7244, 29 USC § 1021, 29 USC § 1132.*

On 22 January 2003, the SEC issued final rules prohibiting insider trading during pension fund blackout periods. See Release No 34–47225. The rules create new Regulation BTR (Blackout Trading Restriction) and are designed to prevent the inequities that arise when directors and executive officers trade in company securities when rank and file employees are prohibited from doing so due to a pension fund blackout period, *17 CFR Part 245.* The regulation covers trades in any equity or derivative security of an issuer, other than an exempt security (as defined in *section 3(a)(12) of* the *Securities Exchange Act of 1934*).

Regulation BTR prohibits trading in issued securities by executive officers (as defined in *Exchange Act Rule 16a-1(f)*) and directors (as defined in *section 3(a)(7)* of the *Securities Exchange Act of 1934*) of an issuer during a pension fund blackout period, *17 CFR § 245.101.* Regulation BTR defines 'issuer' as:

- an issuer whose securities are registered under *section 12* of the *Securities Exchange Act 1934*;
- an issuer that is required to file reports under *section 15(d)* of the *Securities Exchange Act 1934*; or
- an issuer that files, or has filed, a registration statement that has not yet become effective under the *Securities Exchange Act of 1934*, and that the issuer has not withdrawn, *17 CFR § 245.100 (k)*.

Some limitations are imposed where Regulation BTR might otherwise apply to directors and officers of foreign private issuers.

Regulation BTR prohibits directors and executive officers of an issuer from acquiring equity securities or derivative securities of an issuer during a blackout period if the acquisition is in connection with his or her service or employment as a director or executive officer, *17 CFR § 245.100.* In addition, the regulation also prohibits directors and executive officers of any issuer from disposing of equity securities or derivative securities of an issuer during a blackout period if the disposition involves issuer securities acquired in connection with his or her service or employment as a director or executive officer.

The rules define pension fund 'blackout period' as any period of more than three consecutive business days during which 50 per cent or more of the participants in all company pension plans (eg 401(k) plans, profit-sharing and savings plans, stock bonus plans) that permit participants to acquire company equity securities are restricted from trading in company securities held in plan accounts by the company or by a fiduciary of the plan, *17 CFR § 245.100(b)*. Note that certain blackout periods imposed in connection with mergers or acquisitions to allow persons employed by the acquired entity to change their participation are exempted from the rules, *17 CFR § 245.102(b)*. The following types of individual account plans would be included:

- plans that permit participants or beneficiaries to invest their plan contributions in issuer equity securities;
- plans that include an 'open brokerage window' that permits participants or beneficiaries to invest in the equity securities of any publicly traded company, including the issuer;
- plans that match employee contributions with issuer equity securities; or
- plans that reallocate forfeitures that include issuer equity securities to the remaining plan participants, SEC Release No 34–47225.

Companies must provide timely notice of pension fund blackout periods to directors and executive officers and to the SEC. *17 CFR §245.104.* The notice must include:

- the reasons for the blackout period;
- a description of the transactions affected
- a description of the class of equity securities affected;
- the actual or expected dates of the blackout period; and
- the name, address and telephone number of the person designated by the company to respond to inquiries about the blackout period.. Notice is given to the SEC by filing a Form 8-K on the same date that notice is given to directors and executive officers. Finally, advance notice to directors and executive officers is not required if circumstances beyond the company's reasonable control prevent the company from providing such notice, provided that notice is provided as soon as reasonably practicable, *17 CFR §245.104(b)(2)(ii)*.

Violations of Regulation BTR will subject a director or executive officer to potential SEC enforcement actions and possible criminal penalties. The rules also establish a private right of action under which the company is entitled to recover from a violator any profits realised from prohibited trading, *17 CFR, §245.103*.

Implementation of standards of professional conduct for attorneys. 11.16

Section 307 of the Act mandates that the SEC prescribe minimum standards of professional conduct for attorneys appearing and practicing before the SEC in any way in the representation of issuers, *SOX § 307, 15 USC § 7246*. Not surprisingly, *section 307* and the regulations promulgated to implement it have been the subject of intense scrutiny and debate in the legal community. In fact, the controversy surrounding the proposed rule was so intense that the final rule issued by the SEC was significantly modified in light of the comments received by the SEC.

The SEC has made it clear that it believes that the final rules adequately address the 'up-the-ladder' reporting requirements mandated by the Act, but extended the comment period for the 'noisy withdrawal' portion of the rule and have issued a separate release in that regard. SEC Release No 33–8186 (29 January 2003).

Section 205(1) of the Act establishes minimum standards of professional conduct, intended to supplement applicable jurisdictional standards, *SOX § 205(1), 15 USC § 7245(a)*. These standards are not intended to limit the ability of any jurisdiction to impose additional obligations, but this part governs any conflict between the two. Jurisdictions are free to impose stricter rules, but to the extent state ethical rules impose a lower obligation, they will be preempted.

Under the Act, the client of an attorney representing an issuer before the SEC is the issuer as an entity and not the issuer's individual officers or employees that the attorney regularly interacts with. The proposed rule said an attorney 'shall act in the best interest of the issuer and its shareholders', which suggested a duty to shareholders which might form the basis of a private right of action. SEC Release No 33–8150. This language was removed in the final rule.

At the crux of the debate regarding the Act's attorney rules is the imposition of a duty on lawyers to report evidence of a material violation. The Act provides that if an attorney becomes aware of evidence of a material violation by the issuer or any officer, director, employee or agent of the issuer, the attorney shall report such evidence to the issuer's chief legal officer (or the equivalent thereof) or to both the Chief Legal Officer (CLO) and its CEO. Such reporting does not affect privileged or otherwise protected information, *SOX § 307(1), 15 USC § 7245(1)*. The SEC proposed a number of documentation requirements in the proposed rule, but withdrew these in the face of concerns that it would impede an open and candid discussion between attorneys and their clients, and could create a conflict of interest between lawyer and client, SEC Release No 33–8185.

A company's CLO must inquire into the evidence of the material violation, and if he or she determines no material violation has occurred, is ongoing, or is about to occur, he must advise the reporting attorney of the basis for such determination, *17 CFR §205.3(b)(2)*. Unless the CLO reasonably believes that no material violation has occurred, he must take all reasonable steps to cause the issuer to adopt an appropriate response, and shall advise the reporting attorney thereof.. In lieu of conducting an inquiry, the CLO may refer a report of evidence of a material violation to a qualified legal compliance committee (QLLC).

If the reporting attorney does not believe the CLO has provided an appropriate response within a reasonable time, the attorney must report the evidence of a material violation to 'up the corporate ladder'. Specifically, the lawyer must report the evidence to:

- the audit committee of the board of directors;

- another committee of the board consisting solely of independent directors; or

- the entire board, if the board does not have any independent committees, *SOX § 307(2), 15 USC § 7245(1)*.

It should be noted that any attorney retained to investigate a material violation is also appearing and practicing before the SEC and is bound by this rule, *17 CFR §205.3(b)(5)*. Also, the officers and directors who caused the company to retain an attorney to investigate remain obligated to respond to the attorney who initially reported the suspected material violation.

There are certain limited situations in which a lawyer does not have to report a material violation. Specifically, an attorney need not make a report if:

- the attorney was retained or directed by CLO to investigate evidence of a material violation;
- the attorney reports the results of such investigation to the CLO and (unless both the attorney and the CLO reasonably believe no violation has occurred) the CLO reports the results of the investigation to the company's board or an independent committee thereof;
- the attorney was retained by the CLO to assert a 'colourable defence' on behalf of issuer, and the CLO provides reasonable and timely reports on progress to the company's board or an independent committee thereof, *17 CFR §205.3(b)(6)*. If the attorney receives what he or she reasonably believes is an appropriate and timely response, he or she need do nothing more, *17 CFR §205.3(b)(8)*. If, however, the attorney does not reasonably believe she has received an appropriate response, she must explain her reasons to the CLO, the CEO and the directors to whom she reported the evidence of a material violation, *17 CFR §205.3(b)(9)*.

If an issuer has previously formed a qualified legal compliance committee (QLCC), an attorney may report to the QLCC in lieu of reporting to the CLO, *17 CFR §205.3(c)*. The attorney does not need to assess the issuer's response in such a situation. Likewise, a CLO may refer a report of a material violation to a QLCC, and must inform the reporting attorney of such. The

QLCC then becomes responsible for responding to evidence of a material violation. Note that the QLCC must have been previously formed, and not formed simply to respond to a specific incident.

A violation of the attorney reporting requirements subjects an attorney to the civil penalties and remedies for a violation of the federal securities laws available to the SEC, along with other disciplinary measures such as censure or denial of the privilege to appear or practice before the SEC, *17 CFR §205.6*. An attorney who complies in good faith with the regulations will not be subject to discipline because of inconsistent standards imposed by a state or other US jurisdiction. Furthermore, the Act clarifies that the attorney-disclosure provisions do not create a private right of action and that enforcement is vested exclusively in the SEC, *17 CFR §205.7*.

State jurisdictional issues 11.17

As discussed below, the precise impact of the Act on state regulators is unclear. In the Act itself, however, Congress expressed a clear intent that states consider the Act as guidance when considering how to regulate certain accounting firms:

In supervising non registered public accounting firms and their associated persons, appropriate State regulatory authorities should make an independent determination of the proper standards applicable, particularly taking in consideration the size and nature of the businesses of the accounting firms they supervise and the size and nature of the businesses of the clients of those firms. The standards applied by the Board under this Act should not be presumed to be applicable for all purposes of this section for small and medium sized non-registered public accounting firms, *SOX § 209, 15 USC § 7234*. It is still too early to tell how states will react to this 'advice'.

Communicating with auditors 11.18

One common defence raised by auditors in cases of public company accounting fraud is that the officers or directors of the company have purposefully provided false, misleading, or inaccurate information to the auditors. In order to address this perceived problem, the Act specifically makes it unlawful:

> 'for any officer or director of an issuer, or any other person acting under the direction thereof, to take any action to fraudulently influence, coerce, manipulate, or mislead any independent public or certified accountant engaged in the performance of an audit of the financial statements of that issuer for the purpose of rendering such financial statements materially misleading', *SOX § 303(a), 15 USC § 7243(a)*.

Enforcement of this provision is vested exclusively with the SEC, *SOX § 303(b), 15 USC § 7243(b)*.

Standardisation of financial reporting 11.19

While most accounting firms already prepare reports using basic agreed upon principles, the Act went further and codified the fundamental standardisation of accounting reports. Specifically, the Act now requires all financial reports to be prepared in accordance with Generally Accepted Accounting Principles (GAAP), *SOX § 401(a)(i), 15 USC § 78m(i)*. In addition, the Act mandated the preparation of Commission rules designed to cull misleading, untrue, and inaccurate information from pro forma financial information, *SOX § 401(b)(1), 15 USC § 7261(b)(1)*.

Enhanced SEC review 11.20

As discussed previously, the Act arose largely from a series of embarrassing and catastrophic bankruptcies of public companies. Although the companies that declared bankruptcy were 'private' (in that they were non-governmental entities), they were also subject to regulation and oversight by the SEC. The Act increased the SEC's oversight of public companies by requiring a 'regular and systematic' review of public companies' disclosures 'for the protection of investors', *SOX § 408(a), 15 USC § 7266(a)*. The exact scope of the review is unclear from the Act, but it clearly includes the company's financial statements, *SOX § 408(b), 15 USC § 7266(b)*. As an example, the Act requires the SEC to review the periodic reports filed by an issuer at least once every three years, *SOX § 408(c), 15 USC § 7266(c)*.

Real time issuer disclosures 11.21

It is axiomatic that information is power. In the context of publicly traded companies, information is not only power but also, in a very real sense, money. Accordingly, any type of timing gap in the communication of relevant information creates the possibility of inappropriate leveraging of that information—of 'anticipating' value changes based upon advance information. To minimise the likelihood of such gaps, the Act requires the dissemination of information 'on a rapid and current basis.[.]', *SOX § 409(l), 15 USC § 78m(l)*.

Whistleblower protection 11.22

One of the lessons of the last ten years in the corporate fraud arena is that, more often that not, corporate fraud is generally detected from within rather than without. So-called corporate whistleblowers are, accordingly, an integral part of maintaining corporate integrity and such activities must be encourage and protected. *Section 806* of the *SOX* protects whistleblowers from termination and allows them a right of action if they are discharged, demoted, suspended, threatened, harassed, or discriminated against, *SOX § 806, 18 USC § 1514A*. It should be noted that the Act does not provide for the award of punitive damages to a successful whistleblower. *Title XI* of the *SOX*, which deals with corporate fraud and accountability, goes further than *section 806 of the SOX*, imposing criminal penalties on anyone who retaliates against a whistleblower who provides truthful information to the government, *SOX § 1107, 18 USC § 1513*.

Environmental issues under SOX

Introduction 11.23

Before getting further into the specifics of the *SOX* and its impact in the environmental arena, it is probably helpful to summarise briefly how environmental issues were handled pre-*SOX*. Under prior SEC guidance, environmental liabilities were generally addressed under three specific rules:

- *Item 101 of Regulation S-K, 17 CFR. Section 229*;
- *Item 103 of Regulation S-K*; and
- *Item 303 of Regulation S-K*.

Regulation S-K, of which all three of the items listed above are a part, provides the basic requirements applicable to the content of the non-financial statement portions of certain securities filings made by public companies with the SEC (eg, Form 10-K, Form 10-Q, Registration Statements, etc).

Item 101 of Regulation S-K (*Reg 229.101*) requires companies to provide a narrative description of the company's business which must include:

> 'the material effects that compliance with Federal, State and local provisions which have been enacted or adopted regulating the discharge of materials into the environment, or otherwise relating to the protection of the environment, may have upon the capital expenditures, earnings and competitive position of the registrant and its subsidiaries. The registrant shall disclose any material estimated capital expenditures for environmental control facilities for the remainder of its current fiscal year and its succeeding fiscal year and for such further periods as the registrant may deem material.' *Section 229.101(c)(xii)*.

Item 103 of *Regulation S-K* (*Reg. 229.103*) addresses generally the disclosure of 'Legal Proceedings'. The basic rule of *Item 103* is that 'material pending legal proceedings, other than ordinary routine litigation incidental to the business, to which the registrant or any of its subsidiaries is a party or of which any of their property is the subject' must be disclosed in the filing. Moreover, *Instruction 5* to *Item 103* specifically addresses environmental matters and states that:

> 'an administrative or judicial proceeding (including, for purposes of A and B of this Instruction, proceedings which present in large degree the same issues) arising under any Federal, State or local provisions

> that have been enacted or adopted regulating the discharge of materials into the environment or primarily for the purpose of protecting the environment shall not be deemed 'ordinary routine litigation incidental to the business.'

In other words, if the criteria set out in *Instruction 5* are met, the public company must disclose the environmental proceeding to which it is a party.

Finally, *Item 303* of *Regulation S-K* (*Reg. 229.303*), which addresses Management's Discussion and Analysis of Financial Condition and Results of Operations, also can be impacted by environmental issues. Although *Item 303* is not directed specifically at environmental matters, *Section (a)(1)*, which addresses liquidity generally, requires the public company to:

> '[i]dentify any known trends or any known demands, commitments, events or uncertainties that will result in or that are reasonably likely to result in the registrant's liquidity increasing or decreasing in any material way.'

The same concept – trends, demands, etc – recurs a number of times in *Item 303*. The SEC has specifically indicated an increased emphasis on disclosure of known trends.

SOX implications 11.24

Although the SEC has not yet amended any of the Items described above, the *SOX* will still have a significant impact on how public companies deal with potential environmental liabilities.

In addition to the Certification requirements, the SEC has:

- proposed rules which would require 'upstream' reporting of material violations of law (including environmental law) by company lawyers'; and
- proposed rules addressing various disclosure rules which could impact the disclosure of environmental violations.

Outside of the SEC, signs of change abound – in 2002 the American Society of Testing and Materials adopted two new standards for assessing and describing environmental liabilities, and the Senate Committee on Environment and Public Works has asked the General Accounting Office to review current environmental disclosure practices.

While the basic impact of the *SOX* on environmental disclosures is clear, the precise steps a company should take in response to the impact is somewhat less clear. The nature and extent of the changes which need to be implemented by a particular company in response to the Act depend upon the size of the company, the nature of its operations, and how it currently handles environmental issues. Nevertheless three basic principles can be enunciated:

- Companies should be mindful of how new regulations or specific compliance deadlines may impact the company's overall financial performance. One way of handling this issue is to specifically delegate it to a particular individual or committee. By doing this, the company will be able to disclose information about up-coming regulatory impacts and, from an operational perspective, prepare the company for those impacts;
- Ensure that lines of communication are established and fully functional. As noted above, with regard to the Section 906 Certificates, lack of knowledge is no excuse. Accordingly, it is critical to make sure that information known by environmental professionals is being communicated to the officers who will be signing the appropriate certificates. The SEC has suggested the use of a disclosure committee.
- To the extent that potential impacts are being quantified, make certain that a member of management is participating in those discussions to ensure that the estimates are being done correctly.

Conclusion

Due diligence issues **11.25**

The impact of the *SOX* on due diligence depends greatly upon the nature of the company being investigated. Where a public company due diligence effort is being planned, an important part of the process will be identifying what steps the company has taken to comply with the Act and what violations of the Act the company may have committed historically. Among the questions which should be considered are the following:

- What loans has the company made to its officers and directors? What are the terms of those loans? What documentation exists documenting those loans? Have the terms of the loan been amended?
- Does the company have an 'employee hotline' or similar mechanism whereby concerned employees can report problems to management? Have any such complaints been received? What follow-up actions were taken by the company to investigate those complaints?
- Does the company have a written code of conduct or code of ethics? If so, provide a copy.
- Identify the company's auditor (including audit partner). How long has the audit partner served in that capacity?
- Identify whether any officer or director of the company has been employed by the company's auditor within the last three years.
- Identify the exchange which the company is listed on. Confirm compliance with the policies of that exchange.

- Provide copies of all Section 302 and Section 906 Certificates.
- Describe any fees (consulting, advisory, or otherwise) paid by the issuer to any director of the company or any member of the company's audit committee.
- Identify any financial expert on the audit committee. Provide documentation or a narrative explaining the basis for the designation of any financial expert. Review board determination of financial expert status.
- Does the company have a disclosure committee?
- Provide a copy of any audit committee minutes along with any policies and procedures adopted by the audit committee.
- Provide a copy of documentation regarding any internal investigations conducted by the company.
- Provide copies of all written charters for each committee of the board of directors.

While the list above is not intended to be exhaustive, it does highlight some of the critical information that should be obtained during the course of due diligence.

As more fully discussed above, private companies are not directly subject to the provisions of the Act. Nevertheless, investigation of the issues listed above may well provide a window into the company's internal control mechanisms. Accordingly, some limited form of the *SOX* due diligence may be appropriate even if the company being investigated is not a public company.

Corporate governance issues 11.26

From a corporate governance perspective, the mandate of the Act is clear: it will no longer suffice for high level corporate executives (particularly CEOs and CFOs) to simply adopt an ostrich approach whereby they remain aloof from corporate malfeasance. Indeed, such upper level executives now bear an affirmative burden to investigate the corporations that they serve. From the top down, each corporation must ensure that:

- it has in place a framework designed to gather the information covered by the Act;
- its audit committee is empowered to probe possible problems which may need to be addressed in publicly filed reports;
- that such committee is, in fact, using those powers; and
- that problems suspected or detected are being correctly communicated within the corporation and swiftly rectified. As of the date of this chapter, each public corporation should have conducted a comprehensive internal review to ensure that it is not in violation of any of the

applicable provisions of the Act. As discussed previously, it is likely that many larger private corporations have undertaken similar review.

Criminal prosecutions 11.27

When the *SOX* was signed into law, the United States Attorney General circulated an internal memorandum acknowledging the powerful new tools for criminal enforcement provided by the Act:

> 'The President this week signed into law the Sarbanes-Oxley Act of 2002 (the 'Act'). The Act provides tough new tools to expose and punish acts of corporate corruption, promote greater accountability by financial auditors, and protect small investors and pension holders.'

Memorandum from the Attorney General (1 August 2002). As of the date of this chapter, however, there have only been a few significant prosecutions involving alleged violations of the Act, most notably:

- the information and subsequent plea by a former officer of HealthSouth Corporation to violations of a number of criminal laws including, but not limited to, the Act; and
- the arrest of an accountant in California for, among other things, violation of the Act and obstruction of justice.

It is likely that more arrests and prosecutions under the Act will follow in the coming years.

International impact of SOX 11.28

From the inception of the *SOX*, United States regulators have been acutely aware that the Act has an inevitable and profound international impact. In recognition of the need to work, in a cooperative way, with the international community, the SEC has consistently provided international authorities the opportunity to comment on proposed regulations. Despite this spirit of cooperation, the international community remains deeply skeptical regarding the long-term impact and repercussions of the Act. Listed below are some of the primary international impacts of the Act:

- Although foreign accounting firms that audit United States public companies must still register with the PCAOB, such firms have been provided a number of accommodations by the SEC such as:
  - not having to provide registration information to the Board where such disclosures would violate home country laws; and
  - having an addition six months to register with the PACOB.
- With regard to audit committee requirements, taking in account 'foreign corporate governance schemes, while preserving the intention of the Act

to ensure that those responsible for overseeing a company's outside auditors are independent of management.'

- Allowing 'foreign accounting firms ... to provide tax services ... despite their local definition as legal services, which are among the enumerated list of services prohibited by the Act'.
- Excluding most foreign lawyers not licensed to practice law in the United States from coverage under the attorney conduct rules.
- Allowing a financial expert to become qualified not just by learning US GAAP, but also by developing 'an understanding of an issuer's home country GAAP[.]'

Indeed, even beyond the direct impact of the Act on international markets, there is increasing indication that some other countries (including the European Union) may be looking to their own corporate laws to determine whether they should be conformed to comply with the United States. The Act clearly will have a ripple effect as the years pass and other companies (and countries) move to comply with the standard set in the Act.

Corporate Governance in Europe: impacts of SOX **11.29**

It was always tempting fate to say that Enron was an American problem, and that European rules would prevent similar corporate collapses here. Now, of course, Europe has seen its own corporate scandals, such as Parmalat, and the European authorities are responding with draft new rules of their own. The difference, say the rule-makers, is that the European proposals are not a 'knee-jerk' response, but the result of careful scrutiny over many years. Knee-jerk or not, some of the proposals bear a striking similarity to those pushed through in the United States in 2002.

The draft rules, issued by the European Commission earlier this year, would underpin the independence of auditors and regulate their fee structures to ensure that they are not dependent on other services provided by the same firm. Companies would also have to disclose any other services provided -although the proposals stop short of banning firms from taking on other engagements for their audit clients. Audit partners would have to change at least every five years, or the firm would have to change every seven. There would also be a legal requirement for publicly traded companies to establish an audit committee, including at least one member competent in accounting or auditing, to select the auditors. Other measures would strengthen the oversight of accountants (ending self-regulation) and require registration of firms from countries outside the European Union to ensure consistently high standards of regulation.

European accounting firms have welcomed some of the proposals, but all are strongly opposed to auditor rotation, arguing that it damages audit quality. Businesses have reacted against a rules-based approach, saying that, for example, the requirement for audit committees should come from codes of conduct and not from legislation. The UK particularly prides itself on outstanding model codes rather than descriptive regulation that can mean more of a checking box mentality.

Whatever the outcome, the Commission's draft rules have fuelled the debate within Europe. They are some way away from finalisation, let alone from becoming law. However it is certain that something will emerge which moves further towards the US approach; indeed, in some countries, changes are proceeding ahead of European-wide reforms. In the UK, for instance, a bill is currently going through Parliament that will strengthen the oversight of auditors and improve their powers of investigation. Whether the *SOX* provided the impetus or not, there is no doubt that stricter regulation is coming. The robust defence of the European system by some immediately after the collapse of Enron may therefore have been premature.

Chapter 12

International Dimensions and Corporate Governance: the Indian Perspective

Introduction 12.1

The legal expression, corporate governance has acquired significance in India during the last few years. After the collapse of the Soviet Union and the end of the Cold War in 1990, it has become the conventional wisdom all over the world, including India, that market dynamics must prevail in economic matters.

This has also coincided with the thrust given to globalisation because of the setting up of the World Trade Organisation (WTO) and every member of the WTO trying to bring down the tariff barriers. Globalisation involves the movement of four economic parameters, namely:

- physical capital in terms of plant and machinery;
- financial capital in terms of money invested in capital markets or in Foreign Direct Investment (FDI);
- technology; and
- labour moving across national borders.

The pace of movement of financial capital has become greater because of the pervasive impact of information technology and the world having become a global village, (see N Vittal, *Ethical issues in corporate governance,* Central Vigilance Commissioner, Inaugural Address delivered at NBCC Seminar on 22 March 2000, New Delhi, source: http://cvc.nic.in/vscvc/cvcspeeches/march2k5.html).

When investment takes place in emerging markets like India, the investors want to be sure that not only are the capital markets or enterprises with which they are investing run competently but that they also have good corporate governance. Corporate governance represents the value framework, the ethical framework and the moral framework under which business decisions are taken. In other words, when investments take place across national borders, the investors want to be sure that not only is their capital handled effectively and adds to the creation of wealth, but the business decisions are also taken in a manner which is not illegal or involving moral hazard, (N. Vittal, *Ethical issues in corporate Governance*, Central Vigilance Commissioner, Inaugural Address delivered at NBCC Seminar on 22 March 2000, New Delhi, source:http://cvc.nic.in/vscvc/cvcspeeches/march2k5.html).

The real genesis of corporate governance lies in the business scams and failures in India and abroad. The junk bond fiasco in US and the failure of Maxwell, BCCI and Polly Peck in the UK resulted in the Treadway Committee in the US and the Cadbury Committee in the UK on corporate governance. After 20 years of such incidents, the corporate world had witnessed similar corporate collapses such as Worldcom, Xerox, Enron et al, leading to the US legislation, the *Sarbanes-Oxley Act of 2002*, (*SOX*), as is discussed in **CHAPTER11**. Similarly, the Kumar Mangalam Birla Report and the Naresh Chandra Report were published in India by way of response to the crises in corporate governance.

History of governance in India 12.2

The concept of governance is not new to India. The principles of governance were discussed at length in 'Arthashastra' which was written in the fourth century BC. This treatise on government is said to have been written by Kautilya, the Prime Minister of India's first great emperor, Chandragupta Maurya.

In the context of Indian management, corporate governance can be drawn from the following: Kautilya's arthashastra and mulyas (values).

There is a famous quotation from Kautilya's Arthasastra, (Sanjiv Agarwal, *Corporate Governance: Concepts and Dimensions*, (1st edn, 2003) Snow White Publications, page 17):

> 'Praja sukhe sukham rajyaha prajanamcha hitehitam
>
> natma priyam hitam rajnaha prajanam cha hitam piryam.'

This means:

> 'In the happiness of his subject lies the king's happiness, in their welfare his welfare. He shall not consider as good only that which pleases.'

There are four principles of governance:

- Raksha – Protection
- Vridhi – Enchancement
- Palana – Maintenance
- Yogakshema – Safeguard

Arthashastra also describes the concepts of 'well being' as: 'it is the duty of the king to protect the wealth of the state and its subjects, to enhance the wealth, to maintain it and safeguard it and the interests of the subjects', (Sanjiv Agarwal, *Corporate Governance: Concepts and Dimensions*, (1st edn, 2003) Snow White Publications, page 17).

It is interesting also to consider the term 'governance' from the philosophical and family wealth perspectives described in **CHAPTER 18**.

Corporate governance in India 12.3

In India, despite a long corporate history the phrase 'corporate governance' remained unknown until 1993. It came to the fore due to a spate of corporate scams and fraudulent practices during the first phase of economic liberalisation in the early 1990's and thereafter in successive recurrence. These led to the prominence of corporate governance within:

- the corporate body/sector;
- financial institutions;
- enlightened business associations;
- the regulating agencies; and
- the Government.

Good governance is imperative for edge in competition and critical to economic and social progress. In an ever-increasing globalised economy, firms need to tap domestic and international capital markets for investment, but capital providers have a choice and the quality of corporate governance is increasingly becoming a deciding factor for investment and lending.

Enlarging and deepening the capital pool for developing or emerging economies requires full attention to corporate governance standards. This sets the imperatives for reform, (excerpts from the *Report on the Committee on the Companies Bill, 1997*, Department of Company Affairs, Ministry of Finance and Company Affairs, Government of India, New Delhi, September 2002).

A new era of good corporate governance has been put in place with the enactment of the *Companies (Amendment) Act, 2000*. The latter is in line with the changing needs of the corporate sector in the wake of globalisation. With the emphasis being on good corporate governance, the *Companies (Amendment)*

Act 2000 aims to make the Indian corporate sector as a whole and its corporate bodies respected globally so that they enjoy the confidence of investors and other stakeholders including consumers, (*Report on the Committee on the Companies Bill 1997*, Department of Company Affairs, Ministry of Finance and Company Affairs, Government of India, New Delhi, September 2002).

The legal and regulatory framework regarding corporate governance in India 12.4

The *Companies Act 1956* is the principal legislation providing a formal structure for corporate governance. Akin to this, the *Monopolies and Restrictive Trade Practices Act 1961*, the *Foreign Exchange Management Act 2000*, the *Industries Development and Regulation Act 1951* (*IDR Act*) and other economic legislation also govern corporate activities. The Securities and Exchange Board of India (SEBI) has assumed a greater role in recent years, particularly after the implementation of the Industrial Policy of 1991.

The government of India, has, in conformity with the above policy, tailored various laws and enacted new legislation with a view to providing a conducive atmosphere for the corporate sector to operate effectively, with the necessary safeguards. Increasing corporate activities, together with an interaction with multinationals, have raised various new issues requiring urgent attention. The legal and regulatory framework in India, pertaining to corporate governance comprises the following laws.

The Companies Act 1956 12.5

This is a monumental piece of legislation which *inter alia* sets out the requirements concerning the board of directors, meetings, management, conduct of meetings, appointment or removal of auditors or directors, directors' responsibility, corporate restructuring, mergers, inter-corporate activities, audit committee, accounting standards, shareholders directors and postal ballots.

The Monopolies and Restrictive Trade Practices Act 1969 12.6

The *Monopolies and Restrictive Trade Practices Act 1969* (*MTRP1969*) regulates the control of monopolistic acts, restrictive and unfair trade practices and protects the interest of consumers. The *MRTP 1969* is gradually in the process of being replaced by the new Competition Law.

The Foreign Exchange Management Act 2000 12.7

The *Foreign Exchange Management Act 2000* (*FEMA 2000*) regulates the control and monitoring of activities concerning the foreign flow of funds, investments and investors. The object of *FEMA 2000* is to consolidate and amend the law relating to foreign exchange with the objective of facilitating external trade and payments and for promoting the orderly development and maintenance of the foreign exchange market in India.

The Sick Industrial Companies (Special Provisions) Act 1985 12.8

The *Sick Industrial Companies (Special Provisions) Act 1985's* (*SICA1985*) main objective is to determine sickness and expedite the revival of potentially viable units or closure of unviable units (the word, 'units' refers to a sick industrial company). It was expected that by revival, idle investments in sick units would become productive and by closure, the locked up investments in unviable units would get released for productive use elsewhere. *SICA 1985* was enacted with a view to securing:

- the timely detection of sick and potentially sick companies owning industrial undertakings;
- the speedy determination by a body of experts of the preventive, ameliorative, remedial and other measures which need to be taken with respect to such companies; and
- the expeditious enforcement of the measures so determined and for matters connected therewith or incidental thereto.

The Securities and Exchange Board of India Act 1992 12.9

The *Securities and Exchange Board of India Act 1992* (*SEBI 1992*) provides for the establishment of a Board to protect the interests of investors in securities and to promote the development of, and to regulate the securities market and for matters connected therewith or incidental thereto. The Rules and Regulations, made thereunder relating to unfair trade practices, insider trading, takeover and mergers, raising money from the market, regulation of secondary market et al all bring within their ambit various aspects of corporate governance.

The Securities Contract (Regulation) Act 1956 12.10

The law governing securities in India is regulated, among others, by the *Securities Contracts (Regulation) Act 1956* (*SCRA 1956*). *SCRA 1956* was enacted by the Parliament of India to prevent undesirable transactions in securities dealings. According to *section 16(1)* of *SCRA 1956*, the central government has the power to issue notifications, preventing or restricting securities transactions, which, in its opinion, are rife with speculation.

The Depositories Act 1996 12.11

The Depositories Act 1996 was enacted to enable the creation of the National Securities Depository Limited (NSDL) to modernise India's antiquated settlement system, primarily by enabling dematerialised equity trading. Measures were taken to promote investor security by establishing a comprehensive system of margins, intra-day trading and exposure limits, capital adequacy norms for brokers, and trade/settlement guarantee funds for each exchange. Unlisted companies with a net profit in each of the last three years were given free access to the market.

Corporate governance: Indian practice 12.12

In India, corporate governance is being observed and implemented in the spirit and the law and corprates have begun realising that it is the observance and implementation of good corporate governance practices which will lead to corporate excellence. This is also evident from the various legislative changes that have been introduced in the last couple of years in relation to corporate legislation and in the law relating to capital markets. Important changes were made in 1999 and 2000 to the *Companies Act 1956* and which are outlined below.

Buyback of shares 12.13

Section 77A of the *Companies Act 1956* (inserted by the *Companies (Amendment) Act 1999* permits companies to buy back their own shares. This has been supplemented by the *SEBI (Buy Back of Securities) Regulations 1998* (published in the Gazette of India, Extraordinary, Part iii – Section 4, Bombay, 14th November 1998), which allows a company to buy back its shares directly from its shareholders or through the stock markets. The extent of shares to be bought back and the buyback price have to be approved by the shareholders. The stringent guidelines set out by the market regulator require a company to maintain an escrow account, besides extinguishing the shares bought back. The rationale behind introducing the *SEBI (Buy Back of Securities) Regulations 1998* was to return the surplus cash to the shareholders, avoid hostile takeovers and promote corporate good governance.

Issue of sweat equity shares 12.14

The Securities and Exchange Board of India has notified the *SEBI (Issue of Sweat Equity) Regulations, 2002,* (published in the Gazette of India, Extraordinary, Part ii – Section 3 – sub-section (ii), Mumbai, 24th day of September, 2002), for the issue of sweat equity by listed companies. A company whose equity shares are listed on a recognised stock exchange may issue sweat equity shares in accordance with *section 79A* of the *Companies Act 1956* (inserted by the *Companies (Amendment) Act 1999*) to its employees and directors. Further, *Section 79A* of the *Companies Act 1956* permits a company to issue sweat equity shares of a company subject to the guidelines to be issued in this regard. Accordingly, the Department of Company Affairs has notified the *Unlisted Companies (Issue of Sweat Equity Shares) Rules 2003*, which have to be complied with by an unlisted company proposing to issue sweat equity shares.

Nomination of shares **12.15**

The Companies (Amendment) Act 1999, has, for the first time, laid down the law in respect of the nomination of shares, debentures and fixed deposits. The relevant sections are *ss 109A, 109B and 58A(11)* of the *Companies (Amendment) Act 1999*. Earlier, the position was that to avoid the problems of succession one had to hold the shares in joint names. On the death of one of the joint holders, the shares continued to be held in the name of the holder who had survived. But the matter did not end there. It transpired that the legal heirs of the deceased joint-holder could make a successful claim on the shares. The matter was finally settled by the Company Law Board in the case of *Kana Sen and others v C K Sen and Co{P} Ltd and another, 1997, 27, CLA 82 CLB* where it was held that in case the articles of association of a company provided that the survivor of joint holding alone will have a right to the shares on the death of the joint-holder, it is only that survivor who will be recognised by the company as having a title to the shares.

Transmission of shares **12.16**

Members can sell their shares to any person, either directly or indirectly. This can be done through a duly executed share transfer deed.

Pursuant to the provisions of *s 108* of the *Companies Act 1956*, a company shall not register any share certificates received for transfer unless a proper instrument of transfer duly stamped is executed. There is a common share transfer form prescribed (Form No 7B in accordance with Rule 5A of the *Companies (Central Governments) General Rules and Forms 1956*) for transfer of securities. This form has to be duly filled in by the transferor and the transferee.

Unpaid dividend account **12.17**

In order to make the legal position beneficial and convenient to the investing public, and also for protecting the rights of the public, *s 206A* of the *Companies Act 1956* (inserted by the *Companies (Amendment) Act 1988*) was made, with effect from 15 June 1988. Clause 28 of the Notes on the Clauses of the *Companies (Amendment) Bill* is appended below:

> 'With a view to providing protection to the investing public, this clause introduces a new section providing for payment of dividend and allotment of bonus and rights shares, to the transferee on a mandate in this regard from the transferor and in the absence of such mandate, also imposes an obligation on the company to transfer the

dividends accruing on such shares to the Unpaid Dividend Account and to keep in abeyance any offer of rights or bonus shares, till the title to shares is decided.'

Investor education and protection fund 12.18

Pursuant to the provisions of *s 205C* of the *Companies Act 1956* (inserted by the *Companies (Amendment) Act 1999*) the Central Government [of India] has notified the establishment of a fund called the Investor Education and Protection Fund with effect from 1 October 2001. The fund shall be credited with the following amounts:

(a) amounts in unpaid dividends accounts of companies;

(b) the application moneys received by companies for allotment of any securities and due for refund;

(c) matured deposits with companies;

(d) matured debentures with companies;

(e) the interest accrued on the amounts referred to in clauses (a) to (d).

Small depositors 12.19

The *Companies Act 1956* contains various provisions for the protection of small depositors/investors who are usually the most vulnerable minority group in a corporate body. The amendments in the *Companies Act 1956* have provided considerable powers to the majority shareholders. However, the principles of good governance have enunciated that every shareholder, whether he is a majority or a minority shareholder has to be protected. The newly enacted *s 58AA* of the *Companies Act 1956* provides guidelines to protect the small depositors.

For the purposes of *s 58AA*, a 'small depositor' (the explanation to *s 58AA* as inserted by the *Companies Amendment Act 2000*) which came in force from 13 December 2000) means a depositor who has deposited in a financial year a sum not exceeding Rupees twenty thousand (approximately $US 445, (the currency conversion rate for the purpose of calculation is $US 1= 45 Indian Rupees)) in a company and includes his successors, nominees and legal representatives.

Section 58AA further requires that every company that accepts deposits from small depositors to inform the Company Law Board (CLB) of any default made by it in repayment of such deposits or part thereof or payment of any interest thereon, within 60 days from the date of default furnishing full particulars of the name and address of each such depositor and the principal sum of deposit and interest thereon due to them. The CLB, on receipt of such information from the company, will exercise on its own motion the powers conferred upon it by *s 58A* [Clause 6 of 58A explains the punishments imposed on a the company in default of inviting deposits without issuing an advertisement] and pass an appropriate order within 30 days from the date of receipt of information. The CLB may also pass the order after the expiry of 30 days on giving the small depositors an opportunity of being heard. For this purpose it is not necessary for the small depositor to be present in person at the CLB hearing.

Under the new amendment, no defaulting company will at any time accept further deposits from small depositors unless each of the small depositors had been paid the amounts due. Where companies that have defaulted in repayment of principal and payment of interest to small depositors, had obtained funds by taking a loan for the purpose of its working capital from any bank, they will first utilise the funds so obtained for the repayment of any deposit and interest to small depositors before applying such funds for any other purpose. Failure to comply with *s 58AA* and any order passed by the CLB will be punishable with imprisonment up to three years and also a fine not less than Rupees 500 (approximately $US 12) a day during the period of non-compliance.

Powers of SEBI 12.20

The amended *SEBI Act* (*Section 11C* of the *Securities and Exchange Board of India Amendment Act 2002*) gives more powers to the market regulator, ie SEBI, including search and seizure power. The decision to give more powers to SEBI flows from the 'limitations' in the stock market regulator's ability to check malpractice and market manipulation. Consequently, apart from expanding the SEBI board to nine from the present six (including the chairman), the regulator will now have powers of 'search and seizure' of company premises and records but only after a magistrate permits it. Secondly, the penalties which can be imposed have been considerably hiked, from Rupees 5 lakhs (approximately $US 11,110) to Rupees 25 crores (approximately $US 5,555,555) or three times the undue profit made, whichever is higher (*s 15G* of the *Securities and Exchange Board of India Amendment Act 2002*). All the money realised by SEBI by way of penalties or fines would go into the Consolidated Fund of India (*s 15JA* of the *Securities and Exchange Board of India Amendment Act 2002*). (NB: 1 Lakh=100,000 (one hundred thousand); 1 Crore=10,000,000 (ten million.))

To enhance the enforcement powers of SEBI, the *SEBI* (*Amendment Act*) *2002* clarifies and defines offences such as insider trading, fraudulent and manipulative trade practices and market manipulation (*Chapter 5A* of the *Securities and Exchange Board of India Amendment Act 2002*). SEBI would also have the powers to call for records from banks and other authorities to facilitate investigations (*Section 11[2][ia]* of the *Securities and Exchange Board of India Amendment Act, 2002*).

Representation of small shareholders on company boards 12.21

The *Companies* (*Amendment*) *Act 2000* has given small shareholders mandatory representation on the board of companies above a certain size, in order to protect minority interests. *Section 252* of the *Companies Act 1956* (Inserted by the *Companies Amendment Act 2000*) for the first time, stipulates the requirement of the small shareholders' director. The amendment provides that a public company having a paid-up capital of Rupees 5 crores (approximately $US 1,111,110) or more, 1,000 or more small shareholders may have a director elected by such shareholders in the manner as may be prescribed.

Introduction of postal ballots 12.22

The *Companies (Amendment) Act 2000* also expands the notion of corporate democracy by permitting postal ballots (*s 192A[1]* of the *Companies Act 1956* was inserted by the *Companies Amendment Act 2000* which came into force with effect from 13 December, 2000). The provisions of *s 192A[1]* of the *Companies Act 1956* provide all listed companies with an option of passing shareholders' resolutions through postal ballot. Further, the listed companies are necessarily required to get certain businesses notified (vide notification number GSR.337 E dated 10 May 2001) by the Central Government to be passed through postal ballot. *The Companies (Passing of the Resolution by Postal Ballot) Rules 2001* (notified on 10 May, 2001), mentions the transactions which have to be transacted through postal ballot.

Director's responsibility statement 12.23

The Board Report should also include the Directors' Responsibility Statement requiring the disclosure of various information and making the directors responsible for the disclosures. *Section 217[2AA]* of the *Companies Act 1956* (inserted by the *Companies Amendment Act 2000* which came into force with effect from 13 December 2000) stipulates that the board's report shall also include the Directors' Responsibility Statement indicating therein that:

- in the preparation of the annual accounts, the applicable accounting standards had been followed along with proper explanations relating to material departures;
- the directors had selected such accounting policies and applied them consistently and made judgments and estimates that are reasonable and prudent so as to give a true and fair view of the state of affairs of the company at the end of the financial year and of the profit and loss of the company for that period;
- the directors had taken proper and sufficient care for the maintenance of adequate accounting records in accordance with the provisions of the Act for safeguarding the assets of the company and for preventing and detecting fraud and other irregularities; and
- the directors had prepared annual accounts on a going concern basis.

Audit committee **12.24**

Section 292A of the *Companies Act 1956* (inserted by the *Companies Amendment Act 2000* which came into force with effect from 13 December 2000) states that every public company having a paid-up capital of not less than Rupees five crores (approximately $US 1,111,110) shall constitute a committee of the Board knows as the audit committee which shall consist of not less than three directors and such number of other directors as the Board may determine, of which two-thirds of the total number of members shall be directors, other than managing or whole-time directors. Every audit committee constituted shall act in accordance with the terms of reference to be specified in writing by the board. The members of the audit committee shall elect a chairman from amongst themselves. The annual report of the company shall disclose the composition of the audit committee. The auditors, the internal auditor, if any, and the director-in-charge of finance shall attend and participate at meetings of the Audit committee but shall not have the right to vote.

The audit committee should have discussions with the auditors periodically about internal control systems, the scope of audit including the observations of the auditors and review the half-yearly and annual financial statements before submission to the board and also ensure compliance of internal control systems. The audit committee shall have authority to investigate any matter in relation to the items specified in this section or referred to it by the board and for this purpose, shall have full access to information contained in the records of the company and external professional advice, if necessary. The recommendations of the audit committee on any matter relating to financial management including the audit report, shall be binding on the board.

If the board does not accept the recommendations of the audit committee, it shall record the reasons therefor and shall communicate such reasons to the shareholders. The chairman of the audit committee shall attend the annual general meetings of the company to provide any clarification on matters relating to audit. If a default is made in complying with the provisions of this section, the company, and every officer who is in default, shall be punishable with imprisonment for a term which may extend to one year, or with fine which may extend to Rupees 50 thousand (approximately $US 1,110) or with both.

Accounting standards **12.25**

Section 211[3A] of the *Companies Act 1956* (inserted by the *Companies (Amendment) Act 1999*) provides that every profit and loss account and balance sheet of the company shall comply with the accounting standards. For the purpose of *Section 211[3A]* of the *Companies Act 1956*, the expression 'accounting standards' means the standards of accounting recommended by the Institute of Chartered Accountants of India (a statutory body constituted by the *Chartered Accountants Act 1949*, as may be prescribed by the Central Government in consultation with the National Advisory Committee on

Accounting Standards (established under *sub-section 1* of *section 210A(1)* as inserted by *Companies (Amendment) Act 1999*).

Secretarial compliance certification 12.26

The amended *Companies Act 1956* has vested various responsibilities on company secretaries in order to maintain good governance of the companies. Every company having a paid up share capital of Rupees twenty five lakhs (approximately $US 11,110) or more shall have a whole time secretary and where the board of directors of any such company comprises only two directors, neither of them shall be the secretary of the company, (*section 383A*(1) substituted by the *Companies (Amendment) Act 1988*). The *Companies Act 1956* makes it mandatory, (*section 383 A(1)* inserted by the *Companies Amendment Act 2000*) for a company having a paid up share capital of Rupees 10 Lakhs (approximately $US 22,220) or more to file with the Registrar of Companies a certificate from a secretary in full-time practice in such form and within such time and subject to such conditions as may be prescribed, as to whether the company has complied with all provisions of the Act and the copy of such certificate shall be attached with Board's report (as referred in *section 217* of the *Companies Act 1956*). On the failure to comply with these rules, the company and the Secretary will be punishable (*section 383A (1A)*, as inserted by the *Companies (Amendment) Act 1988*).

Capital market regulations

Introduction of depositories 12.27

Depositories, ie a system of organisation which keeps records of securities deposited by its depositors was introduced in India in 1996 to increase the investor security and the transparency of the Indian capital markets. The *SEBI (Depositories And Participants) Regulations 1996* (published in the Gazette of India, Extraordinary, Part ii – Section 3 – Sub-section ii, Mumbai on 16 May 1996) provide for the establishment of depositories subject to registration with SEBI. SEBI will prescribe conditions to be met before the 'Certificate of Commencement of Business' can be issued. Investors will have the choice of continuing with their existing share certificates or opting for a depository. Investors opting to hold their securities in the depository mode will be required to get in touch with a 'participant' in the depository system, (who will be registered with SEBI). The 'participant' will be agencies like banks, financial institutions, custodians of securities and stockbrokers. Upon entry into the system, share certificates will be 'dematerialised' and names of beneficial owners would be registered in the books of the depository. The Code of Conduct is prescribed by SEBI for the protection of the investors, (*regulation 20A* of the *Securities and Exchange Board of India [Depositories and Participants] Regulations 1996*).

Regulation of insider trading **12.28**

SEBI has notified the *SEBI (Prohibition of Insider Trading) Regulations 1992* (published in the Gazette Of India Extraordinary, Part II – Section 3 – Sub-Section (II), Bombay, 19 November 1992) [the SEBI (Insider Trading) Regulations) dealing, *inter alia*, with the buying and selling of securities of the company by employees/directors. A plain reading of the *SEBI (Insider Trading) Regulations* appears to indicate that two conditions need to be fulfilled to hold somebody guilty as an insider:

- The 'insider' must be a connected person by virtue of his position in the company such as director, officer or employee with a professional or business relationship with access to unpublished price sensitive information. Otherwise, he is deemed to be a connected person such as a company under the same management or subsidiary, member of stock exchange or merchant banker or others acting in a similar capacity.
- The 'insider' has traded in those securities on the basis of unpublished price sensitive information.

Regulations of collective investment schemes **12.29**

The *SEBI (Collective Investment Schemes) Regulations 1999* (the Gazette of India, Extraordinary Part ii – Section 3 – Sub-section ii, Mumbai on the 15 day of October 1999). Henceforth, no person other than a Collective Investment Management Company which has obtained a certificate of registration under the *SEBI (Collective Investment Schemes) Regulations 1999*, can carry on or sponsor or launch a collective investment scheme. Also, no existing collective investment scheme can launch any new scheme or raise money from the investors even under the existing schemes, unless a certificate of registration is granted to it under the said regulations.

Every Collective Investment Management Company shall be responsible for managing the funds or properties of the scheme on behalf of the unit holders. It should take all reasonable steps and exercise due diligence to ensure that the scheme is managed in accordance with the provisions of these regulations; offer document and the trust deed. A collective management company should exercise due diligence and care in managing assets and funds of the scheme. It should also be responsible for the acts of commissions and omissions by its employees or the persons whose services have been availed by it. It is incompetent to enter into any transaction with or through its associates, or their relatives relating to the scheme (*regulation 14* of the *SEBI (Collective Investment Schemes) Regulations, 1999*).

Takeover code **12.30**

The procedure for takeovers is enshrined in the *SEBI (Substantial Acquisition of Shares and Takeovers) Regulations 1997*, (published in the Gazette of India,

Extraordinary Part ii – Section 3 – Sub-section ii, Bombay on 20 February 1997), as amended in 2002. These make it compulsory to give a public announcement before a takeover deal. No acquirer shall acquire shares or voting rights which (taken together with shares or voting rights, if any, held by him or by persons acting in concert with him), entitle such acquirer to exercise 15 per cent or more of the voting rights in a company, unless such acquirer makes a public announcement to acquire shares of such company in accordance with the regulations (*regulation 10* of the *SEBI* (*Substantial Acquisition of Shares and Takeovers*) *Regulations 1997* as inserted by *SEBI* (*Substantial Acquisition of Shares and Takeovers*) *Amendment Regulations, 1998* published in the Official Gazette of India dated 28 October 1998). The takeover code also contains various disclosure mechanisms like the disclosure of the acquirer to the stock exchanges and continual disclosures (*Regulation 7*(*3*) *and Regulation 8* of the *Securities Exchange Board of India* (*Substantial Acquisition of Shares and Takeovers*) *Regulations 1997* respectively). In addition to the disclosures, SEBI has the power to call for information from the stock exchanges and the company (*Regulation 9* of the *Securities Exchange Board of India* (*Substantial Acquisition of Shares and Takeovers*) *Regulations 1997*).

Fraudulent and unfair trade practices 12.31

SEBI (*Prohibition of Fraudulent and Unfair Trade Practices relating to Securities Market*) *Regulations 1995* has been amended by the *SEBI* (*Prohibition of Fraudulent and Unfair Trade Practices relating to Securities Markets*) *Regulations 2003* (published in the Gazette of India, Extraordinary, Part ii – Section 3 – Sub-section ii, Mumbai, 17 of July, 2003) to control the unfair trade practices in the security market and by this amendment. The policing powers of the SEBI have also been increased as a result of this amendment.

Setting up of the Securities Appellate Tribunal 12.32

Regulation 15K of *SEBI* transferred the appellate function of the Central Government under three different Acts to the Securities Appellate Tribunal (this provision has been inserted by the *Securities Laws* (*Second Amendment*) *Act, 1999* vide Gazette Notification dated December 16, 1999). The objectives of these Acts are to prevent undesirable transactions in securities and depositories by regulating the business of dealing and protecting investors interest. The Securities Appellate Tribunal shall not be bound by the procedure laid down by the *Code of Civil Procedure 1908*, but shall be guided by the principles of natural justice and, subject to the other provisions of this Act and of any rules, the Securities Appellate Tribunal shall have powers to regulate their own procedure including the places at which they shall have their sittings (*section 15U*(*1*) of the *Securities Laws* (*Second Amendment*) *Act, 1999*).

Governance through listing agreement 12.33

SEBI had convened a meeting of all the stock exchanges on January 17 2001 to discuss various issues relating to secondary market including the compliance of the provisions of corporate governance. On the basis of the discussion in the meeting, the stock exchanges are directed to implement the following things outlined below.

Setting up a separate monitoring cell 12.34

The stock exchanges shall set up a separate monitoring cell with identified personnel to monitor the compliance with the provisions of the corporate governance. This cell shall obtain the quarterly compliance report from the companies who are scheduled in the first phase and shall submit a consolidated compliance report to SEBI within 30 days of the end of the quarter, commencing from the quarter ending March 2001.

The companies who are listed in the first phase of the Clause 49 of the Listing Agreement will be required to submit a quarterly compliance report to the stock exchanges within 15 days from the end of the quarter. The report shall be submitted either by the Compliance Officer or the Chief Executive Officer of the company after obtaining due approvals (see the revised Clause 49, Circular dated August 26, 2003 Ref: SEBI/MRD/SE/31/2003/26/08, source: http://web.sebi.gov.in/circulars/2003/cir2803an1.html).

Report on corporate governance 12.35

There shall be a separate section on corporate governance in the annual reports of a company, with a detailed compliance report on corporate governance. Non-compliance of any mandatory requirement ie which is part of the listing agreement with reasons thereof and the extent to which the non-mandatory requirements have been adopted should be specifically highlighted. The companies shall submit a quarterly compliance report to the stock exchanges within 15 days from the close of quarter as per the format given below. The report shall be submitted either by the compliance officer or the chief executive officer of the company after obtaining due approvals (see Part IX of the Revised Clause 49,Circular dated August 26, 2003 Ref: SEBI/MRD/SE/31/2003/26/08, Source: http://web.sebi.gov.in/circulars/2003/ cir2803an1.html).

Listing of Initial Public Offerings 12.36

According to the schedule of implementation of corporate governance specified by SEBI Circular No SMDRP/Policy/CIR-10/2000 dated February 21 2000, all the entities seeking listing for the first time are required to comply with the provisions of corporate governance at the time of listing.

The stock exchanges shall ensure that these provisions have been complied with before granting any new listing. For this purpose, it will be satisfactory compliance if these companies have set up the boards and constituted the committees such as the audit committee, shareholders/investors grievance committee *et al.* Before seeking listing, a reasonable time to comply with these conditions may be granted only where the stock exchange is satisfied that genuine legal issues exists which will delay such compliance. In such cases while granting listing, the stock exchanges shall obtain an undertaking from the company. In case of the companies failing to comply with this requirement without any genuine reason, the application money shall be kept in an escrow account till the conditions are complied with, (Circular No 48/2001 dated March 03, 2001 issued by SMDP, SEBI source: http://web.sebi.gov.in/circulars/2001 /CIR482001.html as on 31–5-2004).

Electronic Data Information Filing And Retrieval (EDIFAR) 12.37

EDIFAR is an Electronic Data Information Filing And Retrieval System. This would involve electronic filing of information in a standard format by the companies. This system has several benefits by way of dissemination of information to various classes of market participants like investors, regulatory organization, research institutions etc. This would also be useful to companies and stock exchanges.

SEBI in association with National Informatics Center (NIC) has set up an Electronic Data Information Filing and Retrieval (EDIFAR) to facilitate filing of certain documents/statements by the listed companies on line on the website to be maintained by NIC. This system is being introduced in a phased manner and would be applicable to 200 companies, (*vide* circular dated July 3, 2002 issued by SEBI, Ref: SMD/POLICY/Cir-17/02, Source: http://web.sebi. gov.in/circulars/2002 /cir232002.h). Initially, the following statements/information would be filed online:

- Financial statements comprising the balance sheet, profit and loss account and full version of annual report; half yearly financial statements including cash flow statements and quarterly financial statements.
- Corporate governance reports.
- Shareholding pattern statement.
- Action taken against the company by any regulatory agency.

Committees on corporate governance 12.38

India has also formulated codes of corporate governance through various committees, more important ones being:

- Confederation of Indian Industries (CII) Code of Desirable Corporate Governance (1998).
- Unit Trust of India (UTI) Code of Governance (1999).

- Kumar Mangalam Birla Committee on Corporate Governance (2000).
- Naresh Chandra Committee on Corporate Audit & Governance (2002).
- N R Narayan Murthy Committee (SEBI, 2003).

Report of Kumar Mangalam Birla Committee on Corporate Governance 12.39

This report should be commented upon as a precedent.

The Securities and Exchange Board of India (SEBI) appointed the Committee on Corporate Governance on May 7 1999 under the chairmanship of Shri Kumar Mangalam Birla, a member of the SEBI Board, to promote and raise the standards of corporate governance. The detailed terms of reference are as follows:

- To suggest suitable amendments to the listing agreement executed by the stock exchanges with the companies and any other measures to improve the standards of corporate governance in the listed companies, in areas such:
  - as the continuous disclosure of material information, both financial and non-financial;
  - the manner and frequency of such disclosures;
  - the responsibilities of independent and outside directors;
- To draft a code of corporate best practices; and
- To suggest safeguards to be instituted within the companies to deal with insider information and insider trading.

The Committee also took note of the various steps already taken by SEBI for strengthening corporate governance, some of which are:

- Strengthening of disclosure norms for Initial Public Offers following the recommendations of the Committee set up by SEBI under the chairmanship of Shri Y H Malegam;
- Providing information in directors' reports for the utilisation of funds and variation between projected and actual use of funds according to the requirements of the Companies Act 1956; inclusion of cash flow and funds flow statement in annual reports;
- Declaration of quarterly results;
- Mandatory appointment of compliance officer for monitoring the share transfer process and ensuring compliance with various rules and regulations;

- Timely disclosure of material and price sensitive information including details of all material events having a bearing on the performance of the company;
- Despatch of one copy of complete balance sheet to every household and abridged balance sheet to all shareholders;
- Issue of guidelines for preferential allotment at market related prices; and
- Issue of regulations providing for a fair and transparent framework for takeovers and substantial acquisitions.

The Committee has identified the three key constituents of corporate governance as the *shareholders*, the *board of directors* and the *management* and has attempted to identify in respect of each of these constituents, their roles and responsibilities as also their rights in the context of good corporate governance. Fundamental to this examination and permeating throughout this exercise is the recognition of the three key aspects of corporate governance, namely; accountability, transparency and equality of treatment for all stakeholders.

The recommendations of the committee **12.40**

This report is the first formal and comprehensive attempt to evolve a code of corporate governance, in the context of prevailing conditions of governance in Indian companies, as well as the state of capital markets. While making the recommendations the committee has been mindful that any code of corporate governance must be dynamic, evolving and should change with changing context and times. It would therefore be necessary that this code also is reviewed from time to time, keeping pace with the changing expectations of the investors, shareholders, and other stakeholders and with increasing sophistication achieved in capital markets.

Corporate governance – the objective **12.41**

Corporate governance has several claimants – shareholders and other stakeholders – which include suppliers, customers, creditors, the bankers, the employees of the company, the government and society at large. This report on corporate governance has been prepared by the committee for SEBI, keeping in view primarily the interests of a particular class of stakeholders, namely, the shareholders, who together with the investors form the principal constituency of SEBI while not ignoring the needs of other stakeholders.

The committee therefore agreed that the fundamental objective of corporate governance is the 'enhancement of shareholder value, keeping in view the interests of other stakeholder'. This definition harmonises the need for a company to strike a balance at all times between the need to enhance shareholders' wealth whilst not in any way being detrimental to the interests of the other stakeholders in the company.

In the opinion of the committee, the imperative for corporate governance lies not merely in drafting a code of corporate governance, but in practising it. Even now, some companies are following exemplary practices, without the existence of formal guidelines on this subject. Structures and rules are important because they provide a framework, which will encourage and enforce good governance; but alone, these cannot raise the standards of corporate governance. What counts is the way in which these are put to use. The committee is of the firm view, that the best results would be achieved when the companies begin to treat the code not as a mere structure, but as a way of life.

The central problem in corporate governance in India today is sociological. Mr Rahul Bajaj, Chairman, Bajaj Auto, epitomised the problem. He said at a seminar on corporate governance in Mumbai in November, 1996 that:

> 'All of us know what boards and managements should do, but are doing what we should not do. We have done things that are questionable – legal but questionable. Why should we need a committee to tell us what to do?'

Mr Bajaj was the chairman of the task force of the Confederation of Indian Industry (CII) entrusted the task of preparing a code of corporate governance. The corporate governance problem in India crystallises into the following questions:

- Why do business leaders do things they know should not be done?
- What are the pressures or fears that force them to do so?
- How can they be helped to be more integral to their own beings?
- How can the board of directors play a more useful role?

The owners have to bring about an attitudinal change in them and identify with the aspirations of each stakeholder. One way in which organisations can improve their standards of corporate governance in the future is by focusing on how they choose external directors who should be selected mainly on the basis of their experience and expertise.

Corporate governance in the 21st Century also emphasises the role of shareholders and financial institutions. Shareholders should be involved in all major decisions and financial institutions should appoint functional experts to represent them on the boards of companies in which they have substantial holdings.

Other corporate laws of India 12.42

The administration of affairs of companies in today's competitive environment entrusted to persons of professional standing. The requirements of public accountability and social responsibility have become a prominent feature of law, making it necessary to pay due attention to the generally recognised principles of equity, shareholders' democracy, prudent governance and transparent accounting.

Apart from the laws that have been analysed above, the Indian regulatory framework concerning corporate governance for companies covers legislation such as the *Banking Regulation Act 1949, the Income Tax Act 1961,* the *Indian Stamp Act 1899, the Insurance Regulatory and Development Act 1999, the Information Technology Act 2000* and various Acts covering the labour laws of India.

Conclusion **12.43**

Although the subject of corporate governance has been receiving explicit attention in India only in the last few years, the institutional and regulatory framework for corporate governance has been in place for a long time. The legal framework for regulating all corporate activities including governance and administration of companies, disclosures, shareholders' rights, ie the *Companies Act* has existed since 1956 and has been fairly stable. The stock exchanges have been executing listing agreements laying down on-going conditions and continuous obligations for companies.

During the last few years, greater attention is being focused on the subject and there has been a discernible growth in awareness about corporate governance being an intrinsic part of companies' best practices and obligations. The lack of it, or, the inadequacy of corporate governance among companies, has been featuring in the media, in the boardrooms of financial institutions who are the block holders in many companies, and in academic debates as one of the reasons for the 'present disenchantment' of small investors. The boards of enlightened companies – even those belonging to the business families – financial institutions and other large institutional shareholders, investors and regulators are increasingly becoming aware of the need for disclosures, open and transparent management concepts, continuing obligation to put out material information, better accounting standards, and for establishing standards of best practices to be pursued by the directors, managers and employees of companies. The protection of the interest of the shareholders and the preservation and enhancement of shareholder value and wealth are concepts, which are being widely recognised in the industry and assuming greater importance.

Several factors have helped drive this change. They are:

- The economic reforms of 1991, which have allowed the growth of free enterprise and given private enterprise its rightful place alongside the public sector;
- The increased domestic and foreign competition to domestic private and public sector companies which has multiplied choices for the consumers, compelled increases in efficiency;

- The growing reliance placed by private and public sector companies on capital markets, underpinning the need for better disclosures;
- The consequential changes in the shareholding pattern of private and public sector companies, Foreign Direct Investment caps and disinvestments;
- The growing awareness of investors and investor groups of their rights;
- The rise of institutional investors, with the public financial institutions gradually asserting themselves and transforming themselves into their new role as active shareholders rather than as lenders;
- The stock exchanges becoming gradually conscious of their roles as self-regulatory organisations and exploring the possibility of using the listing agreement as a tool for raising standards of corporate governance;
- The role of SEBI as the statutory regulatory body for the securities market to protect the rights of investors and to regulate the securities markets.

Chapter 13

International Dimensions: Corporate Governance in Hong Kong Special Administrative Region and the People's Republic of China

Background to corporate governance in Hong Kong 13.1

Since the collapse of the Peregrine Group in 1997 and the Akai Group in 1999, Hong Kong has thankfully been spared from any major corporate failures such as those suffered more recently in the United States with Enron, Worldcom and Tyco, and elsewhere, such as Parmalat in Italy. Hong Kong's public companies, however, continue to be characterised by the predominance of family controlled businesses with their associated problems for minority shareholders and outside lenders.

After a succession of reforms during the 1990's which were mainly in response to the Cadbury Report from the United Kingdom, the pace of corporate governance reform has continued at a steady pace. Many of the ongoing reforms stem from external influences such as the Organisation for Economic Co-operation and Development (OECD) rather than in response to Hong Kong related issues. There is no doubt also that the lessons learnt from the Enron and Worldcom experience have had a profound effect on how the boards of public companies in Hong Kong conduct business, particularly regarding the role of auditors, independent directors and audit committees.

In line with the global trend towards greater corporate transparency, to date four major Consultancy Studies have been commissioned by the Standing Committee on Company Law Reform in Hong Kong relating to corporate governance. Several measures affecting corporate governance have been introduced recently in the *Companies (Amendment) Ordinance 2003* and more are currently being proposed in the *Companies (Amendment) Bill 2003*. Many of their recommendations are still subject to consultation, including those relating to directors' duties, voting by directors in relation to self-dealing, redefining 'associated companies' for the purpose of connected transactions and further expanding the role of independent directors.

The purpose of this section is to briefly highlight the main legislation and codes affecting corporate governance in Hong Kong.

Directors' Duties

Private companies

Standard of care 13.2

Directors should meet the standard of care required by law in performance of their duties, which is one 'an ordinary man might be expected to take in the circumstances on his own behalf' *(Re City Equitable Insurance Co Ltd (1925) Ch 407; Law Wai Duen v Boldwin Construction Co Ltd [2001] HKC 403).*

Duties and requirements 13.3

Directors can make reference to the 'Non-Statutory Guidelines on Directors' Duties' issued by the Companies Registry (January 2004) when acting for the company. These state directors have a duty:

- to act in good faith for the benefit of the company as a whole;
- to use powers for a proper purpose for the benefit of members as a whole;
- not to delegate except with proper authorization and duty to exercise independent judgment;
- to exercise care, skill and diligence;
- to avoid conflicts between personal interests and interests of the company;
- not to enter into transactions in which the directors have an interest except in compliance with the requirements of the law;
- not to gain advantage from use of position as a director;
- not to make unauthorised use of company's property or information;
- not to accept personal benefit from third parties conferred because of position as a director;
- to observe the company's memorandum and articles of association and resolutions; and
- to keep proper books of account.

Directors must ensure that the duty to disclose and prohibitions under the *Companies Ordinance*, (*Cap 32*) (*CO*) are complied with such as *s 129D(3)(j)*, *s 162, Sch 1 Table A, Article 86* where applicable and in relation to interested contracts of directors; *s 157H* which prohibits loans/guarantees to directors or other connected persons; and *s 157HA* requirements in cases where a loan is made to a director for the purposes of the company or in the performance of the director's duties.

Additional requirements for listed companies

Duties and requirements **13.4**

Directors of listed companies must ensure that they are in compliance with the requirements and obligations of Rule 3.08–20 of the Main Board Listing Rules, including:

- Every director must:
  - act honestly and in good faith in the interests of the company as a whole;
  - act for proper purpose;
  - be answerable to the company for the application or misapplication of its assets;
  - avoid actual and potential conflicts of interest and duty;
  - disclose fully and fairly his interests in contracts with the company; and
  - apply such degree of skill, care and diligence as may reasonably be expected of a person of his knowledge and experience and holding his office within the company;
- Every director must satisfy the Stock Exchange that he has the character, experience and integrity and is able to demonstrate a standard of competence commensurate with his position as a director of the company.
- Every director shall comply with the Model Code set out in Appendix 10 of the Listing Rules or the company's own code on no less exacting terms.

Absolute prohibitions **13.5**

The following are a list of absolute prohibitions which directors of a company must comply with:

- A director must not deal in any securities of the company at any time when he is in possession of unpublished price sensitive information in relation to those securities, or where clearance to deal is not otherwise conferred under the Code.

- During the period commencing one month immediately preceding the earlier of the board meeting for approval of the company's results and the deadline for the company to publish an announcement of its results for any year, half-year, quarter or any other interim period, and ending on the date of the results announcement, a director must not deal in any securities of the company unless the circumstances are exceptional.

- Where a director is a sole trustee, the Code will apply to all dealings of the trust as if he were dealing on his own account unless the director is a bare trustee and neither he nor any of his associates is a beneficiary of the trust.

- The restrictions on dealings by a director contained in the Code will be regarded as equally applicable to any dealings by the director's spouse or by or on behalf of any minor child and any other dealing in which for the purposes of Part XV of the Securities and Futures Ordinance (Cap 571) (SFO) he is or is to be treated as interested.

- When a director places investment funds comprising securities of the company under professional management, discretionary or otherwise, the managers must nonetheless be made subject to the same restrictions and procedures as the director himself in respect of any proposed dealings in the company's securities.

Notification **13.6**

The following are a list of circumstances when a director must advise others of his intentions with regard to his dealings:

- A director must not deal in any securities of the company without first notifying in writing the chairman or a director (otherwise than himself) designated by the board for the specific purpose and receiving a dated written acknowledgement. In his own case, the chairman must first notify the board at a board meeting, or alternatively notify a director (otherwise than himself) designated by the board for the purpose and receive a dated written acknowledgement before any dealing. The designated director must not deal in any securities of the company without first notifying the chairman and receiving a dated written acknowledgement.

- The minimum procedure of the company must provide for there to be a written record maintained by the company that the appropriate notification was given and acknowledged pursuant to the Code, and for the director concerned to have received written confirmation to that effect.

- Any director who acts as trustee of a trust must ensure that his co-trustees are aware of the identity of any company of which he is a director so as to enable them to anticipate possible difficulties. A director having funds under management must likewise advise the investment manager.
- Any director who is a beneficiary, but not a trustee, of a trust which deals in securities of the company must endeavor to ensure that the trustees notify him after they have dealt in such securities on behalf of the trust, in order that he may in turn notify the company. For this purpose, he must ensure that the trustees are aware of the company of which he is a director.
- The register maintained in accordance with s 352 of SFO should be made available for inspection at every meeting of the board.
- The directors must as a board and individually endeavor to ensure that any employee of the company or director or employee of a subsidiary company who, because of his office or employment in the company or a subsidiary, is likely to be in possession of unpublished price-sensitive information in relation to the securities of any listed company does not deal in those securities at a time when he would be prohibited from dealing by the Code if he were a director.

Exceptional circumstances **13.7**

If a director proposes to sell or otherwise dispose of securities of the company under exceptional circumstances where the sale or disposal is otherwise prohibited under the Code, the director must comply with the Code, inter alia, regarding prior written acknowledgement. The company shall give written notice of such sale or disposal to the Stock Exchange as soon as practicable stating why it considered the circumstances to be exceptional. The company shall publish an announcement in the newspapers immediately after such sale or disposal and state that the chairman or the designed director is satisfied that there were exceptional circumstances for such sale or disposal of securities by the director.

Disclosure **13.8**

In relation to securities transactions by directors, the company shall disclose in its annual and interim reports: whether the company has adopted a code of conduct regarding securities transactions by directors on terms no less exacting than the required standard set out in the Code; whether its directors have complied with or whether there has been any non-compliance with, the required standard set out in the Code and the company's own code of conduct; details of any non-compliance and an explanation of the remedial steps taken by the company to address such non-compliance.

Independent Non-Executive Directors (INEDs) 13.9

A company must ensure that it has at least three INEDs (Rule 3.10(1)). At least one of the INEDs must have appropriate professional qualifications or accounting or related financial management expertise (Rule 3.10(2)). An INED must also satisfy the Stock Exchange that they have the character, integrity, independence and experience to fulfill their role effectively (Rule 3.12).

Code of Best Practice of the Board 13.10

Directors must ensure that the guidelines under Appendix 14 to the Listing Rules are complied with. These form the skeleton of a code of best practice of the board (including INEDs) and cover the following areas:

- Full board meetings shall be held no less frequently than every six months.
- Except in emergencies an agenda and accompanying board papers should be sent in full to all directors at least two days before the intended date of a board meeting.
- Except in emergencies adequate notice should be given of a board meeting to give all directors an opportunity to attend.
- All directors, executive and non-executive, are entitled to have access to board papers and materials. Where queries are raised by non-executive directors, steps must be taken to respond as promptly and fully as possible.
- Full minutes shall be kept by a duly appointed secretary of the meeting and such minutes shall be open for inspection at any time in office hours on reasonable notice by any director.
- The directors' fees and any other reimbursement or emolument payable to an INED shall be disclosed in full in the annual report and accounts of the company.
- Non-executive directors should be appointed for a specific term and that term should be disclosed in the annual report and accounts of the company.
- If, in respect of any matter discussed at a board meeting, the INEDs hold views contrary to those of the executive directors, the minutes should clearly reflect this.
- Arrangements shall be made in appropriate circumstances to enable the INEDs, at their request, to seek separate professional advice at the expense of the company.
- Every non-executive director must ensure that he can give sufficient time and attention to the affairs of the company and should not accept the appointment if he cannot.

- If a matter to be considered by the board involves a conflict of interest for a substantial shareholder or a director, a full board meeting should be held and the matter should not be dealt with by circulation or by committee.
- If an INED resigns or is removed from office, the Stock Exchange should be notified of the reasons why.
- Every director on the board is required to keep abreast of his responsibilities as a director. Newly appointed board members should receive an appropriate briefing on the company's affairs and be provided by the company secretary with relevant corporate governance materials currently published by the Stock Exchange on an ongoing basis.
- The board should establish an audit committee with written terms of reference which deal clearly with its authority and duties.

Audit committee **13.11**

An audit committee comprising only non-executive directors needs to be established with a minimum of three members, at least one of whom is an INED (Rule 3.21)

The principal duties of the audit committee include:

- assisting the board of directors in providing an independent review of the effectiveness of the financial reporting process;
- internal control and risk management system of the company; and
- overseeing the audit process; and performing other duties and responsibilities as assigned by the board of directors.

The board must approve and provide written terms of reference for the audit committee which clearly establish the committee's authority and duties (Rule 3.22).

The Stock Exchange must be informed immediately and an announcement published in the newspapers containing the relevant details and reasons why the company failed to set up an audit committee or at any time has failed to meet any of the other requirements under Rule 3.21 (Rule 3.23).

Notice of the responsibilities and guidelines for the holding of meetings by an audit committee suggested in *A Guide for Effective Audit Committees* published by The Hong Kong Society of Accountants must be taken.

Transparency and disclosure

Private companies 13.12

Statutory returns on corporate information must be filed with the Companies Registry in compliance with the *CO* and the minimum disclosures required by the *CO* to be made in financial statements needs to be complied with.

The mandatory disclosures as specified by Hong Kong Society of Accountants' Statements of Standard Accounting Practice must also be made.

Additional requirements for Listed companies

Price sensitive information 13.13

The obligations imposed under Paragraph 2(1) of the Listing Agreement to keep the Stock Exchange, their members and the public informed of price sensitive information, eg dividend payments must be complied with

Notifiable transactions 13.14

Disclosure of certain transactions (principally acquisitions and disposals) and arrangements as required under Chapter 14 of the Listing Rules and issuance of circulars to shareholders or obtaining approval from shareholders when required must be carried out.

Disclosure of interests 13.15

Notifications of shareholdings or changes in shareholdings required by *Part XV* of the *Securities and Futures Ordinance* need to be made ie a person who acquires, disposes of, or make changes to an interest in the share capital of a listed company has a duty of disclosure if the interest is notifiable. A notifiable interest exists in any one of the following circumstances:

- interest in the shares comprised in the relevant share capital of the listed company of an aggregate nominal value equal to or more than five per cent of the issued equity share capital; or any change by moving over or below this level;
- any change in percentage level across a whole number percentage; or
- any change in the nature of interest.

Executive pay

Private companies 13.16

Directors must ensure that accounts laid before the company in general meetings contain the aggregate amounts of the directors' emoluments, directors or past directors' pensions, and any compensation to directors or past directors in respect of loss of office (*CO, s 161*)

No payment must be made to any director or past director by way of compensation for loss of office, or as consideration for or in connection with his retirement from office, unless the particulars relating to the proposed payment are disclosed to the members of the company and approved (*CO, s 163*).

The remuneration of directors shall be determined from time to time by the company in general meeting where *Sch 1, Table A, Article 78* is applicable.

Additional requirements for listed companies 13.17

Directors' fees and any other reimbursement or emolument payable to an INED must be disclosed in full in the annual report and accounts.

Shareholders' protection

Approval from shareholders 13.18

Approval from shareholders needs to be obtained on the matters required by the *CO*, including the following eight important matters relating to the company:

- alteration of articles and objects;
- dealing with the company's capital;
- powers to dispose of the company's fixed assets;
- appointment and removal of auditors;
- appointment of directors, directors' salaries and approving payments for loss of office;
- variation of class rights;
- matters relating to dividends; and
- winding up.

Meetings 13.19

The company must hold in each year an annual general meeting within 15 months of the previous annual general meeting (but within 18 months from the date of incorporation in case of the first annual general meeting) (*CO, s 111*). Extraordinary general meetings must be convened when they are called by shareholders. The requirements on notice periods for shareholders' meetings must be complied with in accordance with (*s 114(2), s161BA(6), s 161BA(5), s 114A(2)(a), s 114(3)(a), s114(3)(b)* of the *CO*).

Shareholders' remedies 13.20

Shareholders have personal rights to enforce provisions of Memorandum and Articles of Association of the company (*s 23* of the *CO*).

Shareholders can petition to the court for an appropriate order if the affairs of the company have been conducted in a manner which is unfairly prejudicial to the interests of the shareholders (*s 168A* of the *CO*).

Derivative action can be brought against a wrongdoer who is in control of the company.

Access to information 13.21

Shareholders have a right to inspect the register of debenture holders, register of charges, register of directors and secretaries, minute books of general meeting, register of members, management contracts at general meeting under the CO.

Copy of balance sheet, directors' report and auditors' report should be sent to shareholders not less than 21 days before general meeting (*s 1269G* of the *CO*).

Corporate governance in the People's Republic of China (PRC)

Background to corporate governance in the PRC 13.22

The spectacular collapse of the Guangdong International Trust and Investment Corporation (GITIC) in 1999 leaving over $US 2bn of unrecoverable loans highlighted some of the deep-rooted problems of corporate governance in the PRC and the need for urgent reforms. Among the reasons for GITIC's failure were poor management, rash investments, lack of risk management systems and inefficient operations.

Another shocking story published on 2 August 2001 highlighting that the Guangxia Industry Company Limited (Yinguangxia) had been fabricating profits for several years (see Ling Huawei and Wang Shuo, 'Yinguangxia Trap', *Business & Finance Review*, (2001) August, pp 18–37). The price of Yinguangxia's shares had risen about 800 per cent in one year as a result of:

- falsified contracts;
- exaggerated technologies;
- forged customs returns;
- fraudulent announcements; and
- distorted financial statements.

The following day the China Securities Regulatory Commission (CSRC) suspended dealings in Yinguangxia's shares and began an investigation into the company.

The CSRC has criticised or disciplined several companies for corporate governance violations. Frequent corporate scandals not only brought corporate governance into the newspaper headlines in China – like elsewhere – but also prompted investors and regulators into thinking that the time had come to reform the corporate governance system in China, especially corporate governance in listed companies. This was particularly because of the grave impact of such scandals in the securities market.

Many of the systemic problems came about because most listed companies were re-structured state-owned enterprises or were majority controlled by government agencies. Directors and managers were chosen by bureaucrats who had little incentive to appoint the most suitably qualified candidates and often there was collusion between them to misappropriate funds for their own benefit. There were problems in removing non-performing directors and independent directors were rarely appointed. Because so many shares were non-tradable, there was little incentive to senior management or the board to improve performance or pay attention to the interests of minority shareholders.

In addition to the above matters the PRC corporate governance system has been hindered by various interconnected issues, such as:

- the absence of effective implementation mechanisms of laws and regulations;
- the lack of necessary incentive for management; and
- the absence of appropriate rights protection for small/public shareholders.

The lack of efficient corporate governance that caused corporate failures, financial difficulties, fraud and other scandals meant that reform was a priority for the Chinese legislative agenda. Different aspects were addressed in published CSRC Guidelines, such as 'The Guideline Regarding Independent

Directors in Listed Companies' (The Independent Directors Guideline). Reform was intended to target the following issues:

- a reduction in State-owned Shares;
- reform of the board of directors, including independent directors;
- restrictions on connected transactions;
- information disclosure; and
- reform of the compensation system.

The PRC has so far relied on a legalistic approach to corporate governance reform.

Corporate governance for private companies in the PRC 13.23

The main corporate governance provisions applicable to private companies are set out in the *PRC Company Law* and are summarised below.

Shareholder accountability 13.24

A shareholder must contribute their capital into the company in accordance with the Articles, otherwise they will be held liable for breach of contract (*Article 25*). Capital contributions must be verified by an approved verification authority (*Articles 27* and *91*) and shareholders must not withdraw their capital contribution after the company is set up (*Article 209*).

Board of directors and directors 13.25

The board of directors is accountable to general meeting (*Articles 46* and *112*).

Subject to re-election, the duration of office for a director must not be longer than three years (*Articles 47 and 115*), and a director may not hold a concurrent post as supervisor (*Articles 52 and 124*).

Directors' duties 13.26

Article 57 sets out the circumstances under which persons will be disqualified from acting as a director and a civil servant may not hold a concurrent post as director (*Article 58*). The following are a list of directors' duties:

- Directors must discharge their duties faithfully, uphold the interests of the company, and not seize a company's properties, etc (*Article 59*).
- A director must not misappropriate company's funds or lend company's funds to a third party (*Article 60*).

- A director must not run his own business or work for other businesses which will be likely to compete with the company (*Article 61*).
- Subject to the requirements of relevant laws and the approval of a general meeting, a director must not disclose any confidential information of the company (*Article 62*).

If a director during the course of discharging their duty violates the laws, administrative regulations or the Articles. They are required to compensate the company for any damages suffered (*Article 63*).

Supervisory board and supervisor **13.27**

Articles 57, 58, 59, 62 and *63* quoted above are also applicable to supervisors (*Article 123*). Subject to re-election, the duration of an office for a supervisor must not be longer than three years (*Articles 53 and 125*). A supervisory board must consist of representatives of shareholders and representatives of workers (*Articles 52 and 125*). *Articles 54* and *125* set out the authorities of a supervisory board. A supervisor must discharge his duties faithfully (*Article 128*).

Corporate governance for listed companies in the PRC **13.28**

There are additional requirements as to corporate governance for listed companies. Apart from laws and regulations relating to corporate governance, there are also many regulatory notices, codes and guidelines which have been issued by regulatory bodies, including the National People's Congress Standing Committee (NPCSC), State Council, Ministry of Finance (MOF), the Chinese Securities Regulatory Commission (CSRC), the Chinese Institute of Certified Public Accountants (CICPA), the China Accounting Standard Committee (CASC), the State Economic and Trade Committee (SETC), the China National Audit Office (CNAO), the Shanghai Stock Exchange (SHSE) and Shenzhen Stock Exchange (SZSE).

In particular, based on the OECD Principles of Corporate Governance, the CSRC and SETC jointly issued the 'Codes of Corporate Governance for Listed Companies' in January 2002.

Key principles **13.29**

From the various published laws, regulations and other regulatory documents, the following can be summarised as the five key principles for corporate governance in the PRC:

- shareholder accountability;
- directors' duties;
- the external audit must be independent and penetrating;

- disclosure and transparency are crucial to market integrity; and
- there must be an appropriate regime of regulatory discipline to back these obligations.

The article numbers quoted below refer to the 'Codes of Corporate Governance for Listed Companies' unless otherwise stated.

Shareholders **13.30**

Corporate governance must ensure equal status amongst all the shareholders, especially the minority shareholders (*Article 2*).

Shareholders are entitled to rely on legal channels to protect their lawful interests (*Article 4*).

Guidelines for general meetings **13.31**

Listed companies must ensure that they provide in their Articles for the principles of authority from members in general meeting to the board of directors (*Article 7*). Listed companies must also provide in their Articles procedures for general meetings, including convening meetings, voting, notices, registration and putting motions, etc (*Article 5*).

Associated transactions **13.32**

Rules on the disclosure of associated transactions are set out in *Article 12*. Listed companies must use effective measures to prevent shareholders and other associated parties from taking possession of and transferring company's funds and assets etc (*Article 14*).

Guidelines for controlling shareholders **13.33**

Controlling shareholders owe other shareholders a duty of loyalty (*Article 19*). The controlling shareholders must support labour reforms and improvements of human resources and remuneration systems of listed companies (*Article 18*). Major issues of the company must be decided by general meeting and the board of directors. Controlling shareholders must not intervene in decisions of companies, their lawful operation and business activities (*Article 21*).

Directors' selection process **13.34**

Listed companies must provide at their general meeting detailed particulars of candidates to be elected as directors (*Article 29*). During the process of electing directors, the opinions of minority shareholders must be fully considered (*Article 31*). Listed companies must enter into contracts with the directors clearly setting out their rights and obligations (*Article 32*).

Directors' duties **13.35**

Directors must act in the best interests of the company and the shareholders and have duties of honesty, loyalty and diligence (*Article 33*). Directors must ensure they have sufficient time and commitment to discharge their duties (*Article 34*). Directors involved in any decision-making process which is in violation of laws, regulations and Articles of the company are required to compensate the company for any damage suffered (*Article 38*). Listed companies may buy indemnity insurance for the directors if this is approved in general meeting (*Article 39*). The board of directors must consist of a reasonable percentage of persons from different professions (*Article 41*). The board of directors is accountable to the general meeting (*Article 42*).

Rules of procedure for board meetings **13.36**

Listed companies must provide in their Articles for the rules of procedure for board meetings (*Article 44*), and directors of listed companies must observe the rules of procedure strictly (*Article 46*). Minutes of board meetings must be complete and accurate (*Article 47*).The board of directors may delegate some of its functions to the chairperson of the board when the board is not in session (*Article 48*).

Independent directors **13.37**

Listed companies must provide for the system of independent directors (*Article 49*). Independent directors owe duties of loyalty and diligence to all shareholders (*Article 50*).

Independent director means a person who holds no position in the company other than the position of director. His relationship with the company itself and the major shareholders must not undermine his objective judgment (*Article 1(1)* Guiding Opinions concerning establishment of Independent Directors System in Listed Companies (Guiding Opinions) issued by CSRC on 16 August 2001).

Independent directors owe duties of loyalty and diligence to all shareholders (*Article 1(2), Guiding Opinions*). All listed companies within the PRC must amend their Articles to employ appropriate persons as independent directors including at least one professional accountant (*Article 1(3), Guiding Opinions*). and all independent directors and all prospective independent directors must attend a training course in accordance with the requirements of CSRC (*Article 1(5), Guiding Opinions*).

Independent directors must possess a basic knowledge of the operations of listed companies (*Article 2(3)*, Guiding Opinions) and must have at least five years working experience in the legal profession, accounting profession or other requisite experience as an independent director (*Article 2(4)*, Guiding Opinions).

Article 3, Guiding Opinions sets out the circumstances under which persons will be disqualified from acting as independent directors

Material associated transactions (over RMB300 million) are required to be sanctioned by the independent directors and submitted to the board of directors for discussion (*Article 5(1) 1, Guiding Opinions*). Independent directors are entitled to express their independent opinions on material matters (*Article 6, Guiding Opinions*).

Listed companies must set up an appropriate insurance system for the independent directors in order to lower the risks of independent directors being exposed to litigation (*Article 7(6), Guiding Opinions*).

Supervisor and supervisory board 13.38

A supervisor is entitled to know the operations of the company and owes the company a duty of confidentiality (*Article 60*). A supervisor must report to the board, general meeting and securities regulatory authorities the acts of directors, managers and other senior officials which are in violation of laws, regulations and the Articles (*Article 63*).

A supervisor must possess professional knowledge and relevant working experience such as legal and accounting knowledge (*Article 64*). Further Articles set out the duties of the supervisory board.

Executive incentive compensation 13.39

Listed companies must establish a mechanism linking the managers' remuneration with the company's performance (*Article 77*).

Corporate disclosure and transparency 13.40

In case of serious loss of a company's assets, the investigation and responsibility of relevant personnel shall be disclosed (*Article 36*, Contents and Format of Public Disclosure by Listed Companies (Standard No 2) – Contents and Format of Annual Report (Rev 2003)).

The full names of the ten largest shareholders, their shareholdings and types of shares (A, B, H shares) held at the end of each year shall be disclosed (*Article 25*, Contents and Format of Public Disclosure by Listed Companies (Standard No.2) – Contents and Format of Annual Report (Rev 2003)).

Conclusion 13.41

Hong Kong remains an important gateway to China and the Far East region with its highly developed financial infrastructure, common law tradition, independent judicial system, freedom of movement of capital and strong professional support services. Corporate governance standards in Hong Kong have definitely improved and such improvement will no doubt continue steadily.

The PRC has enormous economic potential, with significant business developments continuing to unfold as its accession to the World Trade Organization forces the pace of creating new opportunities for global players in many areas previously closed to or highly restricted for foreign investors.

The PRC Government has clearly demonstrated its awareness of the need to improve the climate of transparency and certainty for foreign investors, particularly those involved in foreign direct investment, joint ventures and privatisation of state-owned enterprises. Corporate governance reforms in the PRC have already been an essential part of the PRC's rapid legal development recently and are expected to result in more legislation and guidelines over the next few years as its legalistic approach to the issue continues.

Chapter 14

International Dimensions: Corporate Governance in Australia

Introduction 14.1

In Australia corporate governance has become a major issue in the last couple of years and to a large extent, has displaced the quest for quality accreditation under ISO and similar standards during the 1990's.

The reasons for this change in focus are obvious from the discussion in earlier chapters regarding the headline stories of corporate collapse, such as Enron and Worldcom in the US, HIH Insurance in Australia and the Parmalet scandal in Italy. The HIH Insurance case that has had such a dramatic impact in Australia is explained in more detail below (see **14.11**). While this chapter explains the position of corporate governance in Australia the lessons learned from this discussion and the practical hints and tips set out have international relevance.

While the direct causes of these corporate crashes continue to be the subject of extensive investigations by liquidators, regulatory authorities and have provided a catalyst for focus on governance, the collapses themselves are a symptom of poor governance practices rather than the cause.

As has been mentioned in earlier chapters, the recent corporate collapses have provoked a considerable amount of activity by governments with the *Sarbanes-Oxley Act* in the United States, the Corporate Law Economic Reform Program (CLERP) in Australia and similar legislation in many other western countries, as well as in Asia.

There has been a heavy focus on increasing penalties applied to directors and officers of companies and clearly a desire to prove that the legislative and regulatory structures are effective by meeting out harsh punishment to those involved in the collapses.

Whereas the boom and bust cycle of business has been around since human beings first began to trade, it was probably the celebrated collapse of the South Sea bubble in the 18th century that first saw some attention being focused by government to regulate business practices to protect stakeholders.

The initial approach to regulation and issues of governance was a minimalist one. It was argued that the inherent risks in any business helped to create an environment where large profits could be generated and that no government can regulate risk out of business activities, as regulation would stifle economic growth. This was largely the approach in most jurisdictions.

The attitude of the courts tended to support this minimalist approach. In 1892 the Marquis of Bute was sued by a liquidator for neglecting his duties as the President of the Cardiff Savings Bank.

The Marquis' family had a long history with the Bank that was originally established by his father in 1819 who became the President of the Bank. After he died in 1848 his son, then only six months old, inherited the position of President of the Bank. Notwithstanding the family's close association with the Bank, in the period from when he inherited the Presidency of the Bank until it collapsed in 1886, a period of 38 years, he only attended one Board meeting.

Justice Stirling absolved the Marquis from any responsibility holding the 'neglect or omission to attend meetings is not, in my opinion, the same thing as neglect or omission of a duty which ought to be performed at those meetings', clearly, times have changed.

The expectations of directors by shareholders and creditors and the obligation imposed on them by regulatory authorities and underlying legislation would not enable the Marquis of Bute to escape personal liability today by simply ignoring the affairs of the Cardiff Savings Bank.

Spectacular corporate failures make it easier for governments and regulators to add legislation to increase controls and supervision of corporations and those who govern them.

A very large proportion of the electorate in many countries have become shareholders often through privatisation and the sale of state owned utilities including telephone companies, roads, airports and railway lines. This adds to the pressures on governments to regulate business more closely.

Undoubtedly, governments see some shareholders as consumers in need of many of the traditional consumer protection rights and remedies, more commonly involved in the purchase of goods and services.

The heavy focus on governance has seen the end of multi-disciplinary practices (combining accounting and legal service providers under the one roof), a greater role for auditors and in some jurisdictions a positive obligation on legal and accounting advisers to become whistle-blowers.

What is good corporate governance? 14.2

It has been noted in **CHAPTER 10** that the corporate governance framework of the UK has influenced developments elsewhere. In Australia, Royal

Commissioner Justice Neville Owen in his report on the collapse of HIH suggested that corporate governance is not a term of art and his Honour preferred the principles set out in the Cadbury Report (*Committee on the Financial Aspects of Corporate Governance Report*, (1992)) which refers to characteristics of good governance using words such as 'openness', 'integrity' and 'accountability'.

A similar approach was taken by the Core Group on Corporate Governance established by the Pacific Economic Co-Operation Council (PECC), in their Guidelines for Good Corporate Governance launched at the 14th general meeting of PECC in Hong Kong in November 2001.

That committee dealt with the issue in the following way:

> 'Corporate governance principles indicate that business enterprises seeking to remain competitive in free and open markets should ensure that the powers, rights and resources invested in them as corporations are exercised for the benefit, profit and sustained competitiveness of the corporate enterprise.
>
> Moreover, as corporations, with their autonomous standing and separate personality before the law, they are set apart and distinguished from their owners, directors and managers.
>
> There is a need for some distance, defined and protected by professional ethics, to be kept between the corporation on one hand, and the owners, directors and managers on the other.
>
> The relations between owners, directors and managers and with the corporation should be governed by principles whereby each of these groups, in the proper exercise of their rights and duties, promote the best interest of the corporation as a whole.'

In May 1999, the Australian National Audit Office published a paper on *Better Principles and Practices – Corporate Governance in Commonwealth Authorities and Companies*. The ANAO paper noted:

> 'Definitions of corporate governance are many and varied. Broadly speaking, corporate governance refers to the processes by which organisations are directed, controlled and held to account.
>
> It encompasses authority, accountability, stewardship, leadership, direction and control exercised in the organization. For CAC bodies, key elements of corporate governance include the transparency of corporate structures and operations; the implementation of effective risk management and internal control systems; the accountability of the Board to stakeholders through, for example, clear and timely disclosure; and responsibility to society.'

Concepts of good corporate governance **14.3**

As has been noted corporate governance is not a new concept; indeed, the underlying principles of good corporate governance have been long established through the Westminster system of government.

The principle of separation of power between the executive government, the judiciary and the Parliament underpin many of the concepts being put into place to improve the quality of corporate governance in many jurisdictions around the world.

The convention that the Prime Minister of the day should surrender the seals of office at any time when that person loses the confidence of the Parliament is a practice, which can and should translate easily into the corporate world.

The basic concept of ministerial responsibility for matters within a minister's portfolio also sits comfortably with basic principles of good governance and can also be translated into the corporate environment.

As has been seen in **CHAPTER 10** many of the definitions and models put up to improve the quality of governance in the corporate world seek to define corporate governance by reference to structures and processes. For example, it has been said that good governance requires the establishment of an audit committee and once this is done, a corporation can tick the box and feel comfortably satisfied that it has met one of the criteria for good governance.

While undoubtedly legislators and regulators may seek to define corporate governance by reference to form and structure to make it easier to set benchmarks for accountability and prosecution of wrongdoers, this does not necessarily lead to good governance in a real and substantial way.

In Australia recently, a major scandal occurred within the National Australia Bank when traders were operating outside guidelines, resulting in a loss of over $AUD 300m to the bank from these trading activities.

The bank had in place the structures for corporate governance including audit and risk management committees. The independent review however, found that the processes and structures failed to prevent and detect these serious problems, the forms and structures said to be a sign of good governance had in fact, no real substance.

The chairperson and the chief executive officer resigned from the Bank to atone for the failure in governance. The chair of the Audit Committee refused to resign. Clearly, if this scandal has occurred in a government operating under the Westminster system, the chairperson of the Audit Committee as a minister, would have resigned immediately, even though under the Westminster system there is no written rule or legislated process to remove a minister in those circumstances.

Good governance in substance and in fact is defined by reference to the culture of a corporation based on the principles referred to in the Cadbury Report of openness, integrity and accountability. The adoption of such core principles at every level in a corporation will then ensure that the form and structures of the governance model within the corporation have real substance.

The establishment of a prison with the highest level of security can never absolutely prevent an escape and corporations with the strongest culture supporting good governance, cannot guarantee that there will not be some person or persons who will seek to subvert the system and damage the corporation.

As has been mentioned in **CHAPTER 8** the culture of the organisation is very relevant to good corporate governance. The existence of a strong culture supporting good governance, principles of integrity, openness and accountability will however, make it harder for those who wish to subvert the process, more likely that they will be detected and ultimately minimise the damage such persons can cause to a corporation.

Corporate governance and risk **14.4**

One complaint from the corporate world is that corporations, their directors and officers must take risks if they are to deliver growth and profits to shareholders and that an undue focus on good corporate governance may cause a corporation to become risk adverse and ultimately to stagnate and fail. People who advance such arguments misunderstand concepts underlying the principles of good governance.

Good governance is not about avoiding risks or stifling ideas and strategies. Rather, good governance ensures that risks, challenges and strategies are properly evaluated and assessed rather than driven forward by self-interest under a veil of secrecy with little accountability.

While perhaps simplistic good governance requires those responsible for managing corporations to ask the question: 'Why are we doing this, is it within our power and authority, is this the right thing to do?' Of course the lawyers will immediately respond by asking what 'right' means. The oath of office (prepared by lawyers) taken by judges in New South Wales is:

> 'I will do right to all manner of people, after the laws and usages of the State of New South Wales, without fear or favour, affection, or ill-will.'

'Right' in that context has been clearly understood to include transparency, integrity, openness and accountability having regard to all stakeholders and people that may be affected by the actions of the Judge.

Corporate governance makes good economic sense 14.5

There are many issues being debated in contemporary society which people complain are influenced by 'political correctness'.

One such example is the concept of affirmative action as it relates to increasing the number of women in key positions in the workplace. There is much more to the concept of affirmative action than simply political correctness.

A strong economic argument and business case can be made for a policy which encourages using to the full, the intellect, training and experience of women in the workplace. There is a substantial cost to businesses and the community if policies are not developed to create work practices around family responsibilities and these resources are lost.

The same is the case in relation to corporate governance and the economic consequences resulting from any failure or mismanagement of or by large corporations with the inevitable loss of confidence. The economic repercussions creates a very strong business case for corporate governance, apart from the fact that good governance is required simply on ethical or moral grounds. Good corporate governance therefore makes good economic sense.

The ramifications in Australia of the collapse of HIH spread far beyond the creditors and shareholders of HIH, they were felt in remote communities who could no longer obtain public liability insurance to operate a pony club gymkhana or to allow an ANZAC day march.

The wider ramifications of poor governance practices were recognised by the Core Group on Corporate Governance in preparing their guidelines for good corporate governance for the PECC.

The chairperson of that committee noted in the foreword to the guidelines:

> 'Developments in East Asia since the 1997 financial crisis have underscored the critical importance of structural reforms in the governance of the business enterprise. PECC noted that these reforms are necessary for the strengthening of the micro-economic base for the economies in the region.'

A key driver for economic development and investment, more so than the strength of any corporate balance sheet is a feeling of confidence, engendered by transparency and those three principles enunciated in the Cadbury Report openness, integrity and accountability.

In Asia, good governance is not just an interesting legal theory for businesses in developed economies but is a critical factor necessary to ensure sustainable development and social stability, and can assist in reducing regional tensions by improving living standards through economic growth.

The absence of good governance and institutions to ensure enforcement creates a high level of risk and clearly limits or discourages investment and sadly, this often occurs in regions and countries that most need economic growth.

The importance of good governance was recognised and was a key item on the agenda of the Asia Pacific Economic Co-operation (APEC) ministerial meeting in Bangkok, Thailand, in October 2003.

Every government represented at the APEC meeting recognised that irrespective of the state of economic development in their own economy, they must strongly support the principles of good governance for both government and business, as this will make their markets more attractive, increasing the number and size of investment opportunities available and the pace of development.

The importance of good governance was also recognised at a key conference of senior government, judicial and business leaders in Hyderabad, India in November 2002. The lack of good governance practices was seen as being a significant factor slowing the process of economic reform and development in India. Good corporate governance therefore, is not an intellectual luxury but an economic necessity.

Key issues needed to establish good governance 14.6

Good governance is not a matter of form and structures. If it is to be effective it must have substance by driving/encouraging/facilitating a culture within an organisation that is determined to do the right thing.

This requires a process of education and leadership, which involves those in senior positions living good governance principles rather than simply delegating the task of establishing structures to create an appearance of good governance, ie 'walk the talk'.

Regulation and punishment may drive out those who are dishonest but only after the event and they represent a small part of the corporate world. Most corporations, directors, CEOs and mangers are not dishonest but they still need to ensure they have a culture of good governance in their organisation. There are many elements that need to be considered in relation to governance and these are outlined below.

Checks and balances – separation of power and responsibility 14.7

As discussed earlier, see **14.3** above, a tried and tested model of good governance is the Westminster system, which has at its core, the separation of powers.

In a corporate sense, this means there have to be clearly defined roles and responsibility for the board, the chairperson, corporate executives and the auditors.

The *Sarbanes-Oxley Act of 2002* in the United States and similar legislation in countries such as Australia, have reinforced the need to ensure that auditors are independent to avoid the debacle at HIH where one former partner of the audit firm (who had also been the audit partner) was on the board, another was the chairperson of the company and the financial controller had also previously been a partner of the audit firm. See further discussion of the *Sarbanes-Oxley Act of 2002* in **CHAPTER 11**.

HIH is an example where it became impossible to create any separation of power because of the dominant personality of the chief executive who was also founder of the company.

Clearly, this can be a dilemma for any corporation where there is a strong individual who is a driving force in creating a successful company but who forgets that the company is not their own private domain, they are not an absolute monarch and must be held to account in terms of transparency, reporting and decision making.

It is one thing for the legislators to impose obligations on directors but this may not overcome the real-world challenges in a boardroom debate heavily dominated by one or two individuals. A characteristic of most corporate failures has been the involvement of one or two dominant personalities in key positions.

There is no doubt, given the complexity of corporate life that it is important to have specialised committees to focus on specific areas of a company's operation in order to ensure accountability transparency and openness and that the members of those committees should have sufficient expertise to ensure that they can be effective.

The example of National Australia Bank demonstrates that the existence of structures alone is not enough. In that case the Bank did have audit and risk management committees but plainly, those committees were ineffective and apparently had no idea of the problems of the rogue traders, did not know that the regulator had already warned the company of the problem and clearly were faced with a culture of concealment rather than openness.

One has to be cautious in making committees responsible for governance issues to ensure they have sufficient authority, autonomy and resources to carry out their responsibilities.

The PECC guidelines (see **14.2** above) in relation to the separation of powers are a useful model particularly as they have been developed in a region where some may have thought that the diversity of culture and legal systems would render it impossible to develop any guidelines for use throughout the region.

It is important that in the separation of powers and setting up board committees that one does have regard to the realities of the particular corporation. Consideration must be given by boards to ensure that the structure not only addresses the needs of good governance but is also workable and practical, having regard to the needs and risks inherent in that corporation.

Conflict of interest **14.8**

Some conflicts of interest are obvious and easily recognised, some however may be difficult to identify and may be rationalised as no more than a perception of conflict of interest.

Having been both chair and managing partner of a very large law firm for nearly ten years, the issue of conflicts of interest and how to deal with them has regularly arisen for the contributor because of the nature of the special responsibilities lawyers have with their client's to ensure that their counsel is impartial.

Plainly, in a corporate sense, there is a clear conflict of interest to have independent auditors represented on the board and legislation such as the *Sarbanes-Oxley Act* in the United States has recognised that there is also a clear conflict of interest if the audit firm is providing other services to the corporation whether they be accounting, consulting and more recently, legal services.

Some would suggest that the management of conflicts of interest is a complicated and difficult task. This doesn't have to be the case. People generally recognise that there is an actual, potential or perceived conflict of interest by simply asking the question or trying to rationalise a way a conflict of interest situation may arise.

The contributor has had partners speak to him on many occasions when wanting to take on a matter and the conversation will open with the line '*I don't think this is a conflict of interest but..*' The response given is usually along the lines that having asked the question, you obviously feel that there may be a conflict, if not actual at least perceived, and in those circumstances, really need to back out.

Another simplistic test used has been the Sydney Morning Herald test. That test is to ask oneself the question when considering whether there may be a conflict as to how it would look on the front page of the morning papers.

As has been discussed above, one of the reasons for having good governance is to engender confidence that there is transparency and openness. The failure to properly manage situations where there is an actual, potential or perceived conflict of interest, strike at the heart of engendering confidence in the operations of any corporation.

Some corporations seek to deal with the issue of conflicts of interest by establishing rules excluding people from voting, requiring them to leave meetings and so on. Such rules often do not really address the issue of substance in relation to a conflict of interest.

Just because a director withdraws from a meeting at which there is a discussion about engaging a company in which that director has an interest, does not resolve the conflict of interest problem. The question still remains whether or not the director participates in the discussion as to whether it is the right thing to do to give a significant contract to an entity, which will give a director, directly, or indirectly a significant financial advantage.

More significantly as good governance is about confidence and transparency how will such a decision look on the front page of the morning papers or the nightly television news and impact a feeling of confidence in the governance and operations of the corporation.

The issue of conflict of interest can often come up without thinking when there is discussion about appointing a particular person who is a director of a corporation. The issue is not necessarily whether it creates an actual conflict of interest, but rather, whether it creates a potential or perceived future conflict of interest, which diminishes a feeling of confidence in the corporation.

Often, boards may debate the appointment of a person and seek to rationalise away potential or perceived conflicts of interest having regard to some extraordinary skill that the individual may have only to face a barrage of criticism five years later for the appointment, a fall in the share price because of lack of confidence in the stock market and possibly an investigation by one of the regulators.

Some argue that if one adopts a policy of absolutely avoiding potential conflicts of interest on appointment of directors it may reduce the available pool of people with relevant experience and expertise, in my view, this argument is overstated.

One advantage of absolutely avoiding conflicts of interest would be to push the search process for new directors out beyond the usual 'network' from which directors are obtained which of itself adds new and independent thinking to a board and reinforces good governance.

Changes in accounting practices **14.9**

Whilst the causes of corporate failures are many and varied they usually emerge because of a financial crisis and the company cannot pay its debts.

The classic defence by most directors following a serious corporate collapse is that there were misled by the information provided to them, often the most critical information about which they complain is the accounting or financial information.

Legislators and regulators have endeavoured to address this issue by working with the accounting profession to tighten up the use of international accounting standards and practices and to ensure that the need for compliance with internationally accepted accounting standards is more than simply to ensure compliance with rules of an accountant's professional body but has become law and the failure to comply with accounting standards renders officers of corporation and the auditors who acquiesce in non-compliance liable to prosecution.

In almost every large corporate collapse, one can identify problems in the accounting systems of the corporation and manipulation of financial material either to conceal fraud from being detected or simply to enhance the financial justification for a particular strategy or proposal and to make things look better than they really are.

There are often good reasons within accepted accounting standards for changing the way in which a corporation may account for a transaction and accounting concepts are not static and will continue to evolve just as the corporate world itself evolves and becomes more complex.

There are the traditional scams as evident in companies such as Enron and Worldcom where expenses are capitalised to improve the bottom line or assets are included in balance sheet inflated values to preserve the perception of underlying value for the corporation.

There are also more subtle approaches such as avoiding writing down assets, which may have entered the accounts of a corporation at a real figure but which have lost value over time.

In the HIH Royal Commission Justice Owen made reference to 'Aggressive Accounting Practices' within HIH (volume 1 at xivii) he commented:

> 'Put bluntly, HIH management recognised that the group was under performing at a level that could not be sustained. But it failed adequately to respond to the underlying causes of poor performance. Instead, it used and relied on questionable accounting transactions giving rise to doubtful accounting entries, which disguised the seriousness of the situation and the consequences of leaving it unchecked. The process was fatally flawed.'

A common theme in relation to major corporate failures in recent years is that they have all occurred after a period of significant growth (or at least perceived growth). Often these corporations move into new markets, which create difficulties for the corporation. The directors then find it hard to control or manage that growth process, and combined with a lack of accurate financial data don't know whether the growth is producing real profits or whether there is sufficient underlying funding to sustain it.

AMP is one example where it sought to rapidly grow its business in the United Kingdom. It entered into a process of acquiring a number of businesses utilising initially high levels of cash reserves. The process ultimately proved a disaster and only the considerable underlying strength of the company enabled it to survive.

Most growth initiatives are announced by boards and management with great gusto. Problems of management and control are often swept aside or concealed to avoid loss of face. Sometimes management will simply change overhead allocations in management accounts or expense the costs of the new venture against an existing profitable division when providing information to Audit Committees and Boards.

While one cannot expect an audit committee to carry out the role of the external auditors, one area where an audit committee must be extremely vigilant, particularly during a period of rapid growth, is to ensure that it obtains clear and transparent explanation for any changes in accounting practices.

It is often the case when an enthusiastic new chief executive officer comes to a company that after embarking on the usual cost cutting regime which seems common to most new chief executives, they will then put to the board a strategy for growth and put in place a number of new senior executives who that person perceives will be their allies in supporting the new CEO strategy and providing information which justifies that strategy.

This issue particularly arises where the new CEO comes from outside the organisation, often with a brief from the board to implement a change of direction. That person will report to the board a need to change senior executives because they may not be aligned with what the new CEO perceives to be the brief.

A new CEO shortly joined by a new chief financial officer appointed by that CEO and changes in the management accounting system and reports is something, which an audit committee and board should treat with extreme caution.

The audit committee may lose the capacity and information because of changes in the management accounts and reporting systems to compare financial data under the new regime with what they received before. Often a new CEO will seek to justify changes, including changes in a chief financial officer on the basis that they are necessary to move the organisation forward and existing senior management is unwilling to embrace change.

An audit committee and a board will often have some reservations to see the departure of a long standing chief financial officer known for producing timely, informative and accurate information but will be faced with a new CEO who they have hired and who, if they resist the proposed changes will assert:

> 'These are management issues. If you tie my hands there is no point in giving me the job or the brief set for me on my original appointment.'

Often new CEO's will, through this pressure, secure acquiescence of the board and the consequence will be that the first step has been taken which causes the Board and Audit Committee to lose the capacity to properly compare the results of the corporation under the new strategy with the old.

These issues are difficult to deal with. Boards must ensure that the corporation does not stagnate, that it is infused with new people and new ideas. In this process, some will not embrace change, and there will be staff turnover often at levels close to the new CEO.

Boards, consistent with the concept of separation of powers, must avoid becoming embroiled in day to day management, the new CEO, if that person is to succeed must be given an opportunity to bring about change and put forward new strategies.

A board and particularly an audit committee in order to effectively deal with these matters need to have people with appropriate technical experience if they are to manage the risks inherent in the change process or will lose control over their capacity to understand the financial position of a corporation during the period of change.

It is not good enough for an audit committee to simply leave this task to the external auditors. Many of the changes in accounting and reporting which can cause future problems will be perfectly acceptable under ordinary accounting standards (for example changing overhead allocation) and it should be noted that in all cases of a major corporate failure there was a signed audit report.

The most devastating comment one can hear from an Audit Committee or a director of any corporation when considering management reports is that they don't understand the report.

As has been mentioned in **CHAPTER 8**, good governance requires that boards of directors insist on management making themselves understood so that the board can make an informed and considered decision.

It is appropriate for a board to send back to management any material they do not understand, which is given to them for approval or to justify any strategy or proposal put up by management particularly the financial aspects of the proposal.

External advice to assist in good governance **14.10**

As with organisations elsewhere, many corporations in Australia, as part of their governance process, specifically provide for directors to obtain external advice and assistance when examining material put to them. This is a good concept to incorporate into any governance structure but careful thought must be given as to how it is put into place in practice, in particular, the problems created by the individual directors all seeking independent advice on issues put before them.

As indicated above, it is important that boards ensure that what is put to them for consideration is presented in such a way, so that they can understand the issues, the financial data and the risks.

External advice should not become a substitute for directors exercising their own mind and coming to their own views or worse a device to pass their obligations to a third party. A board of directors is a body not unlike a cabinet and cabinet solidarity is important. There is nothing more destructive to any organisation both internally and from an external perception than to have a divided board.

The chair has a heavy obligation to ensure that there is a free flow of discussion and that the chair is not seen as a servant of the CEO or management or taking the place of a very dominant CEO, as occurred in the case of HIH.

Undoubtedly, there will be issues, which come before a board from time to time where there is very vigorous debate and strongly held views by directors. At the end of the day, notwithstanding the personal views any particular director has, it is critical that the board comes to a decision and any individual director whose views have not been accepted, must either live with the decision or resign from the board.

A process where disaffected directors as individuals can use the capacity to obtain external device to overturn what otherwise may be a clear majority decision is bad in terms of governance, bad for the corporation and ultimately bad for all stakeholders including creditors and shareholders.

On the other hand, if a governance policy which enables directors to obtain external advice is to have any substance, there must be a process for examining issues carefully and preferably for the board, even those members of the board that may disagree with a director, to come to a view that the director's wishes to obtain independent advice should be respected. In that case, it would be the board as a whole which seeks the independent advice before moving forward.

Plainly, the role of the chairperson in facilitating this process is critical and it is vital that the chairperson exercises their authority in a way consistent with the judicial oath which requires judges to determine cases without fear or favour, malice or ill-will and most importantly, having regard to the interests of the corporation and all stakeholders as a whole.

HIH – A case study of a catastrophic failure in corporate governance

Brief history of HIH 14.11

HIH began in 1968 with Ray Williams and Michael Payne setting up an agency arrangement for two Lloyds of London Syndicates. The main business was workers' compensation, particularly in Victoria followed by Tasmania and other States.

A pattern of growth began in 1971 when the company then known as M W Payne Liability Agencies Pty Limited acquired C E Heath PLC, a public company based in the United Kingdom and the group's name changed to C E Heath Underwriting Agencies Pty Limited, Ray Williams continued as a director and chief executive, becoming a member of the Board of C E Heath, PLC in 1980.

Core business was workers' compensation insurance until changes in the statutory insurance schemes in the States of Victoria and South Australia reduced indeed, in many ways virtually eliminated that class of insurance in those States and it was then decided to try and enter off-shore markets in Hong King in re-insurance and workers' compensation in California.

The Australian company acquired all of C E Heath, PLC's Australian business in 1989 and as part of the arrangement, ten per cent of shares in the Australian company were held by management including Williams. In 1992, additional non-executive directors were appointed to the Board, which included a former partner at Andersen's, then auditor of the company becoming chairman.

The company was listed on the Australian Stock Exchange in 1992 and became a public listed general insurer with local executives holding approximately eleven per cent of the issued capital of the company and the balance being held by C E Heath, PLC, and the public holding about 45 per cent of the issued shares.

In 1993 the new listed company sought new business in the United Kingdom. Payne, one of the original founders having been appointed as chief executive of a newly established UK subsidiary, Heath International Holdings (UK) Limited.

After 1995, there was a change in the company's core business to public liability and professional indemnity insurance and the period of rapid growth began with the acquisition of CIC. Winterthur became major shareholder in place of C E Heath PLC; but Winterthur did not exercise its entitlement to take control of the board.

In May 1996, the company changed its name to HIH Winterthur International and it had become the second largest general insurance underwriter in the Australian market and moved ahead with further acquisition including Colonial Mutual.

In January 1998, Winterthur disposed of its interest in HIH, which saw a dramatic change in the profile of the HIH Share Registry, by October 2000, the vast majority of HIH shares were held by people with a holding of fewer than 5,000 shares, which meant that the only dominant block of shareholders was the senior executives.

In 1998, HIH acquired FAI (with virtually no due diligence) and overseas operations in the US and the United Kingdom continued to expand.

In March 2001 HIH went into liquidation with a shortfall of between $3.6bn and $5.3bn. So great where the consequences the Australian Federal Government had to establish a support scheme for hardship cases which by February 2003 had received 11,400 applications for assistance.

HIH lessons learned – key factors and risk indicators 14.12

There are many lessons that can be learned from the collapse of HIH and applied by boards and board committees to ensure good governance. Some are blindingly obvious and have been the subject already of action by the government and regulators, others more subtle noting HIH had all the 'forms' of governance but they had no 'substance'.

The underlying causes of the collapse of HIH are not unique and can been seen in nearly every major corporate failure and if recognised and understood do provide practical guidance for those responsible for corporate governance to ensure the same mistakes are not repeated they include:

- Do not be seduced by ambitious proposals that focus on growth, growth is important for the survival and prosperity of every business but it must be manageable and profitable or it will kill the company.

  Warning! – Many CEO's and executives have contracts, which focus heavily on bonuses for growth. Proposals must be examined with rigor by boards and audit committees to ensure they are good for the company and not just the executives ego or remuneration package.

- Be wary of growth outside core business. HIH's move into different areas of the insurance market and different geographic markets took it outside its existing skills and knowledge base.

  Warning! – Boards must test proposals and be satisfied that any move outside core business or geographic markets is supported by analysis from people with the requisite knowledge and expertise which will often have to come from outside the company.

  If the project proceeds the board need to be satisfied that the company can put in place the required skills and systems to manage the project and the costs of these additional resources are properly included in any cost-benefit analysis and not left out to make the project appear more attractive.

- A board which comprises executives who can dominate the decision making process is high risk from a governance perspective.

  Warning! – The role of the board is to provide the primary checks and balances on the CEO and senior executives. It must be robustly independent. It is high risk to have a board, which can be dominated by a strong-willed CEO, backed up by senior executives.

- The FAI acquisition by HIH was an example of a total failure in governance. Neither the board or the audit committee played any significant role in laying down due diligence requirements. The reality is that there was no proper study of FAI and the CEO and senior executives pushed it though the board.

  Warning! – Management may pressurise a board on an acquisition because of the threat someone else will get the deal and put aside due process. Often this factor is exaggerated.

  Boards will inevitably be asked to consider acquisitions. It is vital for good governance to establish basic processes and procedures to evaluate future acquisitions before being placed under pressure by management to approve a specific deal.

- Boards should set clear guidelines, procedures and information required for management, which must be complied with to when bringing proposals to the board. Whilst there may be a need and good reason to make an exception then at least everyone knows to be more careful because it will be treated as an exception. Directors after the event will claim that when a decision was made there was a lack of information or they did not fully understand the deal.

  Warning! – Most bad decisions that haunt boards and destroy companies do not involve day to day matters but acquisitions or deals, which turn an otherwise successful company into a financial wreck.

- If the Board does not understand, it must not allow itself to be bulldozed by threats of resignation by CEO's. The proposals must be sent politely back with a request that it be reviewed and management must explain so that the Board does understand and is convinced. This only has to be done a couple of times to firmly establish the right culture and is a good discipline for management and readily accepted by good quality executives. HIH had a high risk that it could not develop a culture of good governance as a public company because those who really controlled the company had no checks or balances to ensure that the changes were made from when it was essentially a private company owned and run by Payne and Williams both of whom were strong personalities.

The core team remained the same and the addition of several external directors by the invitation of those in control entrenched this problem, which was confirmed as early as 1995 in an independent due diligence report on HIH. An opportunity for change was missed when Winterthur did not take a role on the board as a major shareholder and when they sold out any prospect for the public shareholders of someone outside the management team exerting any influence or control was lost and from then until liquidation the worst decisions were made.

Warning! – Whilst good governance is vital, in every organisation it is a critical issue to consider when a closely controlled company becomes a public company.

- Governance was hopelessly compromised in HIH in relation to the only potentially independent check on the activities of the company by appointing partners of the audit firm (originally appointed before listing in 1973) to the board and one becoming the group finance director. The Royal Commissioner heard evidence of how attempts were made by management to influence the work of the auditors.

  Government and regulators have addressed the most obvious conflict of interest issues preventing auditors from being appointed to boards and key positions but this does not remove the more subtle actions which can compromise independence.

  Warning! – Boards and their audit committees must be uncompromising in ensuring that the auditors are not only independent but see themselves as reporting to the board and audit committee and guard against the prospect influence by meeting with the auditors without management and changing auditors on a regular basis or appointing another firm to review the auditors from time to time.

- At the heart of the operation of any company is it's financial and reporting systems and their capacity to report in an accurate, understandable and timely fashion on how the company is going today and where it is travelling. The integrity of this information is critical to the board and the audit committee.

  These systems are used to generate management accounts and reports used to manage day-to-day operations and are in many respects much more important than annual accounts whose value is largely an historical perspective. Good previously stable accounting, reporting systems and processes can often fail during periods of rapid growth particularly geographic growth remote from head office and require special attention during times of rapid change. There were no controls on or questions asked as to whether the process of growth was profitable, management accounting was unreliable and the company was adopting a fire-fighting approach in dealing with cash flow and day-to-day operations, at the same time trying to absorb new acquisitions.

Whilst accounting standards exist for annual accounts in many companies' management may not follow these standards on day to day reporting (sometimes for very sound and legitimate reasons).

As accounting practices at HIH became more aggressive to cover up problems and to keep the regulator at bay, some of those involved may be convicted of offences under the Australian Corporations Law. The real problem was not so much one of deliberate fraud, but rather a situation where no one had any idea of the true financial position of the company.

Warning! – Boards must through the audit committee ensure that proper processes and procedures are in place to ensure the integrity of management accounts.

- Board must be vigilant when management makes changes to reporting and timeliness of management accounts. New projects must be evaluated in accordance with existing reporting so the outcomes can be properly compared with current operations or prior years.
- Allocation of overheads in divisional reporting must be watched as it can be used to cover problems or make a new project look more attractive and one must look critically if it is suggested that profitability for a new venture should be examined on a special or different basis from that normally followed in the company.
- Whilst the arrival of a new CFO or CEO will often see suggested changes in reporting, be cautious to ensure comparability is not lost and financial results they claim credit for are not enhanced or exaggerated by the change in reporting.
- Be cautious when suggestions are made to develop or introduce new accounting systems for internal accounting. The introduction by HIH of the GEN+ system introduced in 1997 was said by the Royal Commissioner to have significantly impaired the ability of management and the Board to operate HIH's business.
- In the accounting and reporting area do not be seduced by enthusiastic management pushing 'leading edge' concepts. Like most technology the first user is the guinea pig used to iron out the bugs, it is dangerous especially if such change is made during rapid growth or changes mean the risk of losing the capacity to obtain financial information needed to manage the company.

Whilst it is easy to ascribe all the blame for the collapse of HIH to Ray Williams who was clearly a major force in the company because of his role as a founding father and his strong personality, the failure in governance is also a reflection on those who were supposed to be watching over the interests of all shareholders and policy holders, that is to say the other senior executives and non-executive directors.

The culture of HIH ignored any concept of governance. In spite of having structures creating an appearance of a governance system, the reality is there were no checks and balances at HIH to make those responsible for the operations of the company stop and think about the impact on the company of proposed acquisitions and force them to demonstrate whether the proposed acquisitions would in reality add real profit and growth to the company, rather than simply mass.

Mr. Justice Owen standing back from his work as the HIH Royal Commissioner, expresses the view at page 133 of his report:

> 'For me, the key to good corporate governance lies in substance, not form. It is about the way directors of a company create and develop a model to fit the circumstances of that company and then test it periodically for its practical effectiveness. It is about the directors taking control of the regime they have established for which they are responsible. These concepts do not lend themselves easily to specification in something such as a code of best practice. That is why I have not made any formal recommendations.
>
> One thing is clear, though. Whatever the model, the public must know about it and how it is operating in practice. Disclosure should be a central feature of any corporate governance regime. Shareholders, potential shareholders and the wider public are entitled to real, meaningful detail about the way the directors say they are carrying out their stewardship roles.'

Conclusion 14.13

It is impossible to canvass every possible aspect of good governance in this chapter. The concept of good governance is one, which evolves, as does each corporation on a daily basis. The establishment of forms and structures will not create good governance; there are many examples of organisations that have successfully operated without a set of rigid rules for hundreds of years.

The constitution of the United Kingdom was never written down unlike the constitution of the United States or Australia and yet there is a process of good governance under the Westminster system that has worked effectively for hundreds of years.

Good governance will not prevent cheats and charlatans from taking advantage of an organisation but will discourage them and when they have a go, help to identify the problem faster and minimise the potential damage they can cause to an organisation.

Good governance is not a committee or a set of rules but a culture which works effectively if people responsible for corporations recognise the economic value and basic morality of doing the 'right' thing. The structures of governance will only be effective if this culture or set of basic beliefs is embraced by the corporation.

Chapter 15
Corporate Governance in Japan

Introduction 15.1

In the aftermath of the Second World War, Japan experienced an economic miracle with the economy growing at an average rate of around ten per cent a year from the mid-1950s until the Arab oil shocks of the early 1970s. Following the bursting of the bubble in the early 1990s Japan entered into the period of economic stagnation which some commentators have dubbed the 'lost decade', from which Japan is now rapidly emerging. In the first quarter of 2004 the Japanese economy experienced a year-on-year growth rate of 5.6 per cent, faster than that of the US.

One reason is that Japan has benefited from the rise in intra-Asian trade over the past few years. For example, trade with China expanded by 28.6 per cent in 2003. In addition, however, the economy has been increasingly driven by domestic demand. Business investment has rebounded and household investment spending rose by 7.2 per cent in the year ending April 2004. The banking sector is also returning to health, the level of non-performing loans on banks' books has halved to five per cent, while restructuring has at last engendered a real revival of competitiveness at Japan's biggest firms. In view of the current trends it is important to include an overview of the approach to business in Japan, bearing in mind its traditional culture, as well as the clear needs of due diligence, risk management and corporate governance that the global regulatory framework requires.

The core problems of corporate governance in Japan 15.2

One school of thought has long held that Japan is a trust-based society rather than a contract-based society. One by-product of this is that bureaucratic informalism has often been seen as being supreme over law. Indeed, according to Haley:

> 'The avoidance of legal regulation and coercive state control must be viewed as among the most prominent characteristics of governance in postwar Japan.' (John O Haley, Authority without Power, p166 (1991))

As a result of the system of cross-shareholding among major Japanese corporations, a lack of independent directors and a legal system that discouraged shareholder derivative suits, minority shareholders in Japan had little recourse against directors of corporations until the early 1990's. The collapse of the economic bubble forced Japan to re-evaluate its economic model, which led to several substantial revisions of the *Commercial Code* (*Code*).

In a landmark decision in 2000, the Osaka district court ordered eleven current and former directors of Daiwa Bank to pay approximately $US 775 m in a shareholder derivative action for failing to adequately supervise a rogue trader in the New York office who lost more than one billion US dollars. Of particular relevance however, is the fact that the court disregarded the informal consultations that the bank had had with the government and instead chose to focus on the fiduciary duties of the directors. The decision has had a significant impact on corporate governance in Japan with one long-time Japanese specialist noting that:

> '[t]he resulting 'Daiwa shock' had a far-reaching effect in Japan similar to the combined impact in the U.S. of the leading Delaware cases of *Van Gorkom* and *Caremark*.' (Bruce E Aronson, *Reconsidering the importance of law in Japanese corporate governance: Evidence from the Daiwa bank shareholder derivative case*, 36 Cornell Int'l. LJ 11, 13 (2003))

The remainder of this chapter seeks to provide the reader with a snapshot of the major issues and laws as of early 2004. Readers should note however, that this is a fast changing area of the law and consequently are advised to obtain professional legal advice as to the latest developments on corporate governance. Unless otherwise noted all Articles below refer to provisions of the *Code*.

General meetings of the shareholders 15.3

Whilst the general perception used to be that the general meeting of the shareholders in Japan was a largely ceremonial meeting, this is now no longer the case. In fact, the current trend in Japan is towards more active shareholder participation in general meetings.

Convening general meetings 15.4

An ordinary general meeting should be convened at least once a year at a fixed time within three months of the end of the corporation's fiscal year except where the corporation distributes profits more than once a year, in which case the corporation should hold a general meeting in each accounting period for which it distributes profits, (*Article 234*).

An extraordinary general meeting may be convened as necessary, (*Article 235*). A shareholder who has been the owner of not less than three hundredths of the voting rights for at least the preceding six months may by written application to the directors, demand that an extraordinary general meeting be called. If the directors fail to convene a meeting within a reasonable time of the demand, the shareholder may convene the meeting with the permission of the court. The written application must state the matters to be raised at the meeting and the reasons for convening the meeting, (*Article 237*).

Each shareholder must be notified of the general meeting in writing at least two weeks before the day of the meeting. The written notice should also include a record of the matters constituting the object of the meeting, (*Article 232*) The procedures noted above for convening a general meeting can be waived if all of the shareholders with voting rights agree, (*Article 236*).

Resolutions of the general meeting 15.5

The *Code* provides that the general meeting may only adopt resolutions as to matters provided for in the *Code* or by the articles of incorporation, (*Article 230–10*). Each shareholder is entitled to one vote per share, however, the corporation has no right to vote its shares. Where voting is to be by unit, a shareholder has one vote per unit. A unit may consist of any number of shares. Where a shareholder has less than the prescribed number of shares of a unit, the shareholder is not entitled to a vote, (*Article 241*(*1*)(*2*)). A shareholder may exercise his voting rights even if he does not attend the general meeting, so long as the board of directors passes a resolution allowing him to do so, (*Article 239–2*). Unless otherwise stated in the articles of incorporation or elsewhere in the *Code* a resolution may be passed by the general meeting if the majority of the shareholders present at the meeting and representing at least 50 percent of the total voting shares of the corporation vote in favor thereof, (*Article 239*(*1*)). That is, if the articles of incorporation provide for this and there is no provision to the contrary in the *Code*. A quorum is not necessary for the passing of a resolution.

Under the former provisions of the *Code*, the statutory quorum requirement for a special resolution of a general meeting was a simple majority of total voting shares, and a corporation was not able to relax this requirement even by its articles of incorporation. Under a recent amendment to the *Code*, a corporation may relax this requirement to one-third of the total number of voting shares in its articles of incorporation. In such cases, the relevant provision authorising these lesser quorum requirements must be specifically set out in the articles of incorporation, (*Article 343*(*2*)).

Ordinarily, it was difficult for small shareholders to place issues on the agenda at the general meeting since the agenda was decided by the board. This was addressed by a revision to the *Code* that provided that a shareholder who has for the last six months held at least one hundredth of the voting rights of all of the shareholders or three hundredths of the voting rights, may request that a specific matter be the object of the general meeting in writing at least eight weeks before the general meeting, (*Article 232–2*).

Shareholder's rights 15.6

The directors or auditors are obliged to offer explanations on matters as requested by shareholders at the general meeting. The directors and auditor can refuse to provide such an explanation if there are reasonable grounds for doing so, such as:

- the question bears no reasonable relation to the object of the general meeting;
- the explanation would harm the common interests of the shareholders; or
- the explanation would require further investigation, (*Article 237–3*(*1*)).

However, where the shareholder has given reasonable notice in writing prior to the general meeting as to the matters on which explanation is sought, the directors or auditors cannot refuse to provide said explanation on the grounds that further explanation is needed, (*Article 237–3*(*2*)).

Under *Article 294* of the *Code*, a shareholder who has been in possession of at least three hundredths of the right of voting of whole shareholders may apply to the court for the appointment of an inspector to investigate the affairs of the company and the state of its property where there is cause to suspect that a dishonest act or any grave fact constituting a breach of any law or the articles of incorporation in connection with the administration of the corporation.

Likewise, under *Article 237–2* of the *Code* a shareholder who has held at least one hundredth or more of the voting rights of the whole of the shareholders may request that the court appoint an inspector to investigate the procedure for convening the general meeting and the method of adopting a resolution.

A shareholder has appraisal rights in certain circumstances. Notably, a shareholder who opposes any of the following acts by the corporation may, by written request prior to the general meeting request that the company purchase his shares at the value that the shares would have had had the resolution objected to not been adopted, (*Article 245–2*):

- alteration of the articles of incorporation where the articles are altered to the effect that an transfer of shares becomes subject to the approval of the board of directors, (*Article 349*(*1*));
- a transfer of all or a substantial part of the business of the corporation;
- the entry into, alteration of, or rescission of a contract for the leasing of the whole of the business, or giving a mandate to manage the business or sharing with another the entire profits or losses in relation to the business;
- the taking over of all of the business of any other corporation, (*Article 245*(*1*)), and

- succeeding all or a part of its business to a new corporation or another corporation, (*Articles 374–3(1); 374–31(1)*).

A shareholder may also exercise appraisal rights where they object to the amalgamation of the corporation with another corporation, corporate split or share exchange for another corporation's shares.

Finally, a shareholder who has held a share continuously for at least six months may demand on behalf of the corporation that a director who performs any of the following acts refrain from doing so:

- any act that is not within the scope of the articles of incorporation or;

- any act which is against the law, and which raises the prospect of irreparable harm to the corporation, (*Article 272*).

Directors may not offer benefit to a shareholder for the exercise of a shareholder's rights 15.7

A corporation may not offer a benefit to a shareholder for the exercise of a shareholder's rights for the corporation's own account or for the account of a subsidiary, (*Article 295*), Any director or auditor of a corporation who does so is liable for a penal sentence of up to three years or a fine of three million yen or less, (*Article 497*).

The sokaiya are extorters who buy a shareholding in a corporation and then extort money from that corporation by threatening to disrupt the general meeting unless the corporation makes a special payment. The *Code* has been changed to make it illegal for a corporation to make special payments to any person, including sokaiya in return for their non-disruption of a general meeting. Whilst in the past Japanese corporations often held their annual general meetings on the same day in order to try and spread the sokaiya thinly, thereby preventing them from disrupting the meetings, since the revision of the *Code* making special payments illegal, the situation has changed with general meetings becoming more open and shareholders questions more frequent.

Alteration to the articles of incorporation 15.8

The articles of incorporation can be amended by a resolution of the general meeting, (*Article 342*). In order to do so, an amendment must be adopted by two-thirds of the votes of the shareholders present who hold more than half of the voting rights of all of the shareholders or hold the number of voting rights stipulated in the articles of incorporation, (*Article 343*).

The board of directors 15.9

Until the mid to late 1990s, one of the prominent features of the Japanese corporate system was the lack of independent directors. This situation began to change recently. In most Japanese companies the majority of directors are appointed from amongst the employees. In addition, and unlike in some European countries, there is no director who represents the interests of shareholders.

A corporation must have at least three directors, (*Article 255*). The term of office of a director must not exceed two years, (*Article 256*(*1*)), and the directors are appointed at a general meeting, (*Article 254*(*1*)).

The *Code* prohibits certain classes of persons from being directors, notably persons having been adjudicated as bankrupt but not reinstated and persons who have within the previous two years been convicted for a violation of the *Code*, (*Article 254–2*). A shareholder may demand that the corporation hold an election for two or more of the directors by cumulative voting, (*Article 256–3*(*1*)). The shareholder must notify the corporation in writing at least five days in advance of the general meeting of the shareholders, (*Article 265–3*(*2*)). The quorum of shareholders necessary to elect directors may be set by the articles of incorporation, but in no circumstances may it be decreased to below one third of the voting rights of all of the shareholders, (*Article 256–2*).

A director can be removed from office at any time by a resolution of the general meeting. However, if a director has a fixed term of office and he is removed prior to this and without due cause, he may claim damages from the corporation for early termination, (*Article 257*).

The *Code* charges the board of directors with the running of the corporation and supervision of the directors, (*Article 260*(*1*)). Certain matters are within the remit of the board only and the representative directors can not be delegated authority to address them:

- the disposition and acceptance of major assets;
- taking out loans of large amounts;
- the appointment or dismissal of senior executives such as managers, and
- establishment, change to, or dissolution of branch offices or other important branches of the corporation, such as affiliates etc, (*Article 260*(*2*)).

Auditors are also to be present at the board meetings, (*Article 260–3*), In most corporations there are managing directors' meetings or a management committee which in practice are the bodies that make the decisions.

A resolution must be passed by a majority of the directors present at a board meeting, and the board meeting must be attended by the majority of the directors, (*Article 260–2(1)*). A director who has a special interest in the resolution may not participate in the vote for the resolution, (*Article 260–2(2)*). The board of directors may by a resolution of a board meeting elect a particular director to be a representative of the corporation (representative director), (*Article 261(1)*). Two or more such representative directors may be appointed by the corporation, (*Article 261(2)*)

Article 262 of the *Code* provides that a corporation shall be liable to a bona fide third party for any act committed by a director from which it may be assumed that the director has the power and authority to represent the corporation even where such director does not in fact have the power or authority to do so.

Finally, in a noted decision, the Supreme Court of Japan held that articles of incorporation that specify that the directors and auditors of a corporation be Japanese nationals are valid. (Judgment of the Nagoya District Court, April 30, 1968 (Kaminshu 22–3/4–549)).

Directors' duties **15.10**

Article 644 of the *Civil Code* creates in a director a duty to manage the affairs of the corporation with the due care and skill of a good manager. The *Code* also places an obligation on directors to not only obey the law, the articles of incorporation and resolutions of the corporation but also to discharge their duties faithfully, (*Article 254–3*). As mentioned above, the board has a duty to supervise the directors. (*Article 260(1)*).

Where a director intends to carry on business on his own behalf or on behalf of a third party where such business is of the type usually undertaken by the corporation, the director must disclose all material facts relating to business to the board and must obtain the board's approval prior to entering into business, (*Article 264(1)*). Where a director has already entered into a transaction of the type discussed above, they are obliged to report the material matters to the board immediately. Where a director has entered into business in breach of *Article 264(1)*, the corporation may deem such business to have been entered into on behalf of the corporation, (*Article 264(3)*).

Where a director wishes to acquire property from the corporation or wishes to transfer their own property to the corporation, or wishes to receive a loan from the corporation on their behalf or on the behalf of a third party, the director needs to obtain approval from the board of directors. The same applies where a director wishes to receive a guarantee of their liability from the corporation, (*Article 265*).

Directors' liabilities 15.11

A director may be liable to the corporation for damages in the following instances:

- where they have proposed at the general meeting a distribution of profits in contravention of the *Code*;
- where they have paid out dividends in a manner disproportionate to the number of shares held;
- where they have loaned money to another director;
- where they have entered into a transaction in violation of *Article 265* of the *Code*; and
- where they have done an act in violation of the law or of the articles of incorporation of the corporation, (*Article 266*(*1*)).

Where any of the acts above have been done pursuant to a resolution of the board of directors, any director assenting to the resolution shall be deemed to have committed the act, (*Article 266*(*2*)). Those directors who participated in the meeting but did not assent to the resolution but who did not express their dissent shall also be presumed to have committed the act, (*Article 266*(*3*))

Article 266 of the *Code* provides for a director's liability to the corporation. Director's liability to the corporation can only be released by the unanimous consent of all of the shareholders, (*Article 266*(*5*)). The provisions of *Article 266* also apply to Auditors, (*Article 280*(*1*)), promoters, (*Article 196*) and liquidators, (*Article 430*(*2*)).

Notwithstanding the provisions of *Article 266*(*5*) however, where a director has acted in good faith and in the absence of gross negligence, the *Code* relaxes this strict requirement for waiving the liabilities of directors stemming from violations of laws or the articles of incorporation, of *Article 266*(*1*) see above, (*Article 266*(*7*)). A director may be exempted from liability in the foregoing instance by a special resolution of a general meeting where two thirds or more of the total votes of shareholders attending the general meeting have approved such waiver, (*Article 266*(*7*)).

It should be noted that the unanimous consent of all of the corporation's auditors is required for the corporation to submit a proposal for such a waiver at the general meeting, (*Article 266*(*9*)).

The other way to effectuate a waiver of liabilities is to have the board of directors approve the waiver. This method is available only to corporations that have made necessary provision in their articles of incorporation authorising the use of this method. The board of directors may grant the necessary approval of a waiver, but only if they believe such a waiver is really necessary in view of the totality of the circumstances, (*Article 266*(*12*)). The corporation is obliged to issue a public notice of its intention to waive the liability of a director pursuant to *Article 266*(*12*). Even if the board approves such a waiver in accordance with the above procedures, the waiver will not be effective if shareholders holding three percent or more of the total votes of shareholders give notice of their objection to such waiver to the company within a certain period specified in the articles of incorporation,.the period must be at least one month, (*Article 266(15)*).

Penal provisions **15.12**

The penal provisions of the *Code* provide for up to ten years imprisonment and fines of up to ten million yen for violations of the provisions of the *Code*. The exact maximum penalty depends on the section of the *Code* that has been violated, (*Article 486–492-2*).

Auditors **15.13**

The term of office of auditors has been extended from three years to four years, (*Article 273(1)*). Auditors are now obliged to attend board meetings and may express their views at those meetings if they so wish, (*Article 260–3(1)*).

From May 1 2005, the minimum number of outside auditors will be increased from only one to at least half of the members of the auditors for any large corporation. A Large Corporation is defined in the *Code* as a corporation with a stated capital of five hundred million yen or more, or a corporation with two hundred billion yen or more of liabilities according to its latest balance sheet, (*Article 1–2(1)* of the *Law for Special Provisions for the Commercial Code concerning Audits, Etc, of Kabushiki-Kaisha* (*Special Provisions*). In addition, whilst the law currently requires that an outside auditor be any person who has not held such office in the corporation at any time over the past five years, the law will, from 1 May 2005, require that an outside statutory auditor of a large corporation be a person who has never been a director or an employee of the corporation, (*Article 18(1), Special Provisions*).

The auditors of a large corporation will be granted the right to consent to any proposal by the corporation to appoint a new corporate statutory auditor. In addition, the board of auditors will be entitled to propose an agenda for the appointment of a new auditor to a shareholder' meeting with or without specifying a candidate for such appointment, (*Articles 18(3) and 3(3) Special Provisions*).

Derivative actions **15.14**

Any shareholder who has held a share of the corporation continuously for at least six months may request in writing that the company institute proceedings against a director for said director's liability to the corporation, (*Article 267(1)*). If the corporation fails to commence proceedings against the director within 60 days of such a written request, the shareholder may institute proceedings against the director on the corporation's behalf, (*Article 267(3)*). However, where the corporation may suffer irreparable harm before the 60-day period above expires, the shareholder may institute proceedings against the director on behalf of the corporation before the expiry of the 60-day period, (*Article 267(4)*). When bringing an action under *Article 267(4)* the court may demand, if the corporation makes such a request, that the shareholder make a security deposit, (*Article 267(6)*).

The corporation may make a settlement of the claim for alleged liabilities of a director or auditor at a court proceeding of a derivative suit without the unanimous consent of all of the shareholders, (*Article 268(5)*).

The 2001 amendment to the *Code* makes it clear that a corporation may participate in a derivative suit to assist the defendant director or auditor who are alleged to have breached a duty to the corporation, with the prior consent of all the auditors of such corporation, (*Article 268(8)*).

Although this provision has been in the *Code* since 1950 when it was introduced from the United States, there were few cases until the 1990s. It has been argued that this is because the cost of litigation is prohibitively expensive for individual shareholders to bear and even if the shareholders were to be successful, the damages would be paid to the corporation. However, after the collapse of the bubble economy in the early 1990s directors were blamed for many corporate woes. Furthermore, the number of foreign shareholders in Japanese companies increased. Indeed, the issue was even raised at the Structural Impediments Initiative talks. As a result the law was changed and shareholders who bring a successful derivative action on behalf of the corporation may now demand that the corporation reimburse reasonable expenses associated with bringing the action, (*Article 268–2*).

Another reason for shareholders' reluctance to file derivative actions until comparatively recently was the filing fees associated with them. Until the 1993 revision to the *Code*, the filing fee depended on the amount that the plaintiff's were seeking. This effectively made it prohibitively expensive for the majority of potential plaintiff's to file derivative actions. The *Code* was amended to fix the fee for filing a derivative action of \8,200.

Liability of directors to third parties 15.15

In the event that a director has, in relation to their duties as a director, been found guilty of gross negligence or wrongful intent, the director will also be liable to third parties for damages, (*Article 266–3*). In addition, where a director has made a false entry on material matters in a written application for shares, pre-emptive rights, the right to subscribe to new shares, or in a prospectus, the director will also be liable to third parties for damages, (*Article 266–3(2)*).

Important asset committees and corporations with committees 15.16

The 2002 amendment to the *Code*, known as the *Law for Special Provisions for the Commercial Code Concerning Audits, etc* (*Special Provisions*) added a significant new feature to Japanese corporate governance. Under the Special Provisions, large corporations, may, at their discretion, create special committees for the management of the corporation. A corporation has two options, to establish an Important Asset Committee or become a so-called Corporation with Committees.

In the former, the board of directors may, where ten directors or more are present and one of them is an outside director, establish an Important Asset Committee, (*Article 1–3(1), Special Provisions*). The board may by a resolution of the board, delegate its decision-making authority with respect to the disposition or acquisition of important assets and large borrowings to the Important Asset Committee, (*Article 1–3(2), Special Provisions*). The committee should comprise of three or more directors of the corporation, (*Article 1–3(3), Special Provisions*). A corporation is obliged to register the fact that it has established an Important Asset Committee and who the members of the committee are within two weeks of its establishment in the same district as its headquarters are located, (*Article 1–5(1), Special Provisions*).

Under the latter system, a corporation choosing to adopt the committee system is obliged to establish a number of governing bodies, including:

- a nominating committee;
- an audit committee;
- a compensation committee, and
- one or more executive officers, (*Article 21–5(1), Special Provisions*).

However, such a corporation with committees may not have an auditor, (*Article 21–5(2), Special Provisions*) and the term of office of the directors in such a corporation shall be no more than one year, (*Article 21–6(1), Special Provisions*).

The duties of the board of directors of a corporation with committees include:

- supervising the directors and the executive officers; and
- the basic management policies of the corporation.

The board of directors of a corporation with committees may not entrust directors of the corporation with any decisions regarding the corporate affairs of the corporation, (*Article 21–7(2), Special Provisions*), however, it may properly delegate substantial management authority to the executive officers, (*Article 21–7(3), Special Provisions*). The *Special Provisions* limit the corporate affairs which the board can entrust to the executive officers. limitations include:

- the appointment and dismissal of executive officers;
- the power to convene a general meeting;

- the determination of the items on the agenda of a general meeting (with some exceptions);
- the contents of a de-merger plan or agreement, and
- the contents of a merger plan, (*Article 21–5(3), Special Provisions*).

The purpose of the nominating committee is to determine the contents of the proposals appointing or dismissing directors at the general meeting, (*Article 21–8(1), Special Provisions*). The audit committee is charged with a number of duties, including: the audit of directors' and executive officers' execution of their duties and determining the contents of the proposals appointing or dismissing the accounting auditor at the general meeting, (*Article 21–8(2), Special Provisions*). The remuneration to be received by the directors and executive officers of the corporation with committees is to be determined by the Compensation Committee, (*Article 21–8(3), Special Provisions*). Each of the above mentioned committees must consist of three or more directors, the majority of whom must be outside directors, (*Article 21–8(4), Special Provisions*).

The executive officers have the power to determine those matters that are entrusted to them by a resolution of the board of directors and the execution of the corporate affairs of the corporation. (*Article 21–12, Special Provisions*). An executive officer should be appointed by resolution of the board of directors and the term of office should be longer than one year, but may be dismissed at any time pursuant to a resolution of the board, (*Article 21–13(1)(3), Special Provisions*). Directors may serve concurrently as executive officers, (*Article 21–13(5), Special Provisions*). Executive officers are to report to the board at least once every three months, (*Article 21–14(1), Special Provisions*). A corporation is obliged to also appoint a representative executive officer to represent the corporation, (*Article 21–15, Special Provisions*). Where the corporation has cloaked an executive officer with the appearance of a representative executive office, for example, by giving an executive officer the title of president, senior vice-president or any such similar title, the corporation shall be bound by acts of the executive officer vis-à-vis bona fide third parties, (*Article 21–16, Special Provisions*).

Conclusion **15.17**

A number of corporate scandals, including the Daiwa Bank scandal, the Nomura Loss-Compensation Scandal and the Mitsubishi Motors scandal, have forced a re-examination of the Japanese model of corporate governance. The initial result has been a reaffirmation of Japan's adoption of the US model of corporate governance, although the law is currently in a state of flux and a clearly defined new model has yet to emerge. Nonetheless, the above provides the reader with a basic starting point for an understanding of Japanese corporate governance.

Chapter 16

Environmental Due Diligence and Risk Management

Introduction

Macro and micro issues 16.1

The comments in earlier chapters have already demonstrated that environmental regulations and trends cover many areas of business activities. A comprehensive treatment of this topic would require a separate book, but some of the key issues may be touched on here. It would be an omission to ignore the vital area of environmental issues in a book on corporate governance and due diligence.

In a good practice series published by the European Commission's Directorate-General for Enterprise the publication *Responsible Entrepreneurship* (2003) it was stated (at p32):

> 'Recent decades have seen a marked increase in awareness and public concern about the impact of productive activities on the natural environment. Environmental impacts associated with business operations include:
>
> - Inefficient and unsustainable use of natural resource such as oil, gas and water;
> - Emission of greenhouse gases such as CO2 contributing to climate change;
> - Emission of pollutants contributing to air and water pollution;
> - Long-term effects of hazardous chemicals;
> - The rapid loss of biodiversity; and
> - A high level of waste generation and hazardous waste.
>
> These impacts increasingly result from goods and services rather than production processes. Instruments used by business to address such environmental impacts are manifold and include amongst others:

environmental management systems, both formal (EMAS, ISO 14001) and informal, eco-design tools, cleaner production techniques and technologies and eco-labels'.

In the discussion below some of the instruments are referred to. The main headings in relation to traditional EDD and risk management may be categorised in the following summary:

- air and water pollution;
- hazardous and special waste;
- toxic substances;
- planning;
- pesticides and herbicides;
- radioactive substances;
- employee protection;
- biotechnology;
- statutory nuisances;
- marine pollution;
- environmental information; and
- health and safety matters.

There is no doubt that there are many types of liability and responsibility associated with these activities since environmental law is increasingly having an impact on business transactions. The relevance of the environmental performance of a business to its core business activities is a debate that has been evolving over decades. Moreover, several matters covered by this chapter are business issues that have been on the agenda for many years. Accordingly, specific strategies have been developed to avoid or mitigate environmental damage and consequences. Indeed, in the same way as any business transaction is automatically considered from the tax angle, it is imperative to carry out a comprehensive environmental investigation as a matter of good business practice. In addition, as discussed in **CHAPTER 4**, a business should take account of the important risks of health, safety, social and environmental (HSSE) in the interest of good corporate governance.

A recent survey on EDD by KPMG (*Environmental Due Diligence: A Survey of major UK companies*, May 2004) has provided the following key insights:

> 'EDD has become an important feature of an increasing number of merger and acquisition (M&A) transactions. Interestingly, however, it is not automatically included in the approach many companies take to transaction evaluation, even in sectors at the greatest risk from health, safety, social and environmental (HSSE) issues ...

Legal and financial (capital and operating expenditure and liabilities) consequences of HSSE issues remain fundamental to most EDD investigations. But HSSE issues now also impact other business performance variables such as sales, operations, customer relations and reputation – each of which can directly impact the key transaction 'check points' of: sale and purchase agreement; valuation model (and assumptions); deal breaker evaluation; acquisition accounting; post acquisition action planning; and exit strategy (where applicable).

While some companies have made the transition towards a more commercially oriented EDD process, others are part way through the transition, having adapted their assessment process to take account of the business performance issues, but not yet altering the EDD scope.

One of the critical success factors is for companies to have a commercially driven scoping and assessment approach which integrated EDD findings into the legal, commercial and financial due diligence assessments, underpinned by robust technical competencies. The survey data indicates these are the companies which are suffering fewer material issues arising post acquisition and in the long run are more likely to have their M&A succeed.'

Moreover, as regards management procedures and guidance the survey found that despite the availability if good management procedures and guidance many EDD teams "work without clear internal guidance and procedures". "Improving the management procedures and guidance in place around EDD appears to be a clear opportunity for many companies, and may come under increasing internal scrutiny as broader corporate governance reforms come into play.'

Transactional tools **16.2**

It is useful, to consider the tools that have been available for some time in other jurisdictions. In corporate transactions such as acquisition/disposal of corporate assets and mergers, certain tools are employed in order to discover and make adequate provisions for potential environmental problems. Bearing in mind the above survey findings, however, it is important that the environmental and related issues are fully integrated into the business as a whole.

Pre-contract enquiries **16.3**

The most useful means of creating an environmental picture of the vendor company or business is to make appropriate pre-contact enquiries in the form of an environmental questionnaire (see the example of a simple US questionnaire in the **APPENDIX**). This serves as a prompt to identify and perhaps assess the existence of circumstances which, either individually or in conjunction with other similar or linked claims, could lead to environmental liability. The acquisition or disposal of corporate assets requires a rigorous assessment of the vendor company's environmental position. Information needs to be gathered to enable the purchaser to decide on the environmental warranties and indemnities that may be required.

The environmental questionnaire would normally contain a full set of specific questions, answers to which may give an indication as to whether the land or adjoining property is contaminated or not and whether other environmental concerns are present. The following types of environmental liabilities should, for example, be disclosed in response to the questionnaire:

- liabilities which are historical, existing or potential;
- liabilities which concern civil matters such as damages, injunctions or personal injury claims relating to industrial diseases;
- compliance/regulatory matters including liabilities which concern criminal matters such as fines, prosecutions or enforcement action;
- liabilities which reveal capital or revenue cost, whether of a compulsory or voluntary nature.

In addition, the environmental questionnaire will be directed at specific sites which are to be acquired and in this regard the following matters should, for example, be covered:

- liabilities relating to existing sites, closed sites and operating sites;
- eased sites which are no longer owned or operated by the company;
- sites which may have been used by former owners; and
- closed operations and practices.

It is advisable to ask the vendor to go back as far as possible – at least 15 years and more where earlier liabilities or incidents of which the vendor is aware should be disclosed. The vendor company should also be asked to give an indication of significant changes or new issues of which it is aware and which may emerge between the date of their response and the proposed date of completion.

The subject-matter of the environmental questionnaire should cover:

- details of monitoring and reporting procedures of the vendor company within its own organisation – this should reveal actual and potential problems;
- details of monitoring and procedures for reporting accidents or incidents;
- details of any environmental audit carried out – both internal and external;
- whether the vendor is aware of any health and safety matters affecting its business, plants or sites or industrial diseases caused by its business or operations which could result in environmental liability;

- pending or threatened litigation matters;
- questions concerning disposal of waste produced or otherwise handled by vendor company;
- any internal guidance on landfills or other guidance relating to storage management and handling of waste, whether on or off site;
- details of any known or suspected cases of contamination of land, water or air;
- clean-up obligations in relation to contamination; and
- detailed questions should be asked of the relationship with regulatory authorities and related expenditure.

Sample pre-contract enquiries cover:

- details of any litigation with neighbours; and
- details of environmental liabilities which have or could be incurred under contract for example by way of indemnities or warranties given by the vendor company.

Industrial sites and their associated activities and sites used for waste disposal or waste disposal practices themselves are particularly high risk when considering environmental liabilities. The site specific environmental questionnaire should accordingly concentrate on such matters. This would include detailed information on:

- the operations and activities of the vendor company and whether they caused (or may cause) harm to human health;
- discharges or releases into the air, water and land which could cause environmental liabilities;
- materials and chemicals used, stored or disposed of at the site; and
- the vendor company should be asked to check that all conditions relating to permits licences and consents used for the business are being complied with and whether they may be revoked.

Sometimes it will be possible to issue a single questionnaire if the size and complexity of the assets/shares being acquired are relatively small. The enquiries should cover:

- the property and its historic uses;
- waste disposal policies of the company;
- usage of storage tanks, on or under the property;
- regulatory compliance and notices;
- civil liability; insurance policy on the property;
- health and safety issues; and

- the historic uses of the adjoining land.

Replies to pre-contract enquiries will often indicate that there are environmental concerns that require immediate attention in the transaction. Various matters that might cause concern are discussed below by way of example. In addition suggested guidance is given, although, of course, any business should decide on the basis of the particular circumstances of the case in hand.

The product and raw materials 16.4

Although the vendor may well have a product which does not itself appear to cause pollution, the production process should be checked as it may do so. For example, toxic materials may be used as part of the manufacturing process. In such a case the purchaser is advised to:

- request sight of environmental policy adopted by the vendor;
- require information as to how it is implemented; and
- inquire how the vendor company monitors compliance with its own policy.

In addition it is important to carry out investigations relating to:

- the process and previous use of the site.

The purchaser needs to find out from the vendor whether the process may cause emissions which could impose environmental liabilities. In addition previous processes need to be checked. Although they are no longer used, they may lead to current or future liabilities. Checking out the previous use of the site may alert the purchaser as to the possibility of the land being currently contaminated, with all the attendant costs for clean up that this could bring.

Waste disposal 16.5

Whenever waste is being stored, transported or disposed of, the purchaser must be aware of its waste management responsibilities under the waste management regulations. The vendor should also be aware of its waste management activities and any problems in this connection in order to be in a position to respond to the purchaser's queries.

Environmental consents and licences 16.6

The purchaser should ask to see all environmental consents and licences which are necessary for the business being acquired. In this context, it is particularly important for the purchaser to check carefully the conditions which attach to them and also to take a view as to whether or not these conditions are being complied with, as current liability may arise out of a previous breach of conditions. Sometimes the regulatory authority which has granted a licence may review it, so it is crucial to check whether such a review is taking place. In addition, the purchaser should ask to see copies of correspondence between the vendor company and the regulatory authorities in order to consider what local opinions there are about the industry being carried on by the vendor company.

Site inspection **16.7**

It is, of course, a matter of good practice for the purchaser, as well as relevant advisers, to visit the site, as issues that trigger off environmental enquiries may be discovered. For example, there may be a potential risk to groundwater caused by discharge or loss of fluids resulting in groundwater pollution. This may manifest itself in cracked concrete or soakaways. In view of the extensive impact of the contaminated land regime in the UK since its implementation in 2002, as well as the developments in the EU regarding environmental liability, as well as related developments elsewhere, it is vital to understand the implications of EDD on asset values.

Other significant potential risks for which inspection should be made relate to:

- the position of drains and the route of rain water run-off;
- local sub-surface hydro-geology; and
- uses of groundwater.

The results from the site visit may lead to the appointment of specialists to carry out site inspections and to report to and advise the purchaser.

The vendor would be advised to limit his liability by:

- imposing a financial limit and a time limit on his liability in respect of the warranties or indemnities; and
- demanding that the vendor should control the conduct of any claims which may be made.

Replies to pre-contract enquiries will often indicate that there are environmental concerns that require immediate attention in the transaction.

Searches **16.8**

In addition to making these pre-contract enquiries, it is now common practice for the purchaser to commission an environmental search report. This often identifies the historical uses of the land and the adjoining properties for up to the previous 150 years This may be obtained in certain public registers such as ordinance survey maps, which are often available in the relevant local authority offices. The planning history of the site should also reveal any potential problem areas, as far as the use of the site is concerned.

Where the land is found to have had contaminative uses in the past and there is no evidence that satisfactory remediation has been carried out on the site, then there may be need for a full environmental audit on the site. This would serve to determine the extent of contamination and the appropriate method of remediation.

Environmental audit 16.9

As indicated above, see **16.1**, in today's climate of increased environmental awareness, a company or lending institution involved in a company takeover or merger will ignore the environmental profile of the companies involved at its own peril. Where a comprehensive pre-contact enquiries and search report show that the land had been used for contaminative purposes in the past, the purchaser has to commission relevant environmental consultants to carry out an environmental audit having regard to costs and the probability of the anticipated risks *vis a vis* the intended use of the property. There are environmental audits, of varying degrees of sophistication, which help to ascertain the effects of previous uses; recommend a cost effective remedial strategy and possible alternative uses of the land. It could also be used to ensure that environmental risks are taken into account and evaluated at the stage which their importance to the transaction can be recognised and the information used with maximum effect. Since environmental audits have become increasingly important as regards corporate activities and transactions the background to the development of auditing, as well as the current role, are detailed here.

Environmental auditing and international standards 16.10

There has been a longstanding debate over the role of environmental auditing and international standards. The 1990s were hailed as a decade of environmentalism. In 1990 there was an emphasis on:

- increased environmental awareness;
- access to environmental information; and
- implementation of sound environmental principles.

Both legislative and economic incentives have placed the environment high on the agenda for businesses and government bodies, as well as consumers, shareholders, investors etc. Some organisations have been aware of this trend and have reflected it in their growth from the beginning, others have taken it on board at a later stage. There have been some that have taken their environmental responsibilities very seriously over the years, aware of the fact that the combination of green concerns and business objectives is a must for business. This has been demonstrated in the ways that they do business as well as their approach to transactions. For example, the Chemical Industry has been conducting environmental audits for many years; the Chemical Industries Association has a well established set of guidelines. It is true to say that many believe that commerce can and must grow alongside an expansion in environmental awareness.

The environment is a transboundary global concern as has been vividly demonstrated in the debate over climate change. Yet much of the push towards the 'greening of business' has also come about through national initiatives of different jurisdictions where the competitive edge has been noted as well as through regional initiatives of the EU's Environmental Action Programmes, both as regards legislative and market-based tools. As has been discussed in **CHAPTER 4** what has emerged as a particular area of concern is that of sound environmental management.

The EU's Fifth Environmental Action Programme 'Towards Sustainability', which addressed the years 1993–2000 particularly emphasised the importance of 'shared responsibility' for the environment which involves all sectors of society, public and private sectors, organisations and individuals alike. While accepting the importance of the regulatory approach, this Programme has highlighted a more creative approach to environmental management through the integration of voluntary mechanisms, financial incentives and central funding. Taking up the thread of the well publicised 1987 report of the World Commission on Environment and Development, *Our Common Future*, (the Brundtland Report) the need to protect the environment and preserve it for future generations has been worked upon to take account of the key role that business must assume so that the concept that the 'polluter pays' is complemented by the notion that business can profit positively through an enhanced environmental sensitivity.

While the several hundred legislative instruments dealing with the environment that have emanated from Europe still have a major part to play in protecting the environment, there is no doubt that economic instruments and financial incentives also have a useful role in coordinating the partnership between the public and the private sector.

In this discussion the intention is to consider environmental auditing in the light of the need for enhanced environmental awareness and improved environmental management standards nationally and internationally, bearing in mind the emphasis on corporate responsibility that is vital to sustainable development and having regard to the need to be proactive.

Role and objective of environmental auditing 16.11

Environmental audits are often regarded as the best tool for bringing about environmental reform of industrial practices. This may be true provided that they are one of several mechanisms used in a much larger scheme aimed at significantly improving a company's environmental performance. It is also important that the results of that performance are regularly and readily available to the public. In the United States the US EPA, the Courts and some States have all sought in one way or another to find a future for Environmental Audits. Under the general EPA Policy on Environmental Auditing no company is forced to conduct an Audit while the EU Eco-Management and Audit Regulation (EMAS) has also left the choice of environmental auditing to the companies or organisations who wish to participate in the scheme. Yet it is also true to say that there can be a real pressure on companies to undertake environmental audits both in respect of their success in the market place and in respect of the success of individual transactions.

The process of environmental auditing was first developed in the US in the early 1970's as a method for an organisation to confirm that it was complying with legislative requirements. Specialist environmental auditors checked compliance and examined sites or plants that were being bought or sold to ensure compliance with, in particular, the Superfund legislation.

After the disaster in Bhopal in India in 1984 and the liability issues raised for Union Carbide, companies became anxious to ensure that their overseas subsidiaries fulfilled similar standards, with the result that parent companies and American multinationals began auditing overseas. This brought the practice to Europe where it has assumed a different role largely because environmental liabilities have been less severe. Environmental auditing was considered a way of making a company greener and demonstrating publicly that environmental responsibilities were being taken seriously. More recently, in the US, organisations have considered environmental auditing as a means of protecting themselves from criticism, as well as a marketing support, rather than simply a defence against legal liability. Meanwhile the ongoing discussion in Europe in connection with extending liability to circumstances of impairment or damage to the environment has emphasised the role of environmental audits in due diligence exercises.

Basic objectives 16.12

The basic objectives of an environmental audit are to:

- check a company's performance against its objectives and policies;
- report to management on any environmental concerns with suggestions for modifications or improvements;
- propose programmes for future environmental activity;
- check compliance with company wide or other standards, legislative requirements and regulations;
- specify steps required to achieve total compliance; and
- recommend action for risk management.

By taking a 'cradle to grave' approach an audit can monitor emissions and discharges, waste and recycling efforts, with cost effective results that can improve both the bottom line and achieve quality management in economic and environmental terms. In addition the audit can be a monitoring exercise to:

- verify training, health and safety and environmental procedures;
- check the adequacy of record keeping;
- provide a satisfactory data base for use for any of several purposes; and
- comment upon an analysis of information produced.

A further objective is to test the validity and quality of the audit management system in place and the adequacy of audit protocols, procedures and compliance manuals as well as to review any contingency plans. Environmental audits are often seen as management tools which are an aid to the framing of appropriate policies to support marketing and as indicators of the efficiency of the company's environmental policy. For instance, at the time of insurance review they may be helpful to obtain satisfactory insurance arrangements while compliance audits assess general compliance with existing and proposed regulatory standards, corporate standards and good practice. In addition, they perform a useful role in transactional due diligence (see also the scope of audits at **16.17**).

While environmental audits, and their results, have been voluntary rather than mandatory, EMAS, for example, does requite the independent scrutiny of the environmental impact of a company's activities and does result in a public environmental statement. Similar guidelines were developed through the British Standard 7750, superseded by the ISO 14001. In the United Kingdom both initiatives have had an effect on the level of corporate green activity.

Corporate environmental performance reporting 16.13

The EU's Fifth Environmental Action Programme recognised that many of the present day environmental issues and threats to the world's ecological balance were posed by trends in political, economic and social life. This work has been further developed through the current programme and the developing environmental, energy and related legislative instruments. It has been regarded as necessary to bring about substantive cultural change in these areas, which required a more flexible, imaginative approach to environmental management than had been previously adopted. These initiatives have been developing over the last decades: yet the message has taken time to be received by business overall and therefore these developments are summarised here.

The underlying principle behind this approach was that by increasing public awareness about the environmental performance of organisations, pressure would be exerted on those organisations to ensure that their performance is 'acceptable' to the public. By way of example, in the United States the Valdez Principles, which were evolved following the environmental accident in March

1989 caused by the spillage of an oil tanker Exxon-Valdez in Alaska, heralded a way forward for socially responsible' companies. The clean-up of the Alaska coastline by the company concerned, Exxon, has to date cost over $1bn, consumed thousands of man years of managerial effort and embroiled the company in over 150 complex law suits which, depending upon their outcome, could add hugely to the total cost. Among other things, these Principles called for an environmentalist on each corporate board and an annual public audit of a company's environmental progress.

The message that was being sent was that eco responsibility would be good for business. Leading corporate managers called for 'corporate environmentalism' in America and, for example, the head of Pacific Gas & Electric stressed the importance of dialogue with environmental groups and others as a way forward in an openness that was good for business. This could be seen in the context of corporate environmental policies, published commitment to environmental performance and environmental performance reports. Organisations have in fact seen that a positive approach to environmental awareness is good for business. It has become increasingly recognised that for environmental reporting to be meaningful, it should be done on a site by site basis rather than by providing overall statistics relating to environmental data, such as emissions of different types. This is especially true bearing in mind the possibility of a director being buttonholed and made personally liable for any statement made in an environmental report.

International co-ordination 16.14

There is no doubt that in order to improve the environmental performance of companies and to maximise the effect of this improvement there should be worldwide co-operation by business to uphold the concept of sustainable development. This is the case, whether in developed or developing regions. Areas as diverse as business investment in Eastern Europe and the development of corporate policy in Japan and India have relied on environmental auditing as an international tool.

In order that more businesses should join this effort and that environmental performance should continue to improve, the International Chamber of Commerce (ICC) established a task force of business representatives to create a business charter for sustainable development. This comprised 16 principles for environmental management. It was formally launched in April 1991 at the Second World Industry Conference on Environmental Management (WICEM 2) in Rotterdam. The last of these principles is:

> 'Compliance and Reporting: To measure environmental performance; to conduct regular environmental audits and assessments of compliances with company requirements, legal requirements and these principles; and periodically to provide appropriate information to the Board of Directors, shareholders, employees, the authorities and the public.'

The charter was one of seven projects by European and North American Business Leaders at the Bergen Conference 'Action for Common Future' in May 1990. This conference was generally regarded as the main follow up event to date for the Brundtland Report. The Bergen Conference brought together at ministerial level the 34 Member Countries of the UN Economic Commission for Europe (ECE), including Canada, the US and the Soviet Union. Non Governmental groups also took part and the 'Industry Agenda for Action' was endorsed by business leaders at the Bergen Industry Forum and centred around the preparation of the charter and of the auspices of the International Chamber of Commerce (ICC). The charter was adopted by the ICC Executive Board on the 27 November 1990.

The aims of the charter are still worth citing:

- 'To provide guidance on environmental management to all types of business and enterprise around the world, and to aid them in developing their own policies and programmes.
- To stimulate companies to commit themselves to continued improvement in their environmental performance.
- To demonstrate to Governments and electorates that business is taking its environmental responsibility seriously, thereby helping to reduce the pressure on Governments to over legislate and strengthening the business voice in debates on public policy'

See Dr L Spedding, Environmental Management for Business, Wiley, 1996, p 8.

Many companies saw the charter as a major response to governmental and activist pressures for environmental 'codes of conduct'. In addition the ICC definition of environmental auditing has been accepted generally as:

> 'a systematic, documented, periodic and objective evaluation of how well environmental organisation, management and equipment are performing with the aim of helping to safeguard the environment by:
>
> (i) facilitating control of environmental practices;
>
> (ii) assessing compliance with company policies which would include regulatory requirements.'

EU Eco Management and Audit Regulation (EMAS) **16.15**

EMAS, which was adopted in June 1993 and took effect in 1995 began as the draft *Eco Audit Directive* which proposed mandatory annual audits by large manufacturing firms. It was aimed at major processes as a complement to Eco Labelling which was a scheme concerned with products. As such EMAS was designed to monitor pollution and reduce the impact of large, individual facilities or sites in accordance with strictly drawn guidelines. While it still remains more suitable for manufacturing facilities in view of its site-based approach, a voluntary 'management systems approach' was proposed during negotiations and the resulting regulation establishes a voluntary scheme within which participating companies must establish an environmental management system. The Regulation was extended to enable Member States to include sectors such as the distributive trades and public services on an experimental basis and the negotiations also dove-tailed the efforts of the British Standards Institution (BSI) which had been evolving the new Environmental Standard BS 7750 which was intended to comply fully with the environmental management system (EMS) requirements of EMAS.

Under EMAS, a verifier will review an organisation's environmental impacts, procedures and targets in relation to its environmental policy and EMAS and the process leads to an independent verification statement which is required to confirm compliance with the scheme.

Despite the longstanding history of EMAS many businesses still do not take advantage of it and the EU continues to strive to improve uptake of the scheme.

ISO standards 16.16

The United Kingdom was the convenor of an International Standards Organisation (ISO) International Electro Technical Commission (IEC) working group on environmental management with regard to a draft International Standard. These developments also took place in the last decade. The working group was set up under the aegis of the strategic Action Group on the Government (AGE) which was established jointly by the ISO and the IEC in 1992. The intention was to design an ISO Standard from the outset and a draft document was submitted to the representatives of the working group in June 1992. The New Technical Environmental Committee (which reflected that of BSI) was based upon an official proposal to produce international standards on

- Environmental Management Systems;
- Environmental Auditing;
- Environmental Labelling;
- Environmental Performance Evaluation;
- An Industrial Mobility Plan for Development Industrial Standards; and
- Life Cycle Analysis.

The objective was to produce an ISO Standard some one or two years after the revision of the BS 7750 Standard and the establishment of certification bodies in 1994, provided that there should be agreement on the production of an international standard. The intention was to evolve a systems approach embracing an environmental management system, health and safety management system and a quality management system. The ISO 14001 replaced the

BS7750. In the UK the accreditation service, UKAS, has been concerned with maintaining the implementation of the standard and has recently suggested that the accreditation should be more robust.

Scope of audits 16.17

It is useful to consider the scope of the audits as they have evolved by looking at the broad range of services that has developed and been offered by environmental consultants. Briefly, audits can, as noted above, see **16.9**, be transaction based, site specific or more general. The broad range of issues carried by environmental auditing evolved rapidly and provided a helpful guide to the basic types of audits available, as follows:

- **Corporate Audit Programmes:** to assist international corporations in the development of company-wide monitoring.
- **Due Diligence Audits:** to assess potential liabilities and to provide key information for negotiating parties prior to the transfer of a site as environmental liabilities can have significant impact on the value of a company's assets. The audits frequently save considerable amounts of money by helping to reduce a site's asking price or in securing warranties to cover future environmental expenditure.
- **Waste Audits:** to assess all aspects of waste management from the on-site storage of materials to off-site disposal by licensed operators as required by new, more stringent legislations. Waste audits can identify opportunities for cost savings through recycling or resource recovery.
- **Health and Safety Audits:** to adjust working practices in response to new legislation. H&S audits in the UK and throughout the world are used to assess compliance as the basis for ongoing training and improvement programmes and may be integrated into an health, safety and environment management approach.
- **Site Audits:** A general example of a site audit could include:
  - Site setting and location;
  - Site history;
  - Management systems;
  - Raw material storage and manufacturing processes;
  - Emissions (liquid, air, noise);
  - Waste;
  - Health and safety;
  - Building materials;
  - Energy;
  - Security;
  - Fire Precautions; and
  - Pest Control.

Disclosure and liability under audits 16.18

There has been a longstanding debate over disclosure and liability. While the US deliberately avoided any public disclosure requirement as part of the EPA Policy on environmental audits, EMAS referred to a 'true and fair' disclosure of 'the environmental issues of relevance to activities at the site'. Nevertheless it must be recalled that this disclosure requirement only comes into play if a company voluntarily chooses to participate in the Eco Audit Scheme. As noted above see **16.15**, the EU is still working to encourage improved participation in the scheme, especially as regards SMEs.

In particular in the UK the audit will disclose legal risks and liabilities for which, as a matter of English law, no privilege will apply. The wide scope of the audit under EMAS for example will tend to destroy the cornerstones of privilege where the dominant purpose of the Audit was to obtain legal advice. This will, of course, vary according to the particular jurisdiction when undertaking a global audit.

There has been some concern that by undertaking an audit and having results published a company may find itself vulnerable to attack by voluntary groups and enforcement agencies. Nevertheless, having regard to the general trend in favour of openness, it would be preferable on balance to select the more open approach with regard to environmental matters.

Ongoing issues 16.19

There is no doubt that environmental auditing can assist in the achievement of the goal of sustainable development and good corporate governance. It has been argued that to be fully effective however a proper international standard is needed. One suggestion from the US has been to look at the best features of the methods in Europe and the US.

It has been stated that EMAS is ahead of the EPA Policy on environmental audits in its approach as a market model and its objectives, that is the improvement of environmental performance beyond the minimally required level, together with broad validated public disclosure. In the US, the EPA Policy requires updating since it does not properly reflect the status of environmental audits and public disclosure requirements provided in other US laws and policies. Although the EPA Policy is piecemeal, the strength of the US approach lies in the fact that companies with poor environmental track records have to conduct audits and to disclose results. A 'Euro-US' proposal has therefore been suggested by some practitioners whereby:

- EMAS would serve as the basic 'block model' applying its comprehensive, integrated and standardised approach including public disclosure requirements.
- Authority to compel an environmental audit would be added to EMAS and be required by any organisation that is engaged in a significant violation of a material environmental standard.

- A further provision could be included to make the absence of positive effort a factor to be weighed in assessing criminal and civil liability. EMAS would be specifically identified as a preferred type of positive effort.
- A 'prevention test' could be added to determine corporate officer liability, either as a supplement or alternative to the 'positive effort' provision.

If this proposal were to be implemented globally, it would bring about a practical hybrid that would merge both the regulatory and the voluntary mechanisms.

Environmental accounting and managerial decisions: governance issues

Incorporating environmental values to improve the environmental performance of companies and to evaluate project viability 16.20

It has often been commented that traditional accounting methods do not necessarily reflect environmental costs. Because the link between environment, cost and company performance was not previously fully appreciated, cost accounting systems evolved without creating a method for identifying true environmental costs.

Traditional financial analysis also lacks the capacity to accurately predict environmental costs and savings. It has employed simple point estimates of the outcomes of future uncertain events or, at best, worst case, most likely and best-case analyses. By collapsing continuous probability distributions of possible outcomes into point estimates, the analyses destroy valuable information that might otherwise be accessible to decision-makers. This increases the odds that bad decisions will result. Financial analysis of capital investment alternatives cannot capture the significance of environmental risk mitigation projects. Yet, environmental experts do not always have the ability to communicate to the financial analysts in the appropriate terms and framework. Environmental capital projects often compete with projects that have a more tangible cost savings equation traditionally.

Cost accounting may also be a useful tool in investment decision-making. By identifying up front the full cost of an investment decision, companies and individuals can better select investments not only in terms of short-run profits, but also in terms of long-term sustainability. The case study regarding SERM that was mentioned in **CHAPTER 4** may be recalled in this regard.

Risk assessment and decision-making are another area where full cost accounting can add benefit to corporate planning. By having realistic figures for actual costs as well as potential costs (remediation of environmental accidents, plant closure, fines, penalties, litigation, damages, costs of bad environment decisions and cost savings from good decisions), companies can accurately predict risk, and calculate the return on investment of projects that reduce those risks.

A variety of 'green' accounting methods have been developed over several years which allow for costing out environmental costs. Because this is a different form of accounting, individual companies have to adapt these accounting systems to their own practices in order to implement them with little cost. Revamping an entire accounting system for most organisations would be a major undertaking. Nevertheless, depending on the accounting systems in place, a modified system may pay for itself by identifying costly and needless environmental expenditures.

Benefits of environmental accounting

Implementing environmental accounting makes environmental costs more visible to company managers, thereby making those costs easier to manage and reduce. Environmental accounting gives companies the opportunity to:

- significantly reduce or eliminate environmental costs;
- be transparent in the spirit of good corporate governance;
- improve environmental performance, and
- gain a competitive advantage.

Costs to companies of environmental regulations 16.21

The following section outlines some of the ways in which companies incur environmental costs. A number of the costs, as well as potential cost savings, have traditionally been hidden from managers because of the accounting methods. In order to properly identify potential cost savings, it is important to first identify the actual costs that regulations impose on companies. Thus, once these costs are identified, it is possible to recognise when pollution prevention and environmental management activities result in cost savings.

Clean-up liability and costs 16.22

The history of the well-known Superfund, the main objective of which was to enable the clean up of contaminated sites in the US, provided a vivid illustration of an approach to clean up liability and costs that met with huge criticism by many stakeholders, including industry. It was estimated that if a company was named a party to a Superfund site, the immediate cost was roughly $US 500,000, largely due to legal fees. Once liability was assessed, costs could run well into the millions. In addition to direct costs paid for cleanup, companies had to pay hefty insurance fees as well. These fees have sharply increased due to the costs of liability which has driven many insurers out of the environmental liability insurance market. By way of example, the cost of environmental liability in the United States was estimated over a decade ago, in 1991, to equal $US 500 to $US 1 trillion while the property-casualty insurance industry was only worth $US 158.2bn.

For example, Motorola, an American semiconductor manufacturer, recently worked out a Superfund settlement in Arizona that will add up to £30m. It is worth noting that many of the 'guilty' parties for Superfund contamination in the Silicon Valley were actually complying with technology and regulatory guidelines at the time the facilities were installed. However, the technology used to store hazardous materials, etc clearly did not do the job and contamination occurred. The debate over clean up liability and costs continues to be very sensitive and it is vital that businesses establish their likely exposure having regard to the location. As mentioned, in the UK there is in place a contaminated land regime, the details of which fall outside the scope of this manual.

Environmental programme administration **16.23**

One significant cost associated with compliance comes from the time it takes to understand the regulations and to keep up with the details of compliance such as mandatory report generation and reporting forms. These costs are above and beyond the actual payments made, for example, to treat or dispose of waste.

Research and development for alternative processes and materials **16.24**

A further cost comes from developing alternative materials and processes. Environmental laws may not require companies to invest in the development of alternatives but when a key process material is banned, companies are forced to develop alternatives in order to stay in business. By way of example, over ten years ago in the US, AT&T spent $US 25m on research, development and testing of substitutes for ozone-depleting substances. Moreover, in testimony to the House Armed Services Committee on March 6, 1990, IBM estimated that it would spend $US 70m in corporate-wide capital costs to eliminate use of CFC 113. IBM further estimated that it would spend about $US 140m to research and develop alternatives processes for these products. This point is further discussed in the section on legislative trends below, see **16.35**.

One advantage for manufacturers in some developing countries is that they may be spared some of these costs, because alternatives have already been developed. For companies just getting started, they can design their processes around banned substances. This means that rather than retrofit equipment or amortising costs of equipment made obsolete by bans on materials, companies can begin production with state-of-the-art processes and equipment produce few emissions.

The cost of inputs 16.25

Raw materials generally occupy the largest proportion of total production costs in any manufacturing setting. Environmental pressures that result in higher raw materials costs are generally the result of suppliers' raised costs. Prices for metals for example have been known to increase approximately 10–12 per cent per year due to environmental pressures and related regulation.

Energy costs 16.26

Energy costs are generally a significant production cost. In heavy manufacturing, they have accounted for up to 40 per cent of production costs. High energy users have considered the effects of higher energy costs, having undertaking energy audits. For example some companies have investigated opportunities for switching to more fuel efficient sources, especially for space heating. Plant design has been a particular constraint to moving to more energy efficient operations. In investment decisions, plant design for energy efficiency should be one consideration for any investor. Energy efficiency will reduce production costs.

Environmental pressures impact on the technology, production and pollution control methods of companies. Most companies have had to make investments and incur costs in response to environmental legislation. Investors would be wise to assure that any company they invest in has state of the art environmental technologies in place and are equipped to respond cost effectively to changes in environmental requirements.

Once a company develops a method for allocating hidden environmental costs, it will be able to realise cost savings. The next section will detail different methodologies and projects underway to further outline environmental accounting methods.

Environmental accounting projects 16.27

A number of organisations in Europe and in the United States and Canada have over the years undertaken efforts to develop environmental cost accounting strategies. Investors, as well as business-decision makers, should keep apprised of relevant developments in the environmental accounting area. This is with a view to adopting and incorporating certain methodologies into the practices of their companies in the interests of transparency and good corporate governance. There have been many initiatives globally. For the present purpose a few illustrations may be mentioned.

By way of example the US Environmental Protection Agency instituted the Environmental Accounting Project. The project was started as a response to concerns expressed by the Pollution Prevention Division that pollution prevention (practices that reduce or eliminate pollution) would not be adopted as a viable option for managing environmental concerns until the environmental costs of non-pollution prevention and the economic benefits of pollution prevention could be seen by managers making business decisions. The project had a network of over 600 members who participated and received information from the project. The main outcome of the project was the extensive gathering and sharing of information on environmental accounting. Many of the results of this project were made available through the internet or from the US EPA. The publication of case studies detailing the experience of individual companies in incorporating environmental accounting were useful for any organisation wishing to move forward into environmental accounting. Software packages could also be purchased to help companies to incorporate environmental costs into capital budgeting decisions.

Information on Environmental Accounting from the US EPA

- An Introduction to Environmental Accounting as a Business Management Tool: Key Concepts and Terms.
- Environmental Cost Accounting for Capital Budgeting: A Benchmark Survey of Management Accountants
- Incorporating Environmental Costs and Considerations in Decision Making: A Review of Available Tools and Software.

In the UK the Chartered Association of Certified Accountants (ACCA) has also examined the question of environmental accounting and reporting and pioneered developments over many years, as well as organising annual awards on environmental reporting in an ongoing bid to engage industry broadly.

In the 1990s the *International Network for Environmental Management* developed a method to give weight to the environmental impact of their activities. The tool was a decision-making tool to help management decide how to get the greatest environmental impact reduction per unit of investment. The report, called *Environmental Accounting for Enterprises*, also inspired practical debate. It is important to note that many of these developments originated from the 1990s yet remain foundational to this discussion.

Environmental accounting methodologies and case studies 16.28

Environmental accounting in a company can be divided into two different areas:

- financial accounting; and
- management accounting.

Financial accounting allows companies to prepare reports for outside investors to use in looking at the risk and environmental profile of a company.

Management accounting, which is the subject of this section, is the process of identifying, collecting and analysing information principally for internal purposes. A key purpose of management accounting is to support a business's future management decisions. Management accounting can involve data on costs, production levels, inventory and backlog, and other important business aspects. Most companies develop management accounting practices as a function of their particular business needs. Business type and size, as well as customer base, can influence which methods are implemented

Types of decisions that benefit from environmental cost information

- Product and Process Design;
- Capital investments;
- Cost Allocation;
- Product Retention and Mix;
- Performance Evaluations;
- Purchasing; and
- Risk Management

Environmental insurance: reducing risk and liability

Efforts by the insurance industry to become more green 16.29

The insurance sector is another area where the integration of environmental issues at all levels has received some attention. The insurance industry has long been aware of environmental risks, because insurers have often had to pay for environmental damage. Companies that wish to protect themselves from exposure to environmental liability purchase a pollution insurance policy. Many banks have required such a policy before allowing certain kinds of investments and transactions. In the past, companies did not actually have insurance to protect them against liability, moreover the environmental insurance market has suffered as a result of the potential size of exposure. As a result of stringent environmental laws and the broadening of the market, the environmental insurance market matured into a billion-dollar industry. Certain initiatives that inspired positive change over the years are mentioned below.

Insurance companies make environmental pledge

Five major insurance companies in Europe began an effort to take their customers' environmental performance into account when setting premiums and to improve their own environmental record. These five companies, including National Provident Institution and General Accident of the UK pledged to integrate environmental considerations and a precautionary principle into their business goals. The effort was sponsored by the United Nations Environment Programme (UNEP). The five companies leading the effort worked to recruit other insurance companies world wide to make a similar pledge. Additionally, they drafted a formal Statement by the Insurance Sector on Environmental and Sustainable Development which integrated environmental risk into casualty insurance, life and pension savings, investment management and real-estate management. They also pledged to use environmental checklists in daily operations and work with suppliers and subcontractors that demonstrate sound environmental management.

Products check environment risk

The 1990s also witnessed the arrival of new software products to support change.

With concerns about enforcement of corporate environmental liability disclosure heightening, Dun & Bradstreet Information Services introduced a line of products to help users screen for potential environmental risk associated with more than one million commercial and industrial properties in the United States.

The products, known collectively as Environmental Information Services, were designed to help perform environmental due diligence. The pre-screening was used in the early stages of a potential loan to pinpoint elevated risks, including those which would require further investigation with a more costly on-site technical evaluation. The reports also provided an estimated cost for the physical clean-up of the site based upon experience at sites with similar environmental problems. One of ➡

the reports listed federal and state environmental filings located by Dun & Bradstreet on and surrounding a target property. Another report helped businesses analyse the information, translating the search into a three-level risk code. It could provide an example of the clean-up costs for other sites with similar filings.

The company's Toxicheck software interpreted various sources of environmental information, including government records, questionnaire responses, and visits to specific properties. The information used was provided by Environmental Data Resources, Southport, Connecticut, which maintains a data base of government records from more than 300 different sources.

The products were developed to meet environmental due diligence standards, such as the ASTM Transaction Screen Standard, and support the FDIC's directive that lending institutions put in appropriate safeguards and controls to limit exposure to potential environmental liability.

Over the years many other software products and tools have come onto the market to identify and evaluate environmental exposures.

Environmental insurance policies

Because many general insurance policies had an absolute pollution exclusion, the environmental insurance market developed offering companies, both large and small, policies tailored to their particular operations and risk. Clearly, insurance is only a protection against unforeseen environmental events, and does not mean a company can take environmental risks.

Insurers looking to capture this growing market developed general liability policies and full pollution legal liability. These policies provided coverage for a full range of risks, covering bodily injury, cleanup of incremental pollution as well as accidents or spills. Other policies have become available that allow clients to tailor packages to what they need for their operations.

Ethics in international trade, investment and project finance: shared responsibility 16.30

Regardless of where they operate it is imperative that every organisation, large or small, focus on their individual role in the protection of the environment, irrespective of whether they are involved in a polluting industry or activity. As was noted above, see **16.13**, this ethic has been clearly supported in Europe. In its Fifth Environmental Action Programme 'Towards Sustainability', covering the period 1993–2000, the European Union (EU) targeted the role of industry in implementing the principle of 'shared responsibility' to achieve improved environmental protection and preservation. This has been developed in the Sixth Environmental Action Programme and the many publications such as *EEA Signals 2004* which is a European Environmental Agency update on selected issues. It has stated:

> 'The main barriers to progress in environmental protection and sustainability are the complex, inter-sectoral, inter-disciplinary and international nature of both the problems and the solutions.'

The European Commission has recognised that small and medium-sized enterprises (SMEs) must be equipped to participate fully in this initiative. In the Enterprise report referred to at the beginning of this chapter it is stated:

> 'Important challenges remain not least among SMEs which are often less aware of current and future environmental trends and regulations or the market opportunities available to them. SMEs tend to underestimate their environmental impacts, which may be small on a company – by – company basis but are considerable when looking at the SME sector as a whole. Internal barriers, such as a lack of skills, awareness and (human) resources, further hamper environmental responsibility … regulation and supply chain pressures seem to be the major drivers for environmental engagement by SMEs.'

The EU's Eco-Management and Audit Scheme (EMAS) exemplified the need for a practical approach by offering incentives to SMEs.

It is also clear that in order to make business more profitable and to benefit society as a whole, organisations should aim to protect the environment by taking account of the highest standards in technology and good management practices. This is true of all of their activities, including their choice of partners in trade, investment and project finance. Management theories indicate that this is best achieved by a combination of formal and informal corporate measures including a comprehensive corporate policy on the environment and follow-up procedures such as environmental audits.

An enhanced awareness of the environmental aspects of running a business and expanding overseas entails reviewing procedures and processes not only to make them more environmentally sound but also to improve efficiency. Businesses across the spectrum have been surprised at the cost-effectiveness of measures to reduce energy consumption and minimise waste generation. The use of low-energy technology, improved insulation, the plugging of leaks, recycling of materials and energy can all lead to significant savings. Good businesses appreciate that sound environmental practice is sound business practice. For instance, every car company, every energy supplier and every supermarket group wants to win the loyalty and admiration of its customers because theses are good for sales and because high environmental standards can bring operating and productivity efficiencies which help add to profits.

One major issue is, of course, climate change and the need for business to grasp the practical issues. According to the EEA Signals 2004:

> 'The climate is projected to continue to continue changing, globally and in Europe, over the next 100 years. Evidence is growing of climate change's impacts on human and ecosystem health as well as economic viability. Substantial reductions in emissions of greenhouse gases will be required to ensure that Europe meets its short-term emission targets. Adaptation measures to manage the negative impacts of climate change also need to be put into place.'

The business bottom line 16.31

There is no shortage of reasons why companies should pay attention to the environment. Evidence of continuing enhanced public awareness of the issues, consumer pressure and the interest of other stakeholders, as well as current or potential regulations from the UK and Europe, are factors that are regularly highlighted in surveys and reports. Often environmental initiatives can often save money and improve profitability and competitiveness. This is clear both from the general approach seen in the pioneering Aire and Calder project and its daughter projects in the UK relating to waste minimisation which involved the regional teaming together of businesses into clubs and in many individual case studies. In this project eleven companies along these rivers grouped together with consultants to improve their environmental management. They reduced emissions by 25 per cent and saved three million pounds, representing a major saving at the time.

There are many initiatives which have resulted in benefits to the bottom line, new product opportunities, or increased market share. For some organisations action has been motivated by compliance regarding current or anticipated regulatory requirements. Business now has substantial evidence to show that there is a huge potential to save by, in particular, good housekeeping, especially in connection with waste and energy issues. Whereas the average pay back period is often 18 months which was the target in the Aire and Calder Projects, some initiatives involve no capital outlay at all.

Sustainability: economic and environmental 16.32

Nowadays every business of stature – regardless of size or location – should take the environment very seriously indeed and business ignores environmental issues at its peril. Consumer pressure remains of fundamental importance, mainly because responding to what customers say they want is always good business and wise managers also listen carefully to NGO's. As has been mentioned in **CHAPTER 4** there is a clear link between sustainability and good corporate governance. In addition environmental legislation and regulation around the world continues to become more stringent. Most striking, however is the arrival of corporate peer group pressure. By way of example, the 1990s survey by KPMG demonstrated that 77 of the FTSE 100 companies discussed their environmental strategies in their latest annual reports.

Environmental reporting

This term has generally referred to the reporting of organisations on their environmental performance, typically to shareholders in the annual report, but also by way of separate reports for other groups such as employees and the general public. It is an essential requirement for registration in the EU under EMAS and at international level under the Valdez Principles. A UK initiative to assist SME's under EMAS was set up, known as SCEEMAS.

The principle of sustainable development as endorsed by the Fifth Action Programme of the EU – and further developed by the Sixth Action Programme – should be at the forefront of any corporate environmental policy. This principle recognises that world natural resources are either renewable or finite. Renewable resources should be managed in order to provide a continuous stream for ongoing productivity. Finite resources should be used only in the most productive and efficient ways in order to allow time for the development of alternative sources once existing resources have been depleted and/or for the development of comprehensive recycling schemes to renew these resources if practicable.

The future growth and forward planning of any organisation requires crucial budgeting of available resources, the most important being financial and human resources. Management should consider projections in terms of a 'green budget' in their effort to support a credible environmental protection policy and to maintain a high marketing profile as well as to attract public support for ongoing activities. Management standards are an increasing feature of business life. For many companies, as is mentioned above, see **16.16**, the quality standard ISO 9000 as well as the environment standard ISO 14001.

Pathways to sustainability

Companies which intend to take sustainability seriously should assess their own sustainable goals. These would constitute the long term vision of the company operating in accordance with sustainability.

Environmental Management Systems (EMS) can be converted into sustainable management systems to provide clarity about what are sustainable levels for the company's impacts, and to drive progress towards them.

There will be difficulties, the challenge of identifying sustainability goals for a particular company is complex, and in certain cases the 'sustainable ➡

level' for an existing process or product may be zero. The implication for the company may be that an alternative process or products will be required on the path to sustainability. Flexibility is a vital element. Existing EMS's may improve current products and processes, but they should also identify environmental problems which can only be resolved by a lateral re-think.

The company serious about understanding sustainability will not just have an EMS, it will also have a sustainability benchmark against which to test performance and identify the sustainability gap. Unsustainable practices will be highlighted, and major strategic decisions about product substitution or process change will be planned by pro-active management.

Environmental liability

The US influence 16.33

In developing its environmental legislation, the EU considered established environmental protection legislation and policy in the US. Liability for the clean-up costs of such sites may fall not only on the owner/operator, but also on companies who transported any waste to the site, previous owners of the site, companies whose waste was dumped on the site at any time and even lenders to those companies. The lengthy debate on environmental liability in Europe (the Commission published a Green Paper on the subject in March 1993) has at last led to an environmental liability regime. What is very evident is that large organisations understand that they are judged by the higher standards that exist.

European and International Standards 16.34

While the EU is working towards harmonisation within its borders, jurisdictional differences between Member States do still exist and they maintain separate legal systems, some having imposed stricter standards than those enacted at EU level. The effectiveness of enforcement measures also varies widely. Looking further afield, there is an undoubted need to fulfil international environmental standards in order to compete globally in terms of trade, tenders and the market. Sustainable development requirements have pushed global business towards an enhanced proactive management approach.

Many developing or so-called 'third world' countries have sophisticated environmental legislation on the books and when they commence effective implementation this will have a significant impact on any business operating within their territories. For instance, India has both environmental auditing and impact assessment regulations in place. In order to minimise any impact, it is important not only to be aware of existing regulations but also of forthcoming regulatory developments and their enforcement.

Therefore from an international point of view the ideas behind an organisation's environmental policy are vital. At the very least the aim must be to comply with minimum standards. Ideally the standards maintained should be the strictest possible, bearing in mind the cost-benefit ratio. For international companies, a single policy encompassing all the varying requirements found in the jurisdictions in which they operate demonstrates good corporate practice. The Bhopal incident in 1984 demonstrated that if high standards of management are not maintained, the consequences can be devastating. That particular incident led to American businesses being forced to maintain uniformly strict standards wherever they carry out their operations, irrespective of local regulations. Where financial constraints do not allow such an approach, a single minimum standard to be applied, with stricter standards for those jurisdictions which have enacted such standards, is a viable option for the effective management of an organisation's environmental pollution control efforts. This is true both of the developed and developing world. But by accepting higher standards before they become compulsory, it may be possible to gain market advantage.

Legislative trends **16.35**

Despite some deregulation initiatives, the trend is towards increasing implementation of environmental legislation with stronger enforcement and tougher penalty provisions, including criminal fines imposed on those having control of the polluting operation. By way of example, the *EPA 1990* strengthened the criminal penalty provisions within the UK and gave sweeping powers to the Secretary of State for the Environment to enact comprehensive regulations for the control of everything from air and water pollution through to litter.

The re-enactments by the US Congress of the *Clean Water Act*, the *Clean Air Act* and the enactment of *Superfund Amendments and Reauthorisation Act of 1986* (*SARA*) in 1986 tightened US federal laws, whilst at the same time imposing stiffer criminal and civil penalties. As the earlier discussion indicated, developments in the emerging jurisdictions in the Third World and Eastern European countries are being closely monitored by international organisations, including the UN. The imposition of penalties for environmental crimes, moves by *inter alia* the EU to impose strict liability on companies causing environmental damage, and international monitoring of environmental protection efforts, should all be borne in mind when developing corporate environmental policy. As has also been noted, institutions such as the EBRD take into account the environmental policy of an organisation when considering an application for project finance.

Business ramifications **16.36**

One of the main reasons legislative developments need to be anticipated is the necessity to budget for capital items such as new plant and process equipment.

It is particularly important to keep abreast of international initiatives such as the phasing out of the use of CFCs, that had repercussions at regional and local level, in order for the business to remain competitive. They had to anticipate new measures and act upon them. This was very evident in the very early debate over the refrigeration business, as noted above.

Friendly fridges

Elstar, which supplied refrigeration to pubs and clubs announced that it had switched its entire production to ozone-friendly chemicals. It was believed to be the first maker of commercial fridges in the world to switch to gas-cooled cabinets, some four years after the EU agreed to ban CFC's under the Montreal Protocol, because fridge coolants damage the ozone layer.

More recently the debate has been over electronic goods and legislation such as the EU's *Waste Electrical and Electronic Equipment directive* (WEEE directive). This has focused the minds of industry.

A well thought out public relations approach to increase awareness of a company's green credentials is invaluable. The UK, for example, has already witnessed the impact of organisations such as B&Q. They have effectively dictated environmental management standards to would-be suppliers for many years and, in some cases, have actually sought to assist smaller organisations in their supply chain to meet those standards. It is clear from the comments throughout this book regarding small business that they must be considered in the whole debate. Moreover, as has been mentioned above the supply chain remains a key issue. Giant companies can have as many environmental policy statements as they like, but if their suppliers do not match the same standards and have the same duty of care as the giants the consumer will realise and the credibility of business and industry will be weakened. This is clearly crucial to overseas operations and linkages and has major ramifications in terms of trade, especially in view of the offshoring or outsourcing arrangements that are prevalent in business today.

Greenness often equals quality in the eyes of consumers, but consumer interest in the environmental performance of a business is no longer limited to those living in the immediate area of the company's activities. As is discussed above at **16.32**, many companies now publish details of their environmental policy in their annual report. The European Union believe that all sectors of society must be made to feel a sense of shared responsibility for the environment, and the drive to increase public access to environmental information is seen as fundamental in this context. The EU's 'eco-labelling' scheme had a similar thrust. In the UK the Department of Trade and Industry (DTI) recommended a 'cradle to grave approach' in relation to product stewardship, ie that the supplier of a product which has the potential for contaminating a site or causing pollution should take responsibility for it by developing environmentally sound practices covering the use of that product even when no longer within the supplier's control (eg when the product is being transported, used by the consumer or when it is sent for disposal). Similarly an eco-mark has been developed in other jurisdictions.

Wherever a company operates, has activities and invests, communication is vital. Workers, shareholders, the local community, green action groups and the press should be kept informed of how perceived environmental problems are being solved and, if possible, given a role in helping find solutions. Set against this background, there is a potentially overwhelming array of issues for a company to address, having regard to the need for profitable yet ethical business practice.

Business ethics

As has been mentioned in **CHAPTER 4** there are various methods that have evolved and that are being evolved relating to the analysis of a company's performance in terms of its environmental performance. There is an obvious ethical link between proper environmental performance and good business performance, which is becoming increasingly clear from the point of view of shareholders, investors, banks etc. In order to have a real picture of the assets and liabilities of a company it must be understood that environmental liabilities are very much part of the equation. This factor was very clear in the debate in the UK over the proposed register for contaminated land proposal which was in fact withdrawn as a result of the lobbying of many property owners who considered that transactions and the market would be blighted as a result. Nevertheless the debate did raise awareness and it is true to say that most companies are likely to hold now or in the future, contaminated property. This means that, especially following the UK's *Environment Act 1995*, there may well be repercussions for such organisations in terms of the assessment of their balance sheet. Many aspects of environmental management can affect the finance of a company.

At the international level one recent vivid illustration of the bad effect of perceived improper environmental management is that of the Shell case in Nigeria where business ethics came up as a major concern and involved Shell in a very expensive large scale media exercise to recover its reputation. Shells quoted share price fell dramatically. Shell also suffered the cost of non-compliance by paying the largest national fine of £1m when sued some years ago and this came directly from its 'bottom line'. More recently Shell has again suffered dramatically through the misstated accounts regarding reserves, bringing about critical decisions in respect of its governance.

No activity is without some impact on the environment even if it is not subject to laws and regulations. Therefore organisations, whether manufacturing companies, retailers, service companies, multinationals or SME's have issues that they should address in regard to the environment. In addition, the cost effective objective applies to any business. For many of the companies that do address these issues it has been made clear that unnecessary costs and wastage occur through wastage of energy, water and raw materials and the final disposal of 'waste' products. Thinking efficiently, thinking quality, looking for new uses for materials previously considered waste, taking ideas from people at all levels in the company, working with suppliers – all these characterise the companies that have put themselves in the leadership division. This is why, for example, in the 1990s the UK's National Westminster Bank has coined the expression 'Good business management equals good environmental management' and vice versa. Moreover HSBC has encouraged environmental policies among SMEs.

The environmental balance sheet 16.37

The environmental repercussions of business activity have been considered in a debate that used to be confrontational: economics v environment. This emphasis has shifted. If a business intends to operate a manufacturing process or own, manage such a facility and /or acquire or develop property assets, then the environmental liabilities are increasingly open to question. This applies equally to its business or trade partners or where it seeks project finance.

Largely as a result of the circumstances in the US where the 'deepest pocket' is found to redress environmental damage, lenders, insurers, investors and employees are now much more sensitive to problem. This is true whether they arise from historical or on-going pollution. Such interested parties will scrutinise the health of an organisation's 'environmental balance sheet' much more closely. Unfortunately, as a result of this growing sensitivity, initial estimates for liabilities or potential clean-up costs can be very high. In some cases as has been seen in the US they can also threaten the viability of an entire business, acquisition, divestiture, development or merger. For instance the CEO of Union Carbide felt so strongly about this that he has advocated that the environment should always be considered to be a paramount business issue. He said: that 'speaking from personal experience' he is in no doubt that 'environmental protection has become a survival issue for companies'. In addition, as the insurance industry has experienced to its cost, retrospective measures taken in response to enforcement action by a regulator are rarely cost effective.

Yet, however extensive or substantial liabilities may be, they become considerably more manageable when quantified accurately after thorough investigation and assessment of the issues or impacts involved. Liabilities can also be contained by a pro-active and long term strategy which uses cleaner technology and integrated waste management practices to reduce future pollution, minimise environmental impacts and improve environmental performance.

A carefully monitored and well-reported programme of effective environmental management or remedial action is vital to inspire employees, encourage investors, protect commercial interests and substantially improve the environmental balance sheet of business. Proper environmental performance is not an add-on or an after thought; it requires constant effort, commitment, attention and enthusiasm, in the same way as other key business functions.

Trends in transboundary trade and environmental standards 16.38

By way of concluding remarks certain key trends and issues have been selected to provide a flavour of the extensive nature of relevant developments in environmental management, particularly as they apply to investment, trade and project finance activities. There is no doubt that there will be several areas to watch at International, European and National level which will impact upon how an organisation will consider its approach to environmental management.

It is quite clear from the concept of 'shared responsibility', which has been highlighted by the current Environmental Action Programmes of the European Commission and which promotes the partnership of government and industry as a team working together, that the way forward for the implementation of improved environmental standards will be through a mix of the regulatory, the voluntary and economic instruments. This fact, along with the concern that there should be a level playing field in terms of international trade, has meant that there is an on-going comparative exchange of information both by government and by industry as regards both compulsory and voluntary standards. This is also crucial to the success of corporate governance.

Institutions in the public and private sector are aware of the need for dialogue as has been evidenced in recent talks of the WTO. It is clear from the WTO talks that progress on an environmental understanding is crucial. Indeed, the harmonisation of environmental standards may also occur informally as foreign regulatory models are voluntarily adopted. For example, many developing countries today may admit imported chemical products without national evaluation if the product was duly licensed in the country of origin, thereby relying on the presumed effectiveness of foreign controls. On the other hand, several European countries have borrowed from US Federal or California State standards to update their national legislation, for instance, on automobile emissions. Indeed as with the governance debate US and European environmental legislation has been on a reasonably parallel course for some 30 years. It has been commented that a veritable transatlantic policy dynamic exists.

Similarly, as seen, the concept of a pollution tax or financial charge pro-rated to the volume of pollutant emissions has spread. Its basic idea is to levy a disincentive charge on specified economic activities, depending on the extent of the environmental harm, and to earmark the proceeds of the charge for specific counter-measures in the form of 'effluent charges'. This concept is known throughout Western and Eastern Europe, as well as the US. Moreover the general concept of the 'polluter pays principle' is seen to exist in different context in many jurisdictions, despite varying interpretations.

For some time there has been a debate to promote the 'environmental label', more as discussed, is a system of product labelling and licensing and which was introduced in areas as diverse as West Germany, Japan, Canada and Norway. In view of the European scheme, as well as the information exchange role of the European Environment Agency, there will have to be a practical framework to avoid unfair trade practices in connection with such labels. It has been mooted that arrangements for the mutual recognition of national environmental labels, possibly including harmonised standards and procedures of product selection and identification will become necessary.

Simultaneously, the role of the non-governmental bodies (NGOs) (as noted in **CHAPTER 4**) and the likelihood of complaints as a result of enhanced public information will be another trend that should be noted. It has already been mentioned that the concept of environmental auditing has been taken up directly by non-governmental groups such as Friends of the Earth and by industry. Many trans-national corporations – some partly in response to the Bhopal incident – carry out regular environmental audits to ensure that regulatory requirements and long term environmental liability such as legal waste disposal duties are actively reflected in their subsidiaries' balance sheets. Since the executive board of the ICC adopted its 1988 position paper on environmental auditing for business organisations, reflecting the experience in countries and companies where the practice was already well established, the impetus for the taking up of environmental auditing by government and industry alike has developed over the last 25 years. This has sometimes been referred to as the 'eco-audit trail'.

The eco-audit trail

Throughout the operation of any business, therefore, there will be the need to demonstrate to regulators, investors, lenders, shareholders and consumers the ability of that company to meet the demands placed upon it for environmental protection. Whereas in the past the term 'audit' has referred to ensuring financial probity in business, the term is increasingly being applied in the field of demonstrating environmental probity. For example, the environmental audit is already part of jurisdictions as far apart as India and the US environmental protection legislation. The 'Eco-Audit' is seen as an essential tool to allow the management of a company to be appraised of its continuing environmental performance. The Eco-audit will provide a verifiable trail of the environmental performance of a company from the production of the raw material for its products through the manufacturing process, to the distribution, use and ultimate disposal of that product.

This life-cycle or 'cradle-to-grave' approach demonstrates the commitment of that business to considering the environment in all of its activities. Recent developments, also recognised in jurisdictions as far flung as Canada and India, include schemes for 'eco-labelling', specifying the environmental performance of products to allow an informed choice by the consumer.

However, a fundamental difference remains between the more limited scope of auditing as an internal business management technique and the idea of public review. The latter emerged as the key element of the international environmental audit procedures, which, for example, were followed by the International Labour Organisation (ILO) or in other periodic audits of compliance with agreed upon international standards that are well established. Inherent in these is public disclosure as a means of ensuring democratic control over the implementation of agreed upon international standards.

At the international level it has been mooted that, considering the evident need to make the environmental controls preventative rather than corrective, now may be the time to envisage a global auditing body that would periodically evaluate the performance of states and organisations in complying with their international obligations. There is certainly an argument for more imaginative approaches to compliance control in the field of standard setting and regulations that will require further co-ordination within the United Nations family of organisations. What is increasingly clear is that just as environmental problems have growing transboundary impacts so does the formulation of environmental policy. The internationalisation of policy-making as Governments implement Agenda 21 raised even more concerns that transcend national boundaries, necessitating new rules. UN scientists, like many NGO's, have warned this year that unchecked pollution will cause 'economic collapse'. Moreover environmentalist NGO's increasingly have a place in UN policy development activities, while, for example, the EU actively seeks their input in its policy development activities.

All of these trends, of course, have some impact on the way a company trades, invests, selects partners and generally carries out business activities wherever they operate Those companies that gear themselves up at both compliance and voluntary levels and recognise the value of improved environmental performance in the global market place will have a definite advantage against this background both in private and public sector concerns.

Appendix

Environment, Health and Safety Simple US Company Questionnaire

Please treat this questionnaire as confidential. Once completed this questionnaire should not be reproduced or handed over to third parties outside or even within the company without prior consent. The questions and your answers should only be discussed with other employees directly associated with the project. Responses to these questions can be provided in the space provided or on attached sheets. If the facility maintains a document which addresses these issues, please identify the document, note where the response can be found on the questionnaire, and provide a copy of the document.

The questionnaire is organised in the following format:

1.0 General site information

2.0 Environment, health and safety management

3.0 Production processes and raw materials utilised

4.0 Raw material and waste storage facilities

5.0 Waste materials

6.0 Waste water

7.0 Emissions (air)

8.0 Soil and groundwater quality

9.0 Worker protection for handling of hazardous materials

10.0 Fire protection for hazardous material storage areas

11.0 Safety incidents

12.0 Special health issues

1 General site information

1.1 Company or Facility Name:

1.2 Address (Street, City, Country):

1.3 Is the land or structures leased?

☐ Yes ☐ No

If yes, please identify the name, address and contact of the owner.

1.4 Name of Facility Manager:

1.5 Name of the person(s) in charge of facility environment, health and safety programs and compliance. If more than one person is assigned to these duties, list each person and his/her respective areas of responsibility. Attach the EHS organisation chart.

1.6 Is the facility ISO 9001 registered?

☐ Yes ☐ No

Does the facility plan to be ISO 14001 registered?

☐ Yes ☐ No

1.7 Describe in general terms the products manufactured and/or services performed at this location. Also include whether the facility is primarily used for manufacturing, warehousing, or office use.

1.8 Indicate the number of employees, shifts, and days/week worked at this facility.

1.9 Provide in the spaces below the property size and the appropriate information concerning on-site buildings.

Property Size: acres or hectares

Building Designation	**Building Age**	**Building Area in ft^2 or m^2**	**Primary Use**

1.10 List all former facility owners and/or tenants including type and duration of activities on the site.

Former Owner/Tenant	**Type of Activity and Manufactured Processes Utilized**	**Duration of Activity**

1.11 Within what type of area(s) is the site or facility located? (check appropriate boxes):

☐ Industrial

☐ Commercial

☐ Residential

☐ Agricultural

☐ Water/Groundwater Protection Area

☐ Flood Plain

☐ Earthquake Exposure

Indicate the shortest distance between the facility and other objects (list only if within a mile or 2000 meter radius):

Residential Buildings:	feet/meters
Hospital, School, or Food Processing Company:	feet/meters
Main Highway:	feet/meters
Groundwater Protection Zone:	feet/meters
Surface Water (river, lake, marsh):	feet/meters

1.12 Airport Flight Lanes: feet/meters..

Other: feet/meters..

1.13 List the Governmental authorities who are responsible for overseeing the company's compliance with appropriate environment, health and safety regulations. List the name of the agency, address, telephone number, and responsible individual(s).

1.14 Are there any restrictions to the growth of operations or facility due to these regulations or any other government programs or the land owner, (if applicable)?

1.15 Provide a map of the facility with outlines of all buildings, roads, and other significant site features.

1.16 Provide a topographical map of the site/facility and surrounding area (radius of 1 mile, or 2000 meters).

1.17 Identify the facilities source(s) of potable and process water. If the source(s) is an on-site well(s) mark the location(s) on the map requested in question **1.15**. Also provide other pertinent well information if available (age, depth, screened interval, etc.).

2 Environment, health and safety management

2.1 Describe or provide the facility's EHS policies and EHS responsibilities.

2.2 Describe/list the facility's EHS training programs.

2.3 Describe the facility's EHS emergency procedures.

2.4 Describe the facility's EHS monitoring and record keeping activities.

2.5 Attach a listing of all EHS related reports and documents prepared for the facility. (Include compliance reports, investigation and remediation projects, and permit applications.)

3 Production processes and raw materials utilised

3.1 On the table below indicate if the following processes are in use today or in the past.

PROCESS	IN USE	
	CUR-RENTLY	PAST
CLEANING		
VAPOR DEGREASING		
1,1,1 TRICHLOROETHANE		
TRICHLOROETHYLENE		
PERCHLOROETHYLENE		
WATER BASED		
ALKALI		
SEMI-AQUEOUS		
HANDWIPING		
ALCOHOL		
MEK		
OTHER (LIST)		
ACID PICKLING		
ANODIZING		
CHROMIC ACID		

PHOSPHORIC ACID		
SULFURIC ACID		
OTHER		
PLATING		
NICKEL		
ELECTROLYTIC		
ELECTROLESS		
CHROMIUM		
TRADITIONAL		
HIGH SPEED		

3.1 On the table below indicate if the following processes are in use today or in the past. (Continued)

PROCESS	IN USE	
	CUR-RENTLY	PAST
COPPER		
ACID		
ALKALI		
CYANIDE		
ELECTROLESS		
SILVER		
CYANIDE		
ALKALI		
CADMIUM		
CYANIDE		
ACID		
GOLD		
OTHER (LIST)		

MACHINING OPERATIONS USING COOLANT		
TURNING		
MILLING		
DRILLING		
GRINDING		
OTHER (LIST)		
MACHINING OPERATIONS USING OIL		
TURNING		
MILLING		
DRILLING		
GRINDING		
TAPPING		
BORING		
OTHER (LIST)		
ACID MILLING		
TITANIUM		
NICKEL ALLOYS		
STEEL		
OTHER (LIST)		

3.1 On the table below indicate if the following processes are in use today or in the past. (Continued)

PROCESS	IN USE	
	CUR-RENTLY	PAST
MASS MEDIA FINISHING		
CERAMIC		
PLASTIC		
OTHER (LIST		
SHOT PEENING		
STEEL		
GLASS		
OTHER (LIST)		
STRIPPING		
ACID		
ALKALI		
CYANIDE		
WELDING		
BRAZING		
HEAT TREATING		
PLASMA COATINGS		
PAINTING		
ELECTRODISCHARGE MACHINING (EDM)		
ELECTROCHEMICAL MACHINING (ECM)		
FORGING		

CASTING		
USE OF FIXTURING MATERIAL		
ALUMINUM BASE		
LEAD, CADMIUM, OR BISMUTH BASED		
PLASTIC BASED		
OTHER (LIST)	:	:

3.1 On the table below indicate if the following processes are in use today or in the past. (Continued)

PROCESS	IN USE	
	CUR-RENTLY	PAST
NON-DESTRUCTIVE TEST (NDT)		
FLUORESCENT PENETRATE INSPECTION (FPI)		
OIL BASE		
WATER BASE		
MAGNETIC PARTICLE INSPECTION		
ETCH INSPECTION		
NITAL ETCH		
BLUE ETCH ANODIZE		
X-RAY		
SPECTROSCOPY		
OTHER (LIST)		
OTHER (LIST)		

3.2 List all major raw materials used in the manufacturing process, if not included in question 3.1. Also include maintenance and treatment chemicals.

Commercial Name	Chemical Name

4 Raw material and waste storage facilities

4.1 Underground storage tanks: List and identify the underground storage tanks situated on the property, include removed and abandon UST as well. Assign each tank, or group, a number and mark each tank location on the map requested in **QUESTION 1.15.**

Tank Number	Contents	Vol-ume units	Age	Construc-tion Material of Tank (steel, fiberglass)	Tank Corro-sion Devices

4.2 Aboveground storage tanks: List and identify all aboveground storage tanks on the property, include removed tanks as well. Assign each tank or group, a number and mark each tank location on the map requested in **QUESTION 1.15**.

Tank Number	Contents	Vol. units	Age	Construction Material of Tank (steel, fiberglass)	Tank Under Roofed Structure (Y/N)	Secondary Containment Volume (g/l)

4.3 Drum/container storage: List all the current and former storage locations of drums, including drums with new materials, solid or liquid wastes, and used oils. Also list areas where significant storage of smaller containers is conducted. Assign numbers to each storage area and mark each area on the map referenced in **QUESTION 1.15**.

Storage Location Number	Number of Drums/ Containers	Total Volume units	Contents	Secondary Containment Volume units	Under Roofed Structure? (Y/N)

4.4 Identify and describe any significant releases/spills which have occurred on the site. Include releases/spills in contained areas as well.

5 Waste materials

5.1 Complete the following table concerning wastes generated on-site.

Type of Waste (office, industrial, hazardous)	Monthly Vol- ume	Type of Dis- posal (landfill, incin- eration)	Name of Waste Dis- posal Co.	Disposal Site Location

5.2 **US only**: What is the facility's RCRA status? Attach a copy of the most recent biennial waste generation report.

5.3 Does the facility have any septic systems, drywells, lagoons, or landfills? Include units no longer in use?

☐ Yes ☐ No

If yes, describe material and indicate quantity. Indicate the location of these units on the map requested in **QUESTION 1.15**.

5.4 Has the facility received any Notices of Violation or Administrative Orders pertaining?

☐ Yes ☐ No

If yes, describe briefly:

6 Wastewater

6.1 Does the facility generate and discharge industrial wastewater(s)?

☐ Yes ☐ No

If yes continue, if no go to **SECTION 7.0**. Are permits required for the discharges?

Are permits required for the discharges?

☐ Yes ☐ No

If yes, attach copies of all applicable permits.

Where does the process wastewater drain?

☐ public sewer ☐ surface water or ditch ☐ septic system

6.1 Does the facility generate and discharge industrial wastewater(s)? (Continued)

Where does the sanitary wastewater drain?

☐ public sewer ☐ surface water or ditch ☐ septic system

Briefly describe the wastewater piping and identify the receiving water body, indicate the piping and discharge location on the map requested in **QUESTION 1.15**:

6.2 Does the facility perform any wastewater treatment? (oil/water separator, coagulation/precipitation, filtration, ion exchange, carbon treatment, etc.)

☐ Yes, complete the following table ☐ No

Process Generating Wastewater	Primary contaminants in wastewater	Type of Treatment System

6.3 Has the facility had wastewater contaminant exceedances above appropriate permitted discharge limits?

☐ Yes ☐ No

If yes, describe type, extent and cause of the problems:

6.4 Has the facility received any Notices of Violations or Administrative Orders pertaining to facility wastewater management?

☐ Yes ☐ No

If yes, describe briefly:

7 Emissions (air)

7.1 Describe all existing sources of air emissions from the facility

Source of Emission	Vented yes/no	Type of Emission Controls

7.2 List all permits obtained by the facility for the above referenced emission sources. Attach copies if practical.

7.3 Is there an installation or process on the site which falls under a special law that regulates plants with risk of catastrophic environmental releases?

☐ Yes ☐ No

7.4 Has the facility received any Notices of Violation or Administrative Orders pertaining to facility air emission practices?

☐ Yes ☐ No

If yes, describe briefly:

8 Soil and groundwater quality

8.1 Have any soil/groundwater investigations been performed on-site?

☐ Yes ☐ No

If yes, briefly summarise major conclusions of the investigation:

8.2 If the facility has groundwater supply wells on-site has the well water been analysed?

☐ Yes ☐ No

If yes, list what chemical parameters were tested the results, and in what years.

8.3 Does the facility have any transformers or capacitors that use poly chlorinated biphenyls (PCBs) as a dielectric fluid?

☐ Yes ☐ No ☐ Do not know

If yes, indicate whether the units are located on the map requested in **QUESTION 1.15**.

9 Worker protection for handling of hazardous materials

9.1 Identify the agency which oversees Worker Protection, their address and contact name.

9.2 According to the occupational health authorities, which safety precautions apply to the facility? (check appropriate boxes):

☐ workshop air monitoring

☐ noise protection

☐ respirators

☐ Personal Protective Equipment

☐ other

9.3 Is the facility responsible for regular medical check-ups of the employees?

☐ Yes ☐ No

9.4 Do any building materials at the facility contain asbestos or lead water pipes?

☐ Yes ☐ No

If yes, describe the type of material (ie, floor tiles, pipe wrap, etc).

10 Fire protection for hazardous material storage areas

10.1 How far away is the facility from the nearest fire station?.mile/kilometer

10.2 Does the facility have its own fire response team?

☐ Yes ☐ No

10.3 Notification of a fire occurs through:

Telephone ....

Manually through a fire-alarm key ....

Automatically ....

10.4 How large is the amount of water available for extinguishing a fire:

<1600 liters or <425 gallons/minute

>1600 liters or 425 gallons/minute, but <3200 liter / minute

>3200 liters or 850 gallons/minute

10.5 Is it possible that water which was used to fight the fire can run off into surface waters, into the sewage system or infiltrate into the ground from the property?

☐ Yes ☐ No

10.6 What types of fire safety equipment are installed at hazardous material storage areas?

☐ none

☐ automatic fire detectors

☐ automatic gas detectors

☐ cameras

☐ abundant fire extinguishers

☐ automatic or non-automatic sprinkler systems

10.7 Are employees trained for fire safety in hazardous materials locations?

☐ Yes ☐ No

11 Safety incidents

11.1 Has the facility experienced any employee work related fatalities in the past ten years?

☐ Yes ☐ No

If yes, describe briefly the circumstances of the accident or provide a report.

11.2 Has the facility had any serious employee work related accidents (ie, loss of digit or limb, injury resulting in paralysis or coma, blindness, etc.) in the past five years?

☐ Yes ☐ No

If yes, briefly describe the circumstances of each accident or provide a report.

11.3 Have there been any accidents or inquiries associated with this facility of non-employed people (ie, general public)?

☐ Yes ☐ No

If yes, describe briefly the circumstances of the accident:

11.4 Does the facility keep a log of all work related injuries/illnesses, if yes attach the past five years.

☐ Yes ☐ No

11.5 Has the facility been inspected by any health/safety compliance agency in the past five years? If yes provide copies of any citations and responses.

☐ Yes ☐ No

12 Special health issues

12.1 Are there any known regional or area-wide health issues?

☐ Yes ☐ No

If yes, please describe the circumstances:

12.2 Have there been any nuclear accidents near the facility (within a 500 mile radius).

☐ Yes ☐ No

If yes, please describe the event(s) and any subsequent monitoring activities:

12.3 If available, obtain and attach relevant news articles and documents related to **QUESTIONS 12.1** and **12.2**.

12.4 Does the facility currently or previously used any radioactive source materials? If yes describe the use, time period and disposal location.

12.5 Describe any equipment that generates x-rays which the facility uses.

Chapter 17

Charity Considerations: Issues of Transparency and Governance

Introduction 17.1

Just as business is now operating in an era of exacting requirements of corporate governance, as discussed in earlier chapters (see **CHAPTER 10**), charities are also similarly impacted. Moreover those companies that donate to charities should ensure that the non-profit organisations that they support are following correct procedures and responsive to change. For instance the Royal British Legion has approved radical changes in its structure following recommendations from the Combined Code on Corporate Governance created by the Financial Reporting Council in 2003 (see **CHAPTER 10**) and in anticipation of the legislative changes as a result of the *Charities Bill* (see **17.4** below). Meanwhile charities should be aware that the reputation of large donors can affect their brand, image and reputation too (see **17.2** below). The interaction and influence between the commercial and non-commercial sectors can be witnessed globally.

There are over 20,000 registered charities that are also registered with Companies House. Dual registration enables them to benefit from the limited liability enjoyed by company directors to hold property directly rather than indirectly through individual trustees (see the discussion of corporate vehicles in **CHAPTER 4**). As the governance debate evolves so do the available vehicles or structures in addition to or in place of traditional vehicles and structures. There is a new legal form of incorporation being proposed which is designed specifically for charities, the Charitable Incorporated Organisation (CIO) to take account of this issue (see **17.4** below). Reference may also be made to the proposal for 'community enterprise companies' in the recent *Companies Bill*. Indeed in the UK there is an initiative to modernise charities in many respects in the much anticipated draft *Charities Bill*, which was published on 27 May 2004. The full text can be found at www.homeoffice.govuk/comrace/active/charitylaw/index.html

As has been the case with topics covered in earlier chapters, the discussion of charity considerations can be the subject of another entire manual. Issues of transparency and governance include macro and micro concerns. Developments and comments can extend from the larger organisations to very small organisation in this vast sector. In this chapter, therefore, it is intended to raise awareness of selected key issues relevant to the discussion of due diligence and corporate governance and to highlight related developments. In the main this chapter covers UK developments, with some international comment as appropriate.

Accountability and corporate giving **17.2**

In this age of growing accountability wherever a charity operates clarity over donations is at issue. Corporate support is, of course, vital to many prominent areas of the voluntary sector, including aspects of the art world and other non-profit-making ventures. Meanwhile in the sports world there are often well publicised connections that create issues of governance. For example in the UK the high profile football sector has undergone a recent Inquiry (*English Football and Its Finances* by the All Party Parliamentary Football Committee, February 2004) to address concerns over governance. Moreover there have been questions raised regarding the appropriate vehicle for clubs and associations: the report has welcomed the initiative regarding community enterprise companies referred to above.

In all of these areas support and giving can be highly influential as well as tax efficient. While this has been true for some time in the USA, the tax implications have become increasingly important in the UK more recently. Therefore, as far as concerns both the donor and the beneficiary, transparency is most important. As in the USA, in general since the last collapse of the stock market, charities are finding that there is greater public interest from both taxpayers and shareholders in who is giving to charity and why. Moreover there has been a heightened state of interest as a result of the global concerns regarding terrorism. Trust, once lost, is hard to regain. Just as corporate scandals, such as Enron and Parmalat, have damaged the trust of many investors, impact is felt also on the non-commercial sector (see also earlier discussions in **CHAPTER 7**).

One noteworthy example that illustrates the increasing connections between charities and corporate giving in the USA related to the well documented Enron corporate scandal. The connection between Enron and the foundation associated with Kenneth Lay, the company's disgraced former chairman was controversial. The full extent to which charities favoured by Enron directors benefited from Lay and his company's patronage is unknown, due to the complexity of the arrangements and lack of information. What is clear is that much of corporate philanthropy is strategic. It is usually giving that ultimately affects a company's bottom line. It is not just about being a good corporate citizen. Bearing in mind the significance of governance both for corporates and charities even small business should ensure that it clear about donations In the era of global business and global giving this is a trans boundary concern. Many companies simply do not know exactly how much they give to charity and from where because the giving is diverse – from small donations to local girl guide groups to charity functions used as marketing events. There should be clarity from both sides to ensure the trend in favour of greater transparency, such as that demonstrated by SIR in the UK, is followed (see **17.9** below).

Fundraising and related developments 17.3

One of the most important areas of due diligence, risk management and governance relates to fundraising. There are many aspects that require careful handling. For example one of the crucial matters for the sustainability of a charity is effective fundraising that enables it to get its message across to the correct audience. This is also a sensitive area as complaints about fundraising can affect the reputation of the charity. Neil Morris, deputy chief executive of the Institute of Direct marketing in the UK has been quoted as saying:

> 'Charity fundraisers face really tough targets compared with those faced by commercial marketers.'

Charities rely on accurate lists of potential donors and up to date information: since the *Data Protection Act 1998* the price of such information has risen dramatically. Mailing the right material to the wrong person is as bad as getting the message wrong in the first place. Therefore the building of sustainable relationships and strategic alliances, as well as using the right broker services that are increasingly available, are important as part of the ongoing due diligence and risk management of the charity. As with commercial organisations concerns over the increased reliance on and exposure to IT (see **CHAPTER 9**), operational interruption and disaster management must be addressed (see **CHAPTERS 5** and **9**).For instance, online initiatives related to volunteering are on the increase and they can also be affected by global cyber crime. Charities have to consider:

- How to protect against internet fraud;
- Privacy and security issues;
- Problems of cyber stalking and online harassment;
- Protection against fraudulent emails;
- Different government regulatory frameworks regarding criminal activities; and
- The constitutional issues raised by intellectual property cyber crimes.

Moreover, as with commercial entities, litigation, or the threat of litigation, can be devastating to small charities in particular (see **CHAPTER 3**) in terms of the distraction of resources and damage to reputation.

The draft *Charities Bill* (see **17.4** below) moves towards a unified scheme for regulating public collections and attempts to improve the statements made by professional fundraisers and commercial participators. The pressure is on in today's climate to monitor what they say. For instance NGOs such as

WWF-UK and Friends of the Earth have been criticised for misrepresenting repercussions of climate change and the impact on species to gain more support for campaigns. There is increasing concern over the fundraising ethic and the common technique of using crises to garner support that has become prevalent here and in the USA. Since the 1980s small organisations have grown to become trans national and have adopted corporate styles of performance management. A recent paper produced by Oxford University has highlighted concerns and Paul Jepson, a former programme manager for Birdlife International, its co-author, has stated that much of the sector is still stuck in the 1980s model of fundraising:

> 'You send out a nice emotive letter and you get £10 guilt money. But then you get involved in an arms race and are forced to hype and hype the story. Ultimately this could damage the message of charities and end up with people not believing anything you say. The current state of fundraising is not good. We need a different model.'

The Charities Bill

Charitable purposes and public benefit 17.4

In the UK the draft *Charities Bill* is the first attempt to update charity law since Elizabethan times. It extends the list of charitable purposes to 13 and includes a public benefit test criterion. The Bill has enjoyed widespread support in general from the sector but issues still remain over the list, as well as the requirement relating to public benefit. The new charitable headings are:

- Prevention or relief of poverty;
- Advancement of education;
- Advancement of religion;
- Advancement of health;
- Advancement of citizenship and community development;
- Advancement of the arts, heritage or science;
- Advancement of amateur sport;
- Advancement of human rights, conflict resolution or reconciliation;
- Advancement of environmental protection or improvement;
- The relief of those in need by reason of youth, age, ill-health, disability, financial hardship or other disadvantage (including by the provision of accommodation or care);
- Advancement of animal welfare; and
- Other currently charitable purposes.

As it stands, there will be no statutory definition of 'public benefit' as the Government believes that the non-statutory approach provides flexibility, certainty and the capacity to accommodate the diversity of the sector. Nevertheless the presumption of public benefit for charities for the traditionally recognised charitable objectives for the relief of poverty, the advancement of religion and the advancement of education will be removed. For new charities there will be the need to establish not only the recognised charitable purpose but also the provision of public benefit. It is believed that this will involve the use of case law dating back to the original Elizabethan Act. This has meant some controversy over whether campaigning organisations will want to become charities. There is a trade-off between tax advantages and the potential restrictions of charitable status. Moreover it leaves a great deal of power to the Charity Commission (CC). Indeed the Bill as a whole gives increased regulatory powers to the CC. This has alerted concerns over the potential for over policing, especially as regards the vulnerable smaller charities. The regulatory aspects will be subject to CC guidance, another area of concern due to potential problems of regulatory creep. Moreover several charities threatened to withdraw support unless the position regarding public benefit is clarified.

Independent tribunal — 17.5

The Bill also creates an independent Charity Appeal Tribunal (see the role of tribunals in **CHAPTER 5**). This independent adjudicator could theoretically challenge or reject the CC guidance or rulings. This initiative has been very well received. As has been mentioned in **CHAPTER 5**, the Government is currently working on wider scale tribunal reform and the charity tribunal will have to be consistent with the reformed tribunal system as a whole.

Liability of trustees — 17.6

The proposal relating to the relief from personal liability for trustees in the draft Charities Bill has also been well received. The matter of personal liability has discouraged many from being trustees. The intention is that if a trustee acts honestly and reasonably they will not be personally liable. Nevertheless, as mentioned below, the voluntary sector is generally under pressure to improve governance and, in particular, the role of trustees. There has been some debate regarding the need for a trustee code of practice. The Enron and Worldcom scandals, as well as the subsequent Higgs Review of the role of NEDs in the private sector (see **CHAPTERS 7 AND 10**) gave this debate more urgency. Although the Higgs Review considered corporate boardrooms the Institute of Chartered Secretaries and Administrators (ICSA) has deemed the issues raised to be also relevant to charities. Since trustees are sometimes unclear about their role a single document that clarifies governance is important. Moreover, while some consider that a voluntary sector Enron would not be very likely, organisations are taking the threat seriously.

Some organisations have demanded that every charity should carry out an audit of its management structures in the interest of proper due diligence and good governance. There is no doubt that, as with commercial organisations, non-commercial organisations will be increasingly monitored by stakeholders to ensure that best practice prevails. Whereas there is a financial threshold of £1,000,000 (see **17.9** below) that exempt small charities, practical impacts are being felt on the sector as a whole.

Reporting requirements **17.8**

In the UK since the introduction of the Statement of Recommended Practice for Charities (SORP) accounting framework in the 1990's charities have had to comply with increasingly rigorous requirements in accordance with the *Charities Acts*. SORP, which is reviewed annually, has been updated to cover the need for charities to explain their activities and their achievements. This has of course increased the pressure on charities to be transparent in their objectives and operations. Recently SORP has focused on reporting costs and expenditure, strengthening the emphasis on achievements against objectives and encouraging small equity-holding charities to declare their ethical investment policy. The annual review 2004 has also considered issues raised by the Strategy Unit regarding the completion of a Standard Information Return (see **17.9** below).

The Government does not want charitable companies to opt out of compliance with SORP. Despite anticipated EU legislation to permit charities that are registered as companies to adopt International Reporting Standards rather than those laid down by the CC, from 2005 the DTI has indicated that charity companies should continue to prepare their accounts in accordance with the UK's generally accepted accounting practice. It was thought that international charities with offices in many different places might have adopted the international standard in order to offer a uniform manner of presenting their accounts. The CC, however, was opposed to this, as was the Finance Directors' Group in the UK. They both wanted to maintain SORP, not as a dismissal of international standards but rather to enable consistency while harmonisation of practice occurs. They believed that an opt out by charitable companies would have led to fragmentation. However companies and building societies will be able to switch to the international standards. This area should be considered in the light of the new proposed CIO, whose detailed structure will be dealt with in secondary legislation. The Government will review the need for other forms of incorporation for charities five years after the introduction of CIO.

Standard Information Return **17.9**

Recently a new reporting instrument has been proposed to deal with information from charities This tool is known as the Standard Information Return (SIR). The SIR was proposed in 2002 by the Strategy Unit (SU) in its report on charities with a view to raising transparency and accountability in the voluntary sector. It has been agreed that SIR will be added to the reporting requirements for charities that have an income in excess of one million pounds (£1,000,000). Therefore where small business representatives make a decision to invest in such charities they should ensure that the beneficiary organisation is following the procedure and fulfilling the requirements once the SIR is implemented.

The SIR will be a two page statement that includes a range of topics, such as:

- The charity's achievements over the reporting period;
- Its aims in the forthcoming year;
- Information on the sources of support; and
- Information on the charity's spending.

Purpose of SIR **17.10**

The intention of SIR is that donors, funding organisations and other stakeholders will be able to appreciate quickly:

- What a charity does;
- The financial base of the charity; and
- Whether it is fulfilling its objectives.

The Charity Commission will operate SIR and the information will also feed into the Guidestar online charity information website. It intends that the SIR procedure will be available from 2005. The main idea is that donors, funders and stakeholders will be able to understand clearly and quickly:

- What a charity does; and
- How it is doing.

Considerations of the Strategy Unit **17.11**

The Strategy Unit proposed the SIR essentially on the basis that many of the reports and accounts of charities submitted to the Charity Commission were 'inaccessible and often ill-suited to the public's needs'. The CC has added that it was especially difficult to find credible information about performance or outcomes. It has also expressed concerns that it has been hard to make any meaningful comparisons between similar charitable organisations (see also **17.12** below).

Concerns over SIR **17.12**

Accountants and some charities have, nevertheless, raised concerns over the new tool. Their argument is that it will be very hard to define and accurately set out the work of a charity in the small space available. This could mean rough rankings of charity performance. This could in turn bring about the possibility of league tables that the regulator might also use as a tool in its armoury to assess the performance of charities. Meanwhile a major study to measure charities' financial performance is being carried out by the Charity Finance Directors' Group. The benchmarking study will also compare the voluntary sector with the private and public sectors. The umbrella body says that the study follows a recommendation in the SU 2002 report on charity law and regulation reform that proposed that charities benchmark their performance against their peers in different areas.

The consultation process with key groups of charities and donors has therefore focused on the format of SIR and the extent to which the topics covered can be dealt with in the two page statement. There is no doubt that in practice, this will be a real challenge. The SU proposed nine categories of information, including:

- achieving against objectives;
- stakeholder involvement;
- governance;
- fundraising;
- trading; and
- reserves and investment.

However there remain real concerns, demonstrated by the consultation process that a two-page statement cannot represent the correct position of the charity, thereby going against the objective of transparency.

Another concern is whether the SIR should be audited like the charity's accounts. This has yet to be resolved. The SU has stated that:

> 'in order to confer some external scrutiny, the information provided should be professionally audited and where possible should make use of accredited processes (such as use of accredited quality tools)'.

However the Charity Commission's head of policy, Rosie Chapman, has said that the SIR will be certified rather than externally audited. The lack of audit could mean that there is a risk that an SIR will lack credibility among the public, again rather defeating the object of transparency. There is a risk that the SIR could be considered to be a marketing exercise unless charities are able to support their statements in the SIR with credible evidence.

As regards the future some accountants believe that there will be a compulsory audit. This is largely because of the issues mentioned above and because funders are increasingly requiring detailed reports from charity applicants with verification by accountants. In this regard there is some provision in the *Charities Bill* relating to the sensitive issue of whistleblowing.

Protecting whistleblowers **17.13**

The debate over whistleblowers has affected commercial and non commercial organisations across the world. A worker who blows the whistle on an employer can expect to feel the full force of institutional anger and discrediting including:

- criticism;
- poor performance evaluations;
- punitive transfers;
- job loss;
- ostracism from colleagues;
- blacklisting; and
- stress and health damage.

Historically whistleblowers have found that they have made a decision that entails all risk and no reward. A detailed discussion of whistleblowing is not appropriate in this manual as it is a subject that deserves much more comprehensive treatment. However it has arisen, of course, as a major issue in the well-documented corporate scandals. Moreover it is especially relevant in the consideration of charities, bearing in mind the crucial area of professional ethics. If the public does not protect whistleblowers it tacitly accepts the risks of being denied important information. Although government ministers have been quick to condemn whistleblowers who raise awkward questions – as in the recent case of UN telephone tapping – the regulatory departments have begun to appreciate, and even encourage, whistleblowers.

Following the lead of the UK in 1998 and the USA in 2002, many governments across the world, including the Netherlands, Japan, Korea and South Africa, have drafted legislation to protect whistleblowers. In 2002 the UK's Financial Services Authority established a hotline that received 276 calls from whistleblowers between May 2002 and October 2003. In the US SOX obliges all US-listed companies to establish 'confidential, anonymous' procedures for employees to submit material about 'questionable accounting or auditing practices'. It also protects whistleblowers from retaliation. Impeding a whistleblower from passing information to enforcement agencies now carries a jail sentence of up to ten years, as well as large fines. This demonstrates a major departure for a country whose legal culture has emphasised the duty of loyalty to the employer (see also **CHAPTER 11**). Evidently Ernst and Young 's annual survey of global fraud has rated whistleblowing above external audits as the second most effective means of detecting corruption. People are now prepared to acknowledge that whistleblowing is about good citizenship.

In the *Charities Bill* it is proposed that auditors of charity accounts will be protected from the risk of action for breach of confidence or defamation when they communicate relevant information to the CC. Independent examiners of charity accounts will also be protected. In this respect they are protected since they are often able to identify abuse or significant breaches of trust during the audit process.

Governance issues 17.14

In the UK, governance in the not for profit sector has been under scrutiny. The Charity Commission has deemed existing regulations to be ineffective and has examined the role of trustees. The voluntary sector is under pressure from all sides to consider a code for trustees to bridge the gap between law and good practice.

In March 2004 a final draft for a potential governance strategy framework for charities was produced. Simultaneously ICSA publicised its draft proposals to improve charity governance, focusing on the regulation of trustee boards. The Enron and WorldCom scandals and the subsequent Higgs Review of the role of Non Executive Directors (NEDs) in the UK private sector gave the debate an air of urgency. Although the review considered corporate boardrooms, ICSA has stated that the issues raised are also relevant to charities. Clearly this should not entail a box ticking exercise, practical solutions should also be reached as a result of a code of practice that covers:

- Best practice
- Recruitment; and
- Professional development.

There should be a set of expectations that parallels the code of the private sector, which effectively says comply or explain. In this report, and having regard to US developments (see **17.15** below) the sector is intending to be proactive so that a voluntary sector Enron does not happen. ICSA is aware of the fact that there is still a high level of public confidence in the sector but that it would take just one major scandal to set back the sector and an audit of a charity's management systems would assist.

The SIR is another indication of the growing importance of transparency and governance. While stakeholders have generally welcomed the introduction of SIR in principle, they have indicated that there are issues to be dealt with regarding the use of the SIR as a tool for comparison. It is to be hoped that in time the SIR can be implemented to the positive satisfaction of donors and donees alike. Meanwhile this is a matter of interest to organisations, including small business, and should be monitored carefully.

Trustees duties: governance for non profits in the USA 17.16

The corporate governance debate in the United States is also spreading from the for-profit to the not for profit world. There have been well-publicised controversies at organisations such as The Nature Conservancy, the American Red Cross, and the James Irvine Foundation. This has even caused observers such as Eliot Spitzer, the Attorney General of New York State, to suggest that the *Sarbanes-Oxley Act of 2002* should be applied to non profit boards. Clearly, those non profit boards operate under unusual constraints, for example directors:

- volunteer their time;
- play an important role in raising funds, and
- in some cases are so numerous that board meetings resemble conferences rather than deliberative assemblies.

They also answer to a wide range of stakeholders who may lack a single common goal, such as increasing shareholder value, prevalent in the commercial sector. Thus it is not surprising that a recent McKinsey survey of executives and directors of not for profit social-service organisations found that only 17 per cent of the respondents felt that their boards were as effective as they could be. It has been suggested that to improve the governance of non-profit organisations, their boards must venture beyond the traditional focus on raising funds, selecting CEOs, and setting high-level policy. McKinsey's research indicates that the best boards should also:

- provide professional expertise;
- represent the interests of their nonprofits to community leaders;
- recruit new talent to the organisation; and
- provide the more rigorous management and performance oversight that funders increasingly demand.

In the USA over the longer term, however, not for profit organisations have no choice but to reconsider the way they replace and recruit directors. Meanwhile regular evaluations can help by:

- setting out expectations;
- indicating when a change of behaviour is needed; and
- motivating underperforming directors to leave.

As regards the recruitment of new directors, a standing nominating committee should have the responsibility for creating a board on which each member brings not only the all-important fund-raising capabilities but also necessary skills or relationships with community leaders, politicians, or regulators. The committee should recruit candidates from as wide a range of channels as possible and recognise that sustained cultivation may be needed to get the best possible directors. This should be compared with the Higgs Review and the Tyson Report in the UK.

The Tyson Report 17.17

In the UK the Tyson Report has called for more diversity on company boards so that they included more NEDs from non-traditional backgrounds, such as the voluntary sector. It is evident that, compared with the voluntary sector, listed companies are very rigid about appointments. The voluntary sector is more flexible while commercial organisations are awaiting more guidance in order to implement the Tyson Report.

Tyson's remit was to develop the ideas raised in the Higgs Report on corporate governance. The role of NEDs is regarded as increasingly important given the high profile corporate failures at companies such as Enron. Tyson has argued that companies would benefit from recruiting a more diverse board and that the non commercial sector was a fertile source of NED talent for UK companies.

The Tyson Report also led to hopes that companies would take advantage of specialist registers that include top people form non commercial sectors. Companies need to decide not only whether they want a more diverse board but also whether they would recruit other than by word of mouth or other traditional methods, thereby changing the culture. Such recommendations also emphasis the increasingly close interconnection between the two sectors in relation to ongoing due diligence, risk management and corporate governance. In any event it is essential that the independence of the voluntary sector should not be jeopardised. If involved in company boards any representatives should ensure that they keep their independence. This should not affect the ability to scrutinise companies, nor the corporate sector generally, on social and ethical issues.

Conclusion 17.18

Across the world the debate over governance has focussed the minds of many in all sectors as they grapple with:

- Changing models and practices;
- Practical threats to their sustainability; and
- The need for positive performance for the benefit of all stakeholders.

As this chapter has attempted to demonstrate, since the commercial and non commercial sectors interact, increasingly they must be alert to developments in both sectors. Even in the case of smaller organisations – whether in business or not – there is a duty in the interest of ongoing due diligence, transparency and governance, to meet the challenges and monitor the swiftly changing requirements and expectations of the society and the economy.

Chapter 18

Governance in the Family, the Family Business and Family Trusts

Introduction 18.1

Governance and the family is an appropriate context in which to reflect upon the potential future of governance, as well as its scope, bearing in mind that families involved in business and trusts often seek:

- more unity of purpose;
- an increasing focus on and realisation of the importance of family business; and
- greater emphasis on personal development.

Currently governance tends to be defined as a function of oversight, regulation or supervision. A broader definition concerns how we communicate with one another to regulate our relationships and to make decisions. In recent times, particularly with the corporate scandals that have been exposed, it has come increasingly to mean regulation. So, how could governance look in the context of best practices? Consider that the world has a natural order and that governance is a tool that allows that order to unfold:

> 'Living in accordance with natural hierarchy is not a matter of following a series of rigid rules or structuring your days with lifeless commandments or codes of conduct. The world has order and power and richness that can teach you how to conduct your life artfully, with kindness to others and care for yourself.' (Chogyam Trungpa; The Sacred Path of the Warrior).

Below are some of the questions to ask when looking for new ways of thinking, doing and being:

- How would governance develop as a field based upon the vision in the quote above?
- What can we do to promote the discovery of this order, power and richness founded on a belief in mankind's tendency towards positive

growth whilst at the same time being realistic about human nature with particular regard to our instinctual natures and motivation-driven behaviour such as our need for food, water and space?

- What are the processes, the fields of knowledge that have to be drawn upon, the language, the practices, the structures of governance that promote long-term thinking, planning and action, as well as the sustainability of our World?
- What framework is achievable for individuals to find meaning in their lives, among other imperatives?

In looking for answers some assumptions proposed are that:

- it is impossible to shape governance without input from the behavioural sciences; and
- it is no longer enough to have to look for meaning outside of the workplace.

Leading management consultants have long recognised the need to harness individuals' passions and shortage of talent is becoming a real business issue hence providing a strong incentive to focus on what retains employees in a business.

Having determined such guiding principles the next step is implementation. What often gets overlooked and which leads to failing mission statements, is that even if we appreciate the significance of certain values such as honesty and openness, if the culture of an organisation or the development of an individual is unable to maintain such values they cannot last. This is an essential issue to grasp and is perhaps why even though much is written about values, vision and mission statements, putting them into practice effectively often evades us.

We hear of organisations not 'walking their talk' and out of this arises scepticism and the disregard of the mission statement. Arguably it is more demoralising to have one that is not effective than not to have one in the first place. So, why is it that these mission statements cannot be carried through? One answer I believe is to be found in Robert Kenny's assertion in 'What can Science Tell Us About Collective Consciousness' (www.collectivewisdominitiative.org) that (where we go wrong is that..) 'The external processes and systems of any collective become the sole focus. If we take a group through these processes, or structure an organisation this way, or teach these techniques, then collective consciousness will be ensured', the thinking typically goes. What gets sacrificed are the inner aspects of individual or collective life, issues like psychological or moral development, meditative practice, culture and so on'.If this is so, then here is some guidance as to what is needed. However, making such shifts are difficult because they require radically different ways of being on the part of individuals and corporations.

Governance theories and practice 18.2

As regards some of the theories, practices, ways of being and thinking which the governance theory has to embrace to support positive development some suggestions are outlined below:

From fragmentation to relationship and systems thinking 18.3

This involves seeing ourselves as part of a system; the system of all living things, the system of human beings or the global economy, whichever system we choose.

From 'power over' to 'power from within' 18.4

Governance is also about leadership and, in line with this shift, the old hierarchical ways are being replaced with more inclusive forms of leadership, even forms of leadership that take a passive role rather than an active one.

A new language 18.5

This new language may introduce new words or revive words, placing them in new contexts. Some examples include terms such as sharing power, subsidiarity collaboration, participation, leadership from behind, servant leadership, inner leadership, consensus, creativity, acting for a greater good, compassion, openness, uncertainty, contemplation, self-awareness, self-accountability, collaboration, passion and participation.

Shifts in decision making processes and communication styles 18.6

As well as what we talk about the way in which we talk is shifting. New processes include consensus, collaboration, dialogue, mediation, appreciative inquiry, meditation, storytelling, open space technology, world café and the circle.

A shift from the dynamics of 'groupthink' to harnessing the power of groups 18.7

Groupthink has tended to have a derogatory connotation, meaning that the sum of the parts is less than the whole. We need to understand and counteract its operation and learn what supports collective intelligence, the combined intelligence of a group

Awareness, acceptance, understanding and public acknowledgement that monetary economics drives our society and that there are alternatives 18.8

New measures of progress that incorporate socially and environmentally just measurements already exist and allow greater measurement of the qualitative rather than the quantitative and so support longer term measures. Examples include the Measure of Domestic Progress developed by the New Economics Foundation (www.neweconomics.org), the Gross National Happiness Index adopted in Bhutan and the Genuine Progress Index for Atlantic Canada (www.gpiatlantic.org). There are increasing ways of valuing the intangible assets and liabilities, as well as SERM demonstrated in **CHAPTER 4**.

The other side of this coin is the need to re-balance corporate power. Left to their own devices corporations are structured to pursue their self-interests. Many recognise that the balance of rights and responsibilities has to be restored, that the power of corporations, unaccompanied by social responsibility is unsustainable. Take for example the **purpose** of the World Business Organisation's Conference 2004 (www.worldbusiness.org).

> 'To engage participants in a deep dialogue on the role of business in the creation of a sustainable global future. In such a future, business must adopt a new form of interaction with society by: re-evaluating the responsibilities inherent in corporate governance; encouraging entrepreneurialism; and willingly contributing beyond the enterprise. Together, these activities naturally thrive in a compassionate, empowering environment. Working in such an environment strengthens the commitment of businesspeople to elevate the actualization of these goals to a high priority.'

A shift from short-term thinking to long-term thinking 18.9

An excellent example of where a failure of long-term thinking is going to have fatal effects for all of us is the global issue of climate warming, the reduction in the use of fossil fuels and a switch to what are called 'renewable' sources of energy, such as wind, water and solar power. Failure to take urgent action on this issue will lead to the destruction of our planet so what better example to take to emphasise a need for this shift. However, other areas where short termism has acute impacts are on maintaining favourable environments for short term foreign investment and corporate valuations driven solely by market need to see cash flow.

Many of the above shifts will rely on awareness and understanding of the issues by business, lawyers and other policy makers. Only by changes in the law and policy in the short term will some of these shifts be possible due to the power of factors such as:

- the media;
- the rewards for acting out of self-interest;
- a global lack of integrated, independent thinking and action;
- a belief that education should be structured to provide necessary skills which fails to appreciate that, if we are to create the world we want,

what most fulfils us should drive what we learn and to do otherwise just perpetuates a world lacking in personal meaning or which provides meaning for the minority;

- lack of awareness of the complexity of events; and
- tolerance for a reluctance to honestly and realistically address the 'shadow' side of human behaviour.

Governance, the family and long-term wealth preservation **18.10**

Wealth is about so much more than money. It is often now categorised as human, social, intellectual and financial capital and is used in this sense throughout this chapter. Human capital is arguably the most important. Certainly this is the attitude of James E Hughes (*Family Wealth, Keeping It In the Family: How Family Members and Their Advisers Preserve Human, Intellectual, and Financial Assets for Generations*, Bloomberg Press, 2004) who has pioneered new perspectives on managing, and attitudes, towards wealth to help wealthy families preserve their wealth for reasons which go beyond a goal of simply holding onto the money. For our purposes:

- human capital refers to who the individual family members are, their health and what they are called to do;
- intellectual capital refers to how family members learn, communicate and make joint decisions; and
- social capital is about how family members engage with society at large and would usually cover the philanthropic aspects of a family policy.

Governance is essential to long-term wealth preservation or, put another way, to the preservation of a healthy functioning family unit. There are those who do not want to pass on their money to their children as they do not believe this is the right way to preserve wealth. Many are concerned that they will disempower their children by removing the incentive to make their own way. Is money without meaning sustainable? Can 'wealth without work' be sustained? (Ghandi) We do not know the answers but it is inevitable that as one of the capitals fails so it is likely to bring down the lot. It takes enormous ongoing commitment to the family to preserve the health of all four capitals. Governance is an essential component. It provides the structure within which a family's purpose can be developed, for ongoing education around the governance process and structure itself, financial literacy, education about the nature of trusts and the responsibilities of being a beneficiary, family history, the family legacy discussions about family giving, individual achievements and so on.

The family purpose and vision 18.11

As with any long-term strategy we start with ascertainment of purpose in each of the four capitals. With relation to human capital Hughes proposes that:

> 'the purpose of a family is the enhancement of the individual pursuits of happiness of each of its members in the overall pursuit of the long-term preservation of the family as a whole.'

In helping a family to decide upon its purpose, it will make a difference whether you are talking to the creator of wealth or an inheritor and this needs to be established at the outset. Each family's purpose will vary and will have its own mix of priorities and values attributed to the various forms of capital. However, human and social capital considerations are arguably the most important. Without considerable attention to their development long term wealth preservation has little chance of success.

In working on a family's purpose and vision it is necessary to look at how our relationship with money – how we spend, invest, earn and save it – is an expression of ourselves, a means by which we make meaning with our lives. As such we can learn about ourselves by understanding this relationship. By addressing their human, social and intellectual capital people can change what meaning they create with their money. Financial capital is considered in the context of the other capitals and generally is not the driver of purpose or vision.

In addition to pursuing the happiness of each individual family member what else is relevant to purpose? Why should the family stick together? Research shows that caring for a greater good such as the family, promotes psychological well-being in the form of life satisfaction, happiness, self esteem and a sense of coherence in life. Stewardship of wealth, a term often seen in the field of wealth management means that one views wealth as being for the benefit of a greater purpose beyond oneself, maybe to be preserved for future generations or as a family which benefits the community, say in the case of undeveloped land

Alongside discussion of values, attention needs to be given to the relationship among family members as well as their intellectual capital. Given the rate of change and the volume of information with which we live today it has become even more important to learn how to think and not what to think, to value experience and to understand how we learn so that each can maximise what they learn. Family members have to be educated in all four capitals and much of this is set out in a worksheet developed by Bonnie Brown of Transition Dynamics in the **APPENDIX**. Lifelong learning should be adopted by every family to perpetuate and continuously recreate the family vision, to stay current with best practices.

With regard to financial capital, Hughes stresses the importance of planning for the long term. If one is guided by the great law of the native America Iroquois:

'In our every deliberation, we must consider the impact of our decisions on the next seven generation.'

This would amount to about 200 years, roughly, six or seven generations.

A powerful way to consider families is to use the following balance sheet of family assets and liabilities.

	FAMILY BALANCE SHEET
ASSETS MINUS	LIABILITIES
	Long-term family risks:
Family's total human capital, including:	• Failure of family governance
Each family member's intellectual capital	• Failure to understand that success requires a 100 year plan
Each family member's financial capital	• Failure to comprehend and manage all forms of family capital, human and intellectual as well as financial
Each family member's social capital	
	Intermediate family risks (internal)
	• Death/Divorce
	• Addiction and other "secrets"
	• Malthus's Law (the geometric increase of family members in each generation)
	• Creditors Poor health
	• Poor beneficiary /trustee relationships Investment programs of less than 50 year

	Intermediate family risks (external)
	• Inflation • Inadequate trustee management • Estate and other forms of transfer and wealth taxes • Holocaust • Acts of God • Changes of political system • Lack of personal security
	Short-term family risks
	• Income taxes market inflation • No mission statement • Lack of financial education
EQUALS	SHAREHOLDER EQUITY
	• Are individual family members successfully pursuing happiness • Are the family's human capital and intellectual capital increasing when measure against the family's liabilities? • Is the family as a whole dynamically preserving itself? • Is the family's governance system producing more good decision than bad by taking a seventh generational view

Special thanks to Charlotte Beyer and The Institute for Private Investors for their invaluable assistance in the creation of this work.

In addition to the list of long-term liabilities, such global issues as peace, access to basic resources such as water and food, climate warming and the wealth/poverty divide should be added (see also the environmental considerations in **CHAPTER 16**). A failure to take account of these issues and use what influence each of us has is going to lead to the destruction of our planet and long-term wealth planning becomes superfluous. At a more practical level the greater the pool of assets the greater one's leverage in the financial world to gain access to investment opportunities and to reduce transaction costs. Decisions made by groups of people have the potential to be better, made in the right conditions, than those made by one person alone.

Advisers will serve their clients by understanding and bringing to the wealth creator's attention the implications of failing to prepare heirs, whether individual family members or charitable foundations, to receive substantial wealth, notwithstanding the fears of losing control, death and conflict among heirs. It is a common assumption that beneficiaries know they will inherit substantial wealth; the reality is often quite different. Without the knowledge and the education about managing wealth and the challenges that accompany wealth there is little chance for its long term preservation. Any governance system must be implemented having regard to the family's purpose and vision Socrates suggested that we find happiness by creating a life in which we honour our most cherished values.

To help families with this process some of the 'big' questions surrounding money and its many meanings are listed in various books, (Hughes, Brown, *Unexpected Wealth: A Fire Drill for Building Strength and Flexibility in Families*; Collier, *Wealth in Families*; Needleman, *Money and the Meaning of Life*). There are various assessments, tools and exercises of differing levels of sophistication that can assist with the articulation of values. As a starting point, they need to understand the different values and beliefs and preferably have determined what their values are in order to be able to facilitate such a process. There are also many ways in which the drafting of a mission statement can be facilitated. Stephen Covey in *7 Habits of Highly Effective People* (Simon & Schuster, 1999) says:

> 'the family mission statement is a combined and unified expression of all family members of what your family is about – what it is that you really want to do and be – and the principles you choose to govern your family life.'

His mission statement reads:

> 'The mission of our family is to create a nurturing place of faith, order, truth, love, happiness and relaxation and to provide opportunity for each individual to become responsibly independent and effectively interdependent in order to serve worthy purposes in society.'

Among the processes that should be implemented, succession planning is particularly important. Research shows that failure to properly plan for succession is a key factor in this failure. A conflict resolution process is equally important as we can eliminate the scope for conflict often at times of great anxiety, particularly if a leader has died. The very presence of a process tends to minimise conflict at all other times

Having determined purpose, mission and processes, what are the issues surrounding the governance bodies, the entities that carry out the governance? The types of bodies that may be required will be determined by factors such as:

- the number of family members and the degree of relationship among the various family branches;

- the complexity of assets;
- the health of the human capital, for example the more conflicted a family the greater will be the need for structure; and
- the range of activities such as philanthropic concerns, private foundations, geographical location of assets and family members and public profile of the family.

A family council and a family committee may be formed if a family is sufficiently large, communicative and open to benefit from these more formal arrangements. One body acts like a board of directors, setting policy, facilitating the family governance process and the other acts as the administrator or manager like a managing team in a business. Any council should address at least the following:

- the eligibility for inclusion, that is. age, family members by marriage, advisers;
- the age of inclusion of younger generations:
- a code of conduct to manage process;
- how and who will record and circulate minutes, notices of, agendas for, calling and frequency of meetings;
- whether advisers will be appointed and terms and conditions of their inclusion;
- eligibility for inclusion in and appointment to the family council;
- reporting procedures;
- time of meeting, quotas, procedure for decision-making;
- election, duration of office, re-election, voting, removal and resignation
- compensation;
- officers such as secretary, treasurer
- expenses;
- procedure for dispute resolution such as mediation, consensus, simple or weighted voting e.g. casting votes may be given to elders or wealth creators:
- other decision making procedures;
- what can and cannot be decided upon by the body and what must be referred back to the family;
- member's obligations and responsibilities;
- processes for removal, resignation, appointment;
- disciplinary procedures; and
- procedures for exiting members.

The family in business **18.12**

The importance of family businesses cannot be overstated. Statistics vary but generally in the western world they contribute over 50 per cent of GDP and numerically exceed 80 per cent of businesses of the country. It has also been ascertained that about 30 per cent of businesses succeed into generation two and about 10–15 per cent will reach generation three. This is the same pattern as applies to family wealth, regardless of the family business, which suggests that it is not so much to do with the business itself as the family, whether succession issues, education, proper attention to the creation and nurturing of the human, social and intellectual capitals. European family businesses employ 80 per cent of the workforce (1992). The benefits which a family-owned business can bring to the economy include greater commitment to employees and community, longer range perspective, opportunity for family members and shared venture maintaining family unity (Dashew, *The Best of the Human Side: Managing Ourselves, Our Relationships; and Our Organisations in a Rapidly Changing World*, Human Side of Enterprise, 1997), thereby providing purpose and meaning. Family businesses also run the range from small unincorporated businesses and partnerships to publicly quoted companies.

Governance of a family business requires governance of the business, in addition to the family and also governance of the overlap between the family and the business. Due to the additional systems and increased complexity there is more scope for conflict, lack of cohesion and greater difficulty in developing a shared vision. Often there is no diversification, the main family asset being the business. Ivan Lansberg in *Succeeding Generations*, Harvard Business School Press, 1999 summarises that:

> 'governance structures in family companies must be designed to safeguard the long-term interests of family shareholders by ensuring the growth and continuity of the enterprise and promoting the family's harmony and welfare.'

There are many issues that have to be addressed in family businesses and the following gives only a flavour of what these are. Begin by asking whether the family serves the business or the business the family? Is the family or the business the most important. This will often determine the exit strategy for the business, business strategy and operation and would be important to know to resolve conflict. Historically family business planning centred on estates and management succession. Today, it is recognised that business families have far more needs, particularly where they see that the business can reflect the family's identity, legacy and create unity. In short, that the purpose of the family and business should be aligned. Carlock and Ward in *Strategic Planning for the Family Business*, Palgrave Macmillan, 2001 have developed a very helpful questionnaire for exploring the family business philosophy.

In addition to the foregoing section in wealth in families consider:

- Education of owners and family members about the issues that family businesses give rise to with the overlapping family, ownership and business systems.

- Learning how each system is mutually reinforcing so as one strengthens so it strengthens the other and so on in an iterative process.
- Education about the role of the board of directors, management and shareholders.
- Education about common pitfalls in family businesses, such as the need for family.
- Forum for discussion of the additional values such as those relating to return on investment, growth rates, target Family employment policy
- Conditions for sale or purchase of shares by family members.
- Identify financial issues of ownership, such as return on investment, liquidity and exit strategies.
- Educate, inform and build understanding and consensus around:
  - need for and the role of independent directors;
  - scope, rights and responsibilities of ownership, management and employment;
  - financial issues such as compensation packages; and
  - life cycle of family members and life cycle of business.

The business introduces the need for more process because of factors such as:

- importance of ensuring consistency of purpose, vision and goals of business and family and of addressing family issues which impact on the business and vice versa from damaging either system;
- motivating family members to participate in the family business, participating in the education about the business, its strategy and ownership;
- extended conflict resolution process. Addressing the greater likelihood of family disputes spilling over into the business education;
- greater understanding by the board of directors of family dynamics and stronger relationships with family members
- the different roles a family member may be performing at any one time. Which hat are they wearing for example, mother or boss?
- merging of family and business time.

Succession is once again of critical importance. Ivan Lansberg suggests that directors of a family business may have a greater responsibility to the business to address succession issues. There is increasing emphasis on succession by institutional investors.

There will a greater need for more formality and possibly a greater number of governance structures each having a code of conduct for proceedings and a constitution. It will be useful to represent each body and each system diagrammatically to better appreciate the overlapping areas to ensure information flows where it should and adequate dissemination of information. In addition to the family bodies consideration should be given to

- a shareholders' council to deal with matters such as:
  - the return on investment;
  - management reports; and
  - mechanisms for sale of interests and events that may have a material impact on the value of the business, and on the family's vision and values.
- additional committees for such matters as:
  - audit;
  - compensation;
  - succession; and
  - strategy.

Each body should have a clear mandate and well-defined reporting processes with supporting ongoing education as to the role of each. The existence of a family business provides for more interaction between family members as the number of opportunities for interaction increase. Social capital literature suggests that when the same people are linked together via multiple overlapping roles their social ties are strengthened (Boissevain, Portes).

The additional structure will have the usual business policies and procedures. In addition consideration should be given to such matters as:

- Issues of family employees working with non-family employees.
- Compensation of all employees should be set by reference to objective standards;
- For the non family managers a formal system for defining responsibilities, setting performance objectives reporting and evaluation
- A board of directors which includes outside directors: appropriate outside directors can provide new momentum and a greater sense of direction which in turn can lead to greater buy in by family shareholders.

Family trusts **18.13**

The prior sections covered:

- governance in general;
- governance of wealth and families;
- governance of family businesses; and

- ownership without control.

By legal definition irrevocable discretionary trusts are all about ownership without control. On establishment, assets are legally transferred to the trustees. It is not possible to consider issues about the validity of trusts in this overview. However, a valuable text of common and fundamental issues relating to trusts and trustees is the *Misplaced Trust* (Peter Willoughby, Gostick Hall Publications, 1999).

Discretionary trusts **18.14**

The following is a list of the special issues and challenges that concern discretionary trusts and that should be taken into account in decisions revolving around governance.

- Usually part of the class of beneficiaries is made up of minors or unborn children who have no representation.
- Often trustees are professional and corporate and have no natural connection with the family.
- Beneficiaries generally do not know anything about the nature of a trust, their role and responsibilities as beneficiaries and the responsibilities of trustees.
- The use of trustees as information conduits to other members of family thereby forming triangles. This can happen wherever third party advisers are appointed.
- Trusts are generally established for tax planning reasons, with little thought given to the consequent restrictions on dealings with the assets during the settlor's lifetime.
- Trusts have a long and sometimes perpetual duration (see in particular James Hughes, How Family Members and Their Advisers Preserve Human, Intellectual, and Financial Assets for Generations, Bloomberg Press, 2004, chapter on 'perpetual trusts'). Mechanisms that perpetuate control after the settler's lifetime should be checked carefully;
- The number of relationships involved. Assuming there is a settlor, trustees, protector, an investment committee and a body of beneficiaries there will be relationships to be managed
- Upon creation thought is rarely given to how the trust will play out through the generations in terms of best supporting individuals in their individual pursuit of happiness, (see above). The impact of this can be as follows:
  - the settlor is more likely to defer any communication over the destination of the assets, shifting responsibility onto trustees to bear the varying reactions of the beneficiaries and future decisions as to distribution of the assets;

- beneficiaries inherit assets that they do not control and that they cannot control without breaching the trust. The restrictions on trustees as to how they may invest requires greater prudence than an absolute owner may exercise;

- beneficiaries are often unprepared to know how to 'manage' this wealth, to deal with the trustees or even to know what their options are, none of which augurs well for long-term preservation of the wealth or good relations between the various parties;

- the assets are controlled by people who the beneficiaries may not get along with, yet have no power to change the trustees. This may be mitigated by the existence of a protector but they may be more attuned to the values and drives of the settlor rather than the children; and

- as a result beneficiaries can have destructive attitudes, behaviour and emotions towards the trust assets, ranging from a lack of concern or interest to conflict. This can exhaust substantial parts of the trust fund in litigation.

These challenges should be met by thorough discussion in advance of the trust's formation, good governance, communication and education. Trusts can be vehicles that preserve family unity and values, facilitate succession to assets upon the death of the settlor or principal beneficiary and minimise the likelihood or intensity of potential conflict. Trusts may also serve the valuable role of protecting wealth from events such as dissipation on divorce, taxation and irresponsible beneficiaries. (See the family balance sheet above at **18.11**).The purpose, vision and goals should be addressed as set out under families and wealth and a mission statement is as important and perhaps more so where assets are held in trusts in addition to any letter of wishes (see **18.16**).

In addition to the above there is an increased potential for conflict for such reasons as:

- The introduction of non-family advisers into the system who control the assets;

- A lack of, or poor, communication by the settlor with the beneficiaries;

- The division of trust fund between a life and remainder interest; and

- The disclosure of information concerning the existence of and extent of the trust to younger generations.

Therefore, particular regard should be given to incorporating a conflict resolution process in the family council. Also education should extend to the role and responsibilities of the trustee, the investment manager, protector and others.

Choice of adviser 18.15

As with all aspects of due diligence and governance, the choice of advisers should be given careful consideration depending upon whether a private trust company, or personal or corporate trustees should be involved or any combination. In choosing trustees factors other than those that will be dictated by tax and other legal mandates include:

- interpersonal skills;
- the scope for conflict of interest within the institution;
- an appreciation of common family dynamics; and
- the expertise or network to be able to provide appropriate support and experience of managing the assets.

Peer review, automatic retirement provisions and audits are all tools which can prevent entrenchment of anyone's position. An increasing number of families are forming family offices or using the services of a multi-family office to deal with the administration relating to the investment functions.

One of the critical aspects to address when choosing a trustee is to ascertain their willingness to work with other professionals in the field of wealth management such as family business advisers, art insurance experts, family psychologists, family group facilitators, peer groups for beneficiaries to meet to discuss common issues as wealth inheritors.

Letter of wishes 18.16

A letter of wishes, which usually sketches out when, to what extent and for what purposes a beneficiary may benefit from the trust fund, generally accompanies a trust deed. It should, in general, be supported by a well thought-through and discussed mission statement which builds in flexibility to adapt to unforeseen and changing circumstances.

Protector 18.17

A Protector is usually appointed who is often a close family friend though sometimes a professional, comprises one or more persons with whom the trustee may be obliged to consult, or even be required to, act in accordance with the protector's directions. This arrangement provides an alternative and is the more usual mechanism for collaborative decision-making throughout the life of the trust and, depending on the powers granted to the protector, can limit the trustee's power, thereby acting as a check and balance of power. The Protector will usually have the right to remove trustees. As a matter of good governance, someone, other than the trustees themselves, should be able to remove trustees in the event they do not carry out the purposes of the trust or mismanage the assets. Of vital importance is the privacy of trust affairs and good mechanisms to manage the key relationship of trustee and beneficiaries in the interest of confidentiality.

Any number of other structures such as. investment committee, advisory committee, compensation committee, beneficiary committee can be formed. As in the other areas, governance without a proper education as to the use and benefit will render it ineffective. Effective decision making at the trustee level (by analogy to the board of directors) can strengthen a family system and a well governed family can reinforce the effectiveness of the trustees.

Conclusion **18.18**

It is appropriate to conclude this book on due diligence and corporate governance with some consideration of wealth management from the above perspectives. If we are to review the evolving areas of due diligence and corporate governance in the broader context of the society and the economy this discussion also provides valuable insight into issues that highlight the need for individual responsibility, with repercussions for the quality of decisions and understanding. This should assist with a more enlightened attitude by individuals that they can bring to all aspects of their life – whether at home or in the workplace- and thereby improve the likelihood of overall sustainability and of ongoing due diligence and good governance without the need for excessive regulation.

Appendix

Building competencies information tracking grid

Arenas of Learning	**Formal Education**	**Internships**	**Work**	**Volun-teer**	**Family**
Planning					
Estate					
Financial (Accounting, Taxes)					
Investment					
Legal					
Media and Public Relations					
Wealth Management					
Communication					
Family Well-Being, Roles and Boundaries					

Arenas of Learning	**Formal Education**	**Internships**	**Work**	**Volun-teer**	**Family**
Conflict Management					
Consensus Building					
Family Meetings					
Grieving					
Listening and Negotiation					
Allocation of Family Assets					
Stress Management					
Team Building for Family and Boards					
Family History and Legacy					
Family Stories					
Genogram					
Philanthropic Vision and Practice					
Relationship Management					
Mentor and Coach Relationships					
Professional Adviser Relationships					
Selection of Advisers					
Evaluation of Advisers					
Termination of Advisers					
Trustee/Beneficiary Relationships					

Arenas for Building Competencies	**Formal Education**	**Internships**	**Work**	**Volunteer**	**Family**
	Resource Contact		*Career Path*	*Individual Goals and Evaluation Criteria*	*Meetings and Training*
Financial Capital					
Annual Reports					
Appraisals					
Audits					
Economic Theory					
Financial Statements					
Investment Action Plans					
Stock Market					
Taxes					
Social Capital					
Family Foundations					
Grantmaking					
Legal Guidelines					
Management					
Networking					
Individual Giving Goals					
Other					

Arenas	Formal Education	Intern-ships	Work	Volunteer	Family
Human Capital					
Adviser Relationships					
Facilitators					
Legal, Accounting and Investment					
Media and Public Relations					
Listening and Negotiation					
Use of Family Assets					
Transport (Boats, Planes, RV's)					
Vacation Homes					
Contingency Planning (Fire Drills)					
Career Uncertainties					
Catastrophic Illness					
Sudden Death					
Unexpected Wealth					
Family History and Legacy					
Family Vacations					
Genogram					
Storytelling					
Family Meetings					
Stress Management					
Team Building for Family and Boards					

Arenas for Building Competencies	**Formal Education**	**Intern-ships**	**Work**	**Volunteer**	**Family**
	Resource Contact		*Career Path*	*Individual Goals and Evaluation Criteria*	*Meetings and Training*
Intellectual Capital					
Governance Structures and Processes					
Adviser Management					
Board Management					
Code of Conduct					
Conflict Resolution					
Decision-Making					
Financial Oversight					
Meeting Management					
Mission Statement					
Policies and Procedures for Giving					
Rights and Responsibilities					
Staff Management					
Strategic Resource Management					
Action Plan Goal Management					
Annual Report Preparation					
Adviser Profiles					
Board Chair Profile					
Board Member Profiles					
By-Laws and Board Legal Parameters					
Project Management					
Strategic Planning					
Values and Ethics Management					
Website Creation and Management					

Dr Bonnie Brown, www.cultivatingchange.com

Index

[all references are to paragraph number]

D

K

L

P

V

W